Quality Control and Quality Assurance - Techniques and Applications

Edited by Sayyad Zahid Qamar and Nasr Al-Hinai

Published in London, United Kingdom

Quality Control and Quality Assurance - Techniques and Applications
http://dx.doi.org/10.5772/intechopen.111297
Edited by Sayyad Zahid Qamar and Nasr Al-Hinai

Contributors
Abdulla Jaafar Desmal, Adekunle Yekinni, Amir Nemati, Anthony Said Hasham, Barbara Ciecińska, Chantelle du Plessis, Diego Mantilla, Elena Khunuzidi, Elham S. Said Hasham, Elif Öztürk, Joost van den Berg, Jordy Coffa, Kayode Samuel Adekeye, Khairat Adesina, Mark Southern, Niamh Burke, Pablo Dávila, Raffaele Novario, Rafiu Olalere, Rajenlall Siriram, Rawhi Alrae, Sheriff Lamidi, Steffen Bocklisch, Svetlana Sheremetyeva, Takalani Rhoda Luhalima, Terry Twomey, Vladimir Shper, Vladimir Smelov, Welf-Guntram Drossel, Zainab Merza Madan

First published in London, United Kingdom, 2024 by IntechOpen
IntechOpen is the global imprint of INTECHOPEN LIMITED, registered in England and Wales, registration number: 11086078, 167-169 Great Portland Street, London, W1W 5PF, United Kingdom

British Library Cataloguing-in-Publication Data
A catalogue record for this book is available from the British Library

Additional hard and PDF copies can be obtained from orders@intechopen.com

Quality Control and Quality Assurance - Techniques and Applications
Edited by Sayyad Zahid Qamar and Nasr Al-Hinai
p. cm.

This title is part of the Industrial Engineering and Management Book Series, Volume 4
Topic: Operational Excellence
Series Editor: Fausto Pedro Garcia Marquez
Topic Editor: Stuart So

Print ISBN 978-0-85014-024-8
Online ISBN 978-0-85014-025-5
eBook (PDF) ISBN 978-0-85014-026-2
ISSN 3029-0511

For EU product safety concerns:
IN TECH d.o.o., Prolaz Marije Krucifikse Kozulić 3, 51000 Rijeka, Croatia,
info@intechopen.com or visit our website at intechopen.com.

IntechOpen Book Series

Industrial Engineering and Management

Volume 4

Aims and Scope of the Series

Industrial Engineering and Management (IEM) is a discipline that focuses on optimizing complex processes and systems within various industries. It involves the integration of engineering, business, economics, mathematics, and behavioral sciences to improve efficiency, productivity, quality, and overall performance in organizations. Key aspects of Industrial Engineering and Management include: Process Optimization; System Analysis and Design; Quality Control and Management; Supply Chain Management; Operations Management; Human Factors and Ergonomics; Project Management; Cost Analysis and Financial Management; Decision Analysis.

Overall, Industrial Engineering and Management aims to optimize resources, improve processes, enhance productivity, and ensure the effective and efficient utilization of all elements involved in the production or delivery of goods and services. It is crucial in today's competitive business environment for organizations to stay efficient and competitive.

Production Engineering and Operational Excellence are fields of study and practices that focus on optimizing and improving the manufacturing and production processes within an organization. It combines principles from engineering, management, and operational strategies to enhance productivity, efficiency, quality, safety, and sustainability in the production of goods and services.

Here are the key components of Production Engineering and Operational Excellence: Process Optimization; Operational Excellence; Manufacturing Systems Design; Quality Management; Supply Chain Optimization; Production Planning and Scheduling; Automation and Technology Integration; Health, Safety, and Environmental Management; Cost Management; Performance Measurement and Key Performance Indicators (KPIs); Continuous Improvement and Innovation.

Production Engineering and Operational Excellence are crucial for organizations aiming to stay competitive in the global market by achieving high levels of efficiency, quality, and customer satisfaction while optimizing resources and minimizing waste. It is a multidisciplinary approach that encompasses engineering principles, management strategies, and the effective use of technology to drive operational success.

Meet the Series Editor

Fausto Pedro Garcia Marquez is a Full Professor at UCLM, Spain, with accreditation since 2013. He also holds the position of Honorary Senior Research Fellow at Birmingham University, UK, and serves as a Lecturer at the Postgraduate European Institute. In addition to these roles, Fausto has experience as a Senior Manager at Accenture from 2013 to 2014. He earned his European Ph.D. with the highest distinction. Throughout his career, Fausto has received numerous awards and honors. These include the Nominate Prize (2022), Gran Maestre (2022), Grand Prize (2021), Runner Prize (2020), and Advancement Prize (2018), as well as Runner (2015), Advancement (2013), and Silver (2012) by the International Society of Management Science and Engineering Management (ICMSEM). He was also the recipient of the First International Business Ideas Competition 2017 Award. Fausto's contributions extend to academic publishing, with over 242 papers to his name. Notably, his work has been recognized in journals like "Applied Energy" (Q1, IF 9.746, Best Paper 2020) and "Renewable Energy" (Q1, IF 8.001, Best Paper 2014). His affiliations include the editorial and authorship roles in more than 50 books, with publications through respected publishers such as Elsevier, Springer, Pearson, Mc-GrawHill, IntechOpen, IGI, Marcombo, and AlfaOmega. He has authored over 100 international chapters and holds 6 patents. Fausto serves as the Editor of 5 International Journals and is a Committee Member for more than 70 International Conferences. His research portfolio encompasses being the Principal Investigator in 4 European Projects, 8 National Projects, and participating in over 150 projects involving universities and companies. His areas of expertise and research interests span Artificial Intelligence, Maintenance, Management, Renewable Energy, Transport, Advanced Analytics, and Data Science. Fausto is a recognized Expert in the European Union in AI4People (EISMD) and ESF. He also serves as the Director of www.ingeniumgroup.eu, holds the status of Senior Member at IEEE since 2021, and has been honored as an Honorary Member of the Research Council of the Indian Institute of Finance since 2021. Fausto is also the Committee Chair of The International Society for Management Science and Engineering Management (ISMSEM) since 2020.

Meet the Volume Editors

Dr. Sayyad Zahid Qamar is a professor at the Mechanical and Industrial Engineering Department, Sultan Qaboos University (SQU), Muscat, Oman. He has more than 30 years of academic and research experience and more than 6 years of industry experience. His research areas are applied materials and manufacturing, applied mechanics and design, reliability engineering, and engineering education. He has worked on funded research projects over 4 million dollars. He has more than 230 research/technical publications (books, chapters, publications in refereed international journals and conferences, and technical reports). He is currently editing a volume (*Thermal Engineering of Steel Alloy Systems*) in the thirteen-volume series Comprehensive Materials Processing, 2nd edition, by Elsevier. He has served as an associate editor, guest editor, and editorial board member for different scientific journals.

Dr. Nasr Al Hinai is an associate professor and the head of the Mechanical and Industrial Engineering Department at Sultan Qaboos University (SQU), Muscat, Oman. He received his MSc from the Department of Mechanical, Manufacturing, and Aerospace Engineering, University of Manchester Institute of Science and Technology (UMIST), UK, in 2003 and his Ph.D. in Industrial Engineering from the University of Manitoba, Canada, in 2011. He joined SQU in 2001 after completing his BEng. His research interests lie in production planning and control, metaheuristics in operations research, product design and development, project management, and analysis of bio-composite development processes. He has published several research papers in peer-reviewed international journals and conference proceedings.

Contents

Preface

In today's competitive market landscape, ensuring the quality of products and services is paramount for businesses to maintain their reputation, meet customer expectations, and achieve long-term success. For any engineering sector (manufacturing, construction, transportation, aerospace, food and agricultural products, oil and gas, etc.), product quality is a prerequisite for success. Quality control (QC) and quality assurance (QA) are two aspects of quality management. The American Society for Quality (ASQ) defines QC as "part of quality management focused on fulfilling quality requirements", while it defines QA as "part of quality management focused on providing confidence that quality requirements will be fulfilled".

QC is basically the process of maintaining standards and ensuring product consistency. Its key objectives are detecting defects, identifying deviations, and maintaining quality standards. It comprises different techniques and tools, the main ones being statistical process control (SPC), inspection methods, and Six Sigma. SPC involves the use of statistical methods to monitor and control production processes. Techniques such as control charts, Pareto analysis, and process capability analysis help identify trends, deviations, and areas for improvement. The basic inspection methods in QC are visual inspection, measurement techniques, and non-destructive testing. These methods enable the identification of defects or irregularities in products or processes. Six Sigma is a data-driven methodology aimed at reducing defects and improving process efficiency. The DMAIC (Define, Measure, Analyze, Improve, Control) framework guides organizations through the process of identifying, analyzing, and eliminating variations in processes.

QC finds applications across diverse industries. In the manufacturing industry, QC techniques are integral to ensuring the quality and consistency of manufactured goods. By implementing rigorous inspection and testing procedures, manufacturers can identify and address quality issues before products reach customers. In software development, QC encompasses activities such as code review, testing, and debugging. Techniques like unit testing, integration testing, and regression testing help ensure the reliability and functionality of software products. In the healthcare sector, QC plays a crucial role in medical device manufacturing, laboratory testing, and patient healthcare services. Strict quality standards and regulatory requirements govern processes to ensure the safety and effectiveness of medical products and services.

QA focuses on the systematic planning, implementation, and monitoring of processes to ensure that products or services meet predefined quality standards. Its objectives include preventing defects, continuous improvement, and achieving customer satisfaction by delivering reliable and consistent outcomes. One of its foremost techniques and tools is quality management systems (QMS) that can be used for implementing standards such as ISO 9001 for consistent quality management. It is implemented through audits and reviews for regular assessments of processes, procedures, and documentation. It also involves risk management, which consists of identifying, assessing, and mitigating risks to quality and compliance.

QA applications in the pharmaceutical industry include QA in drug development, manufacturing, and regulatory compliance. In construction engineering, QA processes can help ensure building codes compliance and structural integrity. In the software business, applicable QA methodologies may include agile testing, test automation, and regression testing.

QA provides two types of confidence: internally to the management of the engineering organization and externally to customers, government agencies, regulators, certifiers, and third parties. While QA involves how a process is performed or how a product is made, QC relates more to the inspection of the product. When QA and QC work in tandem, it leads to higher efficiency in the organization and to better products.

As QA and QC are quite interrelated and interdependent, many of the tools and techniques are used in both. These include, but are not limited to, checklists, quality standards and specifications, SPC, control charts, root cause analysis (RCA), Six Sigma, failure mode and effects analysis (FMEA), quality audits, quality management systems (QMS), continuous improvement (CI), Lean manufacturing, testing and inspection, and so on.

This book, *Quality Control and Quality Assurance – Techniques and Applications*, discusses the elements of quality, such as quality planning, quality control, quality assurance, and quality improvement. It includes topics related to QA, such as total quality management (TQM), failure testing, process and product quality assurance (PPQA), and SPC. It also presents QC modules including process control, control charts, acceptance sampling, and product quality control.

Covering some of the latest practices, techniques, and applications in QC and QA, the book is a useful resource for engineering and business professionals and students, managerial personnel, and third-party agencies.

We acknowledge the support of the editorial team at IntechOpen, especially Dominik Samardzija, Patricia Kerep, and Sandra Bolf.

Sayyad Zahid Qamar and Nasr Al-Hinai
Mechanical and Industrial Engineering Department,
Sultan Qaboos University,
Muscat, Oman

Section 1

Basics of QA/QC

Chapter 1

Quality Tools, Technologies, and Techniques: Enhancing Product and Service Excellence

Rawhi Alrae

Abstract

This chapter provides an overview of the significance of Quality Control (QC) and Quality Assurance (QA) in ensuring product and service quality, meeting customer requirements, and achieving organizational objectives. QC involves monitoring and maintaining quality, while QA ensures compliance with specified requirements. The chapter explores various techniques for QC and QA, such as Statistical Process Control (SPC), Six Sigma, Lean Manufacturing/Management, Failure Mode and Effects Analysis (FMEA), Quality Audits, and Inspections, enabling consistent quality, waste reduction, customer satisfaction, and continuous improvement. Additionally, it discusses the applications of computer-aided Quality Control (CAQC) and Quality Assurance (CAQA) technologies in healthcare, manufacturing, service, construction, and food industries, highlighting benefits like real-time monitoring and improved decision-making. Challenges in implementing CAQC and CAQA, including costs and change management, are addressed. Popular techniques and tools in CAQC and CAQA are mentioned, with recommendations for software and tool selection and integration. Future trends in computer-aided Quality Control and Quality Assurance are also explored. Overall, this chapter emphasizes the importance of QC and QA in achieving quality and organizational success across industries.

Keywords: computer-aided quality control, quality management software, quality information systems, quality function deployment, quality assurance

1. Introduction

Quality control and assurance ensure product/service quality, meet customer requirements, and achieve organizational objectives. They prevent defects, support continuous improvement, and comply with standards across industries.

1.1 Introduction to quality control and quality assurance

Quality control and quality assurance are vital for ensuring product/service quality, meeting customer requirements, and achieving organizational objectives. Quality control monitors and maintains quality, while quality assurance ensures products/services meet specified requirements. They prevent defects, detect issues early, support continuous improvement, and comply with standards. Concepts like Total Quality Management, continuous improvement, process control, and compliance enhance these practices. Quality Management Systems (e.g., ISO 9001), Six Sigma, and Lean Manufacturing/Management are interconnected and complement quality control and assurance efforts [1–5].

Quality control is crucial in ensuring products/services meet quality standards and customer requirements. It involves systematic monitoring, evaluation, and rectification of deviations. Elements like inspection, testing, process monitoring, and corrective actions contribute to its effectiveness. Quality control ensures consistent quality, customer satisfaction, and loyalty. It reduces defects, non-conformities, and costs, while ensuring compliance with standards and regulations. By emphasizing quality control, organizations enhance their reputation and establish trust with customers and stakeholders [6–9].

Quality assurance ensures that products/services meet or exceed quality standards through a systematic and proactive approach. It includes elements like quality planning, process documentation, training, and auditing. Quality assurance establishes a culture of quality, prevents defects, and builds trust with customers and stakeholders. It enhances reputation and market competitiveness [10–12].

1.2 Key principles and concepts

Effective Quality Control (QC) and Quality Assurance (QA) practices are built upon key principles and concepts that form the foundation for achieving and maintaining high levels of quality in products or services. Understanding and implementing these principles is essential for organizations seeking to deliver value and meet customer expectations. **Figure 1** shows that the key principles are [13–17]:

Figure 1.
Key principles and concepts of quality control and quality assurance [13].

- Customer focus: Understanding and meeting customer expectations is vital for quality management. Involving customers in the process helps tailor products/services accordingly.

- Process approach: QC and QA emphasize managing interconnected processes to optimize quality and ensure consistent outcomes.

- Prevention over detection: Emphasizing prevention rather than detection reduces costs and fosters a culture of continuous improvement.

- Continual improvement: Regularly analyzing data and customer feedback drives ongoing enhancement efforts.

- Employee involvement: Engaging employees at all levels promotes innovation, a culture of quality, and continuous improvement.

- Data-driven decision making: Relying on accurate data for decision making ensures objective and evidence-based actions.

- Leadership commitment: Strong leadership commitment establishes clear quality objectives, allocates resources, and promotes a quality-oriented mindset.

- Standardization: Establishing standardized procedures and quality standards enhances consistency and process control.

By implementing these principles, organizations can establish robust quality management systems, drive continuous improvement, and meet customer expectations.

1.3 Relationship with other quality management concepts

Quality Control (QC) and Quality Assurance (QA) are interconnected with several other quality management concepts and methodologies. Understanding their relationship with these concepts is crucial for establishing effective quality management systems. The following are some key relationships [6, 15, 16, 18, 19]:

- Total quality management (TQM): TQM is a comprehensive approach to quality management that emphasizes the involvement of all members of an organization in continuous improvement efforts. QC and QA play integral roles in TQM by ensuring that quality standards are met through inspection, testing, and process improvement. They contribute to the overall goal of TQM, which is to achieve customer satisfaction and organizational excellence.

- Statistical process control (SPC): SPC is a technique used to monitor and control processes based on statistical analysis. QC utilizes SPC tools to monitor process performance, detect variations, and take corrective actions. QA, on the other hand, focuses on implementing robust processes that are capable of maintaining stable and predictable performance, reducing the need for excessive QC activities.

- Lean six sigma: Lean Six Sigma combines the principles of Lean Manufacturing (eliminating waste) and Six Sigma (reducing defects and variations) to achieve process improvement and customer satisfaction. QC and QA are essential components of Lean Six Sigma methodologies. QC is utilized to identify defects and non-conformities, while QA focuses on prevention through process standardization, mistake-proofing, and robust quality systems.

- Continuous improvement: QC and QA are closely aligned with the concept of continuous improvement. QC identifies areas for improvement through defect detection, and QA drives the implementation of corrective and preventive actions to enhance processes and systems continuously. Together, they support the culture of continuous improvement by fostering a proactive approach to quality management.

- Risk management: QC and QA contribute to risk management efforts by identifying and addressing quality-related risks. QC activities, such as inspection and testing, help detect potential risks, while QA ensures that appropriate preventive measures are in place to mitigate those risks. By integrating risk management principles with QC and QA, organizations can effectively manage quality-related uncertainties.

It is important to note that the relationships between QC and QA and other quality management concepts may vary depending on the specific context and industry. However, their interplay remains fundamental in achieving overall quality objectives.

1.4 Applications for quality control and quality assurance

Quality Control and Quality Assurance techniques are widely employed in diverse industries to achieve consistent quality, meet customer expectations, enhance operational efficiency, and comply with industry regulations. These techniques involve the use of tools, methodologies, and best practices to monitor, evaluate, and improve processes, products, and services. By implementing effective QC and QA practices, organizations can minimize defects, reduce waste, increase customer satisfaction, and maintain a competitive edge (**Figure 2**) [15, 16, 20–28].

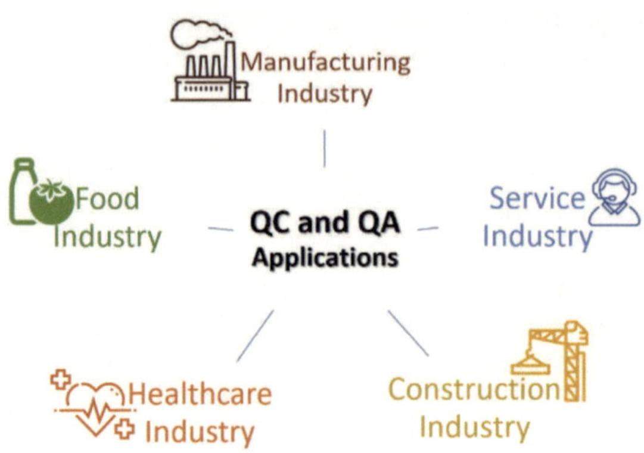

Figure 2.
Quality control and quality assurance applications [15].

1.4.1 Manufacturing industry

In the manufacturing industry, QC and QA techniques play a crucial role in ensuring product quality and reliability. These techniques encompass inspection of raw materials, process control, product testing, and quality audits. By implementing statistical process control, failure mode and effects analysis, and lean manufacturing principles, manufacturers can identify and address defects, reduce variability, optimize production processes, and improve overall quality.

1.4.2 Healthcare industry

In the healthcare industry, QC and QA are vital for patient safety and regulatory compliance. These techniques are applied in clinical laboratories for accurate diagnosis, monitoring medical equipment performance, sterilization processes, and infection control measures. Quality audits, adherence to good manufacturing practices, and risk management are crucial in pharmaceutical and medical device manufacturing. By implementing QC and QA in healthcare, organizations can minimize errors, enhance patient outcomes, and ensure compliance with industry standards.

1.4.3 Service industry

The service industry also benefits from QC and QA practices to deliver high-quality services and ensure customer satisfaction. In sectors such as hospitality, transportation, and banking, QC and QA techniques are employed to monitor service delivery processes, measure customer satisfaction, and identify areas for improvement. Service quality assessments, customer feedback mechanisms, and employee training programs contribute to the consistent delivery of excellent service experiences.

1.4.4 Construction industry

In the construction industry, QC and QA are essential to ensure the safety, durability, and compliance of structures. These techniques involve quality inspections, materials testing, project documentation, and adherence to building codes and regulations. By implementing QC and QA, construction companies can detect and rectify construction defects, maintain quality standards, and deliver projects that meet client requirements.

1.4.5 Food industry

The food industry relies on QC and QA to guarantee food safety, traceability, and product quality. These techniques include rigorous quality inspections, hazard analysis critical control points (HACCP), product testing, and compliance with food safety regulations. By implementing QC and QA practices, food manufacturers can prevent contamination, ensure accurate labelling, maintain product consistency, and protect consumer health.

1.5 Summary of Section 1

This section shows how Quality control and quality assurance are essential for ensuring product/service quality, meeting customer requirements, and achieving organizational objectives. Quality control focuses on monitoring and maintaining quality, while quality assurance ensures products/services meet specified requirements. This section shows how QC and QA can prevent defects, detect issues early, support continuous improvement, and comply with standards. It also shows how these practices are applied in various industries, including manufacturing, healthcare, service, construction, and food, to achieve consistent quality, enhance efficiency, and comply with regulations.

2. Techniques for quality control and quality assurance

Quality Control (QC) and Quality Assurance (QA) encompass a broad range of techniques and practices aimed at ensuring the desired level of quality in products or services. These techniques serve as a systematic approach to monitor, assess, and improve quality throughout various stages of the production or service delivery process. While the specific tools and methods may vary depending on the industry and context, the overarching goal remains the same: to meet customer requirements, enhance customer satisfaction, and achieve organizational objectives.

2.1 Overview of quality control and quality assurance techniques

Quality Control focuses on the inspection, testing, and evaluation of products, processes, or components to identify any deviations or non-conformities from established quality standards. It involves systematic monitoring, measurement, and analysis to ensure that quality requirements are met. The emphasis is placed on detecting and correcting defects, minimizing variations, and maintaining consistent quality throughout the production or service delivery process [6, 15].

Quality Assurance, on the other hand, is a proactive approach that aims to prevent quality issues from occurring in the first place. It involves the establishment of processes, procedures, and guidelines to ensure that products or services meet the specified requirements. Quality Assurance encompasses activities such as quality planning, process documentation, training, and adherence to quality standards and regulations. By implementing robust quality systems and processes, organizations can achieve consistent and reliable quality outcomes [16, 17].

Within the realm of QC and QA, various techniques and practices are employed to achieve the desired level of quality. These techniques can include inspection, testing, statistical analysis, documentation and record-keeping, process control, data analysis, risk assessment, and continuous improvement. The selection and application of these techniques depend on the nature of the product or service, the industry, and the specific quality objectives of the organization.

The overarching purpose of these techniques is to identify, monitor, and improve quality throughout the entire lifecycle of a product or service. They enable organizations to identify and rectify quality issues at an early stage, thereby reducing waste, rework, and customer dissatisfaction. By incorporating these techniques into their quality management systems, organizations can enhance the consistency, reliability, and performance of their products or services [6, 15–17].

Moreover, QC and QA techniques contribute to the overall efficiency and effectiveness of an organization by promoting adherence to quality standards, regulatory requirements, and industry best practices. They facilitate risk mitigation, process optimization, and cost reduction. Additionally, these techniques play a crucial role in building trust and credibility with customers, stakeholders, and regulatory bodies [29].

It is important to note that the selection and application of QC and QA techniques should be tailored to the specific needs and context of each organization. The use of appropriate techniques, combined with a commitment to continuous improvement, allows organizations to deliver products or services that meet or exceed customer expectations, maintain a competitive edge in the market, and achieve long-term success [29].

2.2 Statistical process control (SPC)

Statistical Process Control (SPC) is a powerful technique used in Quality Control (QC) and Quality Assurance (QA) to monitor and control processes based on statistical analysis. It involves collecting and analyzing data to understand process performance, detect variations, and make informed decisions regarding quality control and improvement. SPC enables organizations to achieve and maintain consistent and predictable process outcomes. SPC focuses on assessing process stability, which refers to the consistency and predictability of process outputs over time. It involves monitoring and analyzing data to identify patterns, trends, or shifts that indicate potential variations or changes in the process [6, 15–17, 29].

Control charts are a fundamental tool in SPC. They provide a visual representation of process data over time, enabling the identification of common cause and special cause variations. Control charts have three main components: a centreline representing the process mean, upper and lower control limits (UCL and LCL) indicating acceptable variation bounds, and individual data points or subgroup averages plotted on the chart [6, 15, 29].

Moreover, SPC distinguishes between common cause and special cause variations. Common cause variations are inherent to the process and result from random factors, while special cause variations arise from identifiable and non-random factors, indicating a change or shift in the process. Special cause variations require investigation and corrective action [16, 17].

Furthermore, process capability is another important aspect of SPC. It assesses the ability of a process to consistently produce output within specified limits. Capability indices are used to quantify process performance relative to tolerance limits. Process capability analysis helps identify the need for process improvement and ensures that processes meet customer requirements [6, 15, 29].

Data analysis plays a crucial role in SPC. Statistical techniques are utilized to analyze process data, including measures such as mean, range, standard deviation, and control limits. Hypothesis testing, analysis of variance (ANOVA), and regression analysis may also be employed to further understand process behavior and identify improvement opportunities [6, 15, 29].

SPC is closely linked to the concept of continuous improvement. By monitoring process performance, identifying variations, and taking appropriate actions, organizations can continuously enhance processes and drive improvement efforts. SPC supports the PDCA (Plan-Do-Check-Act) cycle by providing data-driven insights for informed decision making and facilitating the implementation of corrective and preventive actions [6, 15–17, 29].

2.3 Six sigma methodology

The Six Sigma methodology, employed in Quality Control (QC) and Quality Assurance (QA), is a data-driven approach aimed at improving process performance, reducing defects, and enhancing overall quality. It emphasizes the systematic identification and elimination of variation and waste to achieve exceptional levels of quality and customer satisfaction. The methodology follows the Define, Measure, Analyze, Improve, Control (DMAIC) framework, which involves clearly defining the problem, project goals, and customer requirements; collecting and analyzing data to assess the process and identify key factors impacting performance; and developing and implementing solutions to address root causes [15, 16, 18, 30, 31].

In the Six Sigma methodology, key roles play crucial parts in driving quality improvement. The Champion is responsible for aligning Six Sigma efforts with organizational goals and providing necessary resources. The Black Belt, a trained expert, leads Six Sigma projects and applies statistical tools and techniques to drive improvement. Working under the guidance of a Black Belt, the Green Belt is a team member involved in Six Sigma projects. The Master Black Belt, an experienced practitioner, provides coaching and mentoring to Black Belts and Green Belts. These defined roles ensure effective project management and implementation of the Six Sigma methodology [18, 30, 31].

Performance metrics are essential in Six Sigma to quantify process capability and improvement. Metrics such as Defects Per Million Opportunities (DPMO), Yield, Rolled Throughput Yield (RTY), and Process Capability Indices are used to track progress, set improvement targets, and ensure that the process is meeting customer expectations. By utilizing these metrics, organizations can measure the effectiveness of their quality initiatives and make data-driven decisions to drive continuous improvement [15, 16, 18].

The Six Sigma methodology not only focuses on process improvement but also aims to foster a culture of continuous improvement, collaboration, and data-driven decision making throughout the organization. By promoting a culture of quality and encouraging employee engagement, Six Sigma creates a conducive environment for sustained improvements and organizational success. Its successful implementation has resulted in significant enhancements in process efficiency, defect reduction, and customer satisfaction across various industries [15, 16, 18, 30, 31].

2.4 Lean manufacturing/management

Lean Manufacturing, also known as Lean Management or Lean, is a systematic approach to process improvement that focuses on eliminating waste, increasing efficiency, and improving overall quality in manufacturing and service industries. Derived from the Toyota Production System, Lean aims to create value for the customer while minimizing non-value-added activities and resources. Key aspects of Lean include waste reduction, value stream mapping, just-in-time production, continuous flow, kaizen (continuous improvement), standardized work, and visual management. Lean has proven successful in various industries, driving improvements in productivity, lead time reduction, and customer satisfaction [32–37].

Waste reduction is a central theme in Lean, targeting seven types of waste: transportation, inventory, motion, waiting, overproduction, over-processing, and defects. Value Stream Mapping is used to visualize and analyze the flow of materials, information, and activities, identifying areas for improvement. Just-in-Time production ensures products or services are delivered exactly when needed, minimizing inventory and lead times. Continuous flow eliminates interruptions and delays, improving productivity and responsiveness. Kaizen fosters a culture of continuous improvement, while standardized work ensures consistency and efficiency. Visual management techniques enhance communication, transparency, and performance monitoring [35, 36].

2.5 Failure mode and effects analysis (FMEA)

Failure Mode and Effects Analysis (FMEA) is a proactive and systematic technique utilized in Quality Control (QC) and Quality Assurance (QA) to identify and address potential failures or risks in processes, products, or systems. FMEA allows organizations to analyze failure modes, evaluate their effects, and assess their likelihood of occurrence, enabling the prioritized implementation of preventive measures. Therefore, organizations may enhance their quality control and assurance practices. FMEA provides a structured framework for risk management and supports continual improvement efforts. By embracing FMEA, organizations can optimize their processes, minimize failures, and deliver high-quality products and services to customers. The FMEA process begins by identifying all potential failure modes within a process, product, or system. This involves brainstorming, risk assessments, historical data review, and expert consultations. Once failure modes are identified, their potential effects are assessed in terms of severity, considering factors such as safety, quality, customer satisfaction, compliance, and other critical aspects [21, 23, 38–41].

The likelihood of occurrence for each failure mode is evaluated, considering historical data, experience, and expert judgment. Occurrence is rated on a scale to estimate the probability of a failure mode occurring. The effectiveness of existing controls or detection methods in identifying or preventing failure modes is also assessed. Detectability is rated based on the likelihood of detecting failure modes before they cause harm or reach the customer [21, 23, 38–41].

The Risk Priority Number (RPN) is calculated by multiplying the severity, occurrence, and detectability ratings. RPN serves as a prioritization tool, allowing organizations to rank failure modes based on their overall risk levels. High RPN values indicate failure modes that require immediate attention and focused mitigation efforts [21, 23].

Based on the FMEA analysis and RPN values, organizations develop and implement preventive actions to minimize or eliminate identified failure modes. These actions may include process improvements, design changes, enhanced inspections, training programs, or quality control measures. FMEA promotes a proactive approach to risk mitigation, enabling organizations to prevent failures before they occur [21, 23].

FMEA is an iterative process that supports continual improvement efforts. Organizations regularly review and update their FMEA to reflect changes in processes, products, or systems. This ensures that potential failure modes are consistently monitored and addressed. By embracing continual improvement, organizations enhance their quality control and assurance practices, adapt to evolving circumstances, and ensure ongoing risk management [21, 23, 38–41].

2.6 Quality audits and inspections

Quality audits and inspections are essential techniques used in Quality Control (QC) and Quality Assurance (QA) to assess and verify the compliance of processes, products, or systems with established quality standards and requirements. They provide a systematic and objective evaluation of quality practices and help identify areas for improvement. These processes play a crucial role in evaluating and ensuring compliance with quality standards, specifications, regulations, and customer requirements. They serve to identify non-conformities, verify the effectiveness of quality management systems, assess process capabilities, and promote continuous improvement [15, 16, 20, 22, 23, 31].

Internal and external audits are two types of quality assessments. Internal audits are conducted by trained personnel within the organization, focusing on evaluating the organization's own processes. External audits involve independent third-party auditors and are often required for compliance with regulatory or certification standards. Both types of audits provide valuable insights into the effectiveness of quality management systems [22, 23].

Process audits are conducted to assess adherence to defined processes and procedures throughout the organization. These audits examine process controls, documentation, training, and compliance with industry best practices. They help identify areas of inefficiency, non-compliance, or gaps in the quality management system, allowing organizations to streamline operations and ensure consistent quality outputs. Moreover, product audits and inspections involve the examination of finished products, components, or materials to ensure they meet specified quality requirements. Through visual examination, measurements, functional testing, or destructive testing, product audits and inspections help identify defects, deviations, or non-conformities that may impact product quality. Organizations can take corrective actions based on the findings to prevent the release of substandard products to customers [22, 23].

Compliance audits focus on assessing adherence to applicable regulations, standards, and legal requirements. They ensure that organizations operate within the required guidelines and regulations specific to their industry. Compliance audits help identify areas of non-compliance and mitigate associated risks, ensuring that organizations fulfill their legal and ethical responsibilities and maintain customer trust. Alternatively, Supplier audits are conducted to assess the quality management systems and capabilities of suppliers. These audits evaluate supplier performance, quality control processes, and their ability to meet quality requirements. Supplier audits help ensure that the products or services provided by suppliers meet the organization's quality standards. By monitoring and auditing suppliers, organizations can maintain consistent product quality, minimize supply chain risks, and foster long-term partnerships [22, 23].

Findings from quality audits and inspections often lead to the identification of non-conformities or areas for improvement. Corrective actions are then implemented to address these issues and prevent their recurrence. Such actions may include process modifications, training programs, quality system enhancements, or supplier development activities. By taking prompt and effective corrective actions, organizations can continually improve their quality performance and prevent quality-related problems.

Overall, quality audits and inspections contribute to the culture of continuous improvement within organizations. They provide valuable feedback and insights that can be used to refine processes, improve product quality, and optimize the effectiveness of the quality management system. By embracing a culture of continuous improvement, organizations can achieve higher levels of quality, customer satisfaction, and operational excellence.

2.7 Summary of Section 2

This section discusses the techniques for Quality Control (QC) and Quality Assurance (QA) encompass a wide range of practices aimed at ensuring product/service quality. QC focuses on monitoring and maintaining quality by detecting and correcting defects, while QA ensures that products/services meet specified requirements. It highlights the techniques that include Statistical Process Control (SPC) for monitoring and analyzing process performance, Six Sigma for data-driven process improvement, Lean Manufacturing/Management for waste reduction and efficiency, Failure Mode and Effects Analysis (FMEA) for proactive risk management, and Quality Audits and Inspections to assess compliance with quality standards and identify areas for improvement. It shows how these techniques enable organizations to achieve consistent quality, reduce waste, enhance customer satisfaction, and drive continuous improvement.

3. Computer-aided quality control and quality assurance

Computer-Aided Quality Control and Quality Assurance involve the application of computer-based systems, software, and tools to support the planning, execution, monitoring, and evaluation of quality-related processes. These technologies help streamline data collection, automate analysis, improve accuracy, facilitate collaboration, and enable real-time quality control and assurance activities.

3.1 Overview of computer-aided quality control and quality assurance

Computer-Aided Quality Control and Quality Assurance encompass a range of technologies and tools that integrate with existing quality management systems. These tools may include statistical process control (SPC) software, quality management software, quality information systems, computer-aided design (CAD) tools for quality function deployment (QFD), and more. The use of these tools enhances the speed, accuracy, and efficiency of quality-related processes, enabling organizations to proactively identify and address quality issues, improve decision-making, and achieve higher levels of quality throughout their operations.

Computer-Aided Quality Control and Quality Assurance play a crucial role in modern quality management practices. They provide organizations with the means to automate routine quality tasks, capture and analyze large volumes of quality data, visualize quality metrics, monitor process performance in real-time, and identify patterns or trends that may indicate quality issues. By leveraging these technologies, organizations can make data-driven decisions, implement corrective actions promptly, and continuously improve their quality management processes [15, 16, 20, 22, 31, 42–45].

Computer-Aided Quality Control (CAQC) and Quality Assurance (CAQA) technologies have become essential for organizations seeking to enhance the efficiency and effectiveness of their quality management processes. These technologies provide advanced tools and capabilities that enable organizations to monitor, analyze, and improve their quality control and assurance activities. As defined in [42–45], capabilities can be:

- Automation of quality processes: CAQC systems provide data collection, analysis, reporting, and decision-making. By automating these tasks, organizations can reduce manual effort, minimize human errors, and ensure consistent execution of quality control and assurance activities.

- Real-time monitoring and control: CAQC systems enable real-time monitoring and control of quality parameters and processes. Through these systems, organizations can capture data in real-time, monitor process performance, and promptly identify deviations or non-conformities. This enables timely corrective actions, preventing quality issues from escalating and ultimately improving the overall product or service quality.

- Data analysis and insights: CAQC tools offer advanced data analysis capabilities, empowering organizations to extract meaningful insights from large datasets. These tools can perform statistical analysis, generate quality reports, visualize trends, and identify patterns or anomalies. Such analysis aids in proactive decision-making and supports continuous improvement efforts.

3.2 Applications of computer-aided quality control and quality assurance

Computer-aided quality control (CAQC) and quality assurance (CAQA) technologies are widely used in manufacturing, healthcare, service, construction, and food industries. These tools ensure product quality, process efficiency, and compliance with standards. They enable real-time monitoring, statistical analysis, supply chain management, regulatory compliance, process optimization, and customer satisfaction. CAQC and CAQA contribute to improved outcomes, safety, operational efficiency, and innovation in diverse sectors.

3.2.1 Manufacturing industry

Computer-aided quality control (CAQC) and quality assurance (CAQA) are extensively applied in the manufacturing industry to ensure product quality, process efficiency, and compliance with standards. These technologies enable real-time monitoring and control of manufacturing processes, statistical analysis for process improvement, integration of quality processes throughout the product lifecycle, implementation of quality practices across the supply chain, compliance with industry standards, and data analysis for continuous improvement efforts. By utilizing these tools, manufacturers can maintain consistent product quality, optimize processes, and drive continuous improvement [15, 16, 22, 31, 46].

3.2.2 Healthcare industry

Computer-aided quality control (CAQC) and quality assurance (CAQA) technologies play a significant role in the healthcare industry, improving patient care, safety, and overall outcomes. These tools enable the implementation of electronic health records and clinical decision support systems, supporting evidence-based decisions and enhancing diagnosis accuracy. They also aid in patient safety and risk management by facilitating incident reporting and root cause analysis. Healthcare organizations can monitor quality metrics, comply with regulatory standards, optimize processes, and improve operational efficiency using computer-aided systems.

Additionally, these technologies support telehealth and remote monitoring solutions, enhancing access to care and patient engagement [22, 47–50].

3.2.3 Service industry

In the service industry, computer-aided quality control (CAQC) and quality assurance (CAQA) technologies play a crucial role in enhancing service quality, customer satisfaction, and operational efficiency. These tools enable service organizations to monitor and measure service quality, track customer interactions, and personalize service delivery through customer relationship management (CRM) systems. Process automation and streamlining tools improve operational efficiency, while performance monitoring systems help manage service level agreements (SLAs) and track key performance indicators (KPIs). Customer feedback analysis tools provide insights for improvement, and self-service and digital channels enhance accessibility and convenience for customers. Overall, computer-aided technologies in the service industry aim to optimize service quality and deliver an exceptional customer experience [22, 51–55].

3.2.4 Construction industry

In the construction industry, computer-aided quality control (CAQC) and quality assurance (CAQA) technologies have significant applications that contribute to improving construction processes, ensuring compliance with standards, enhancing project management, and achieving high-quality construction outcomes. Building Information Modeling (BIM) tools enable collaboration, clash detection, and optimization of construction processes. Design and engineering analysis tools aid in creating detailed models, performing structural analysis, and ensuring compliance. Quality inspection and testing systems facilitate efficient tracking, verification, and corrective actions. Project management and collaboration tools streamline processes and enhance coordination. Safety and risk management tools identify and mitigate potential risks, improve worker safety, and enhance project quality. Prefabrication and modular construction technologies ensure precision and quality consistency. Overall, computer-aided technologies in the construction industry aim to optimize construction processes, ensure compliance, and deliver high-quality construction projects [22, 56–60].

3.2.5 Food industry

In the food industry, computer-aided quality control (CAQC) and quality assurance (CAQA) technologies have significant applications that contribute to ensuring food safety, enhancing product quality, improving production processes, and meeting regulatory requirements. These technologies support the implementation of Hazard Analysis and Critical Control Points (HACCP) systems, facilitate quality inspection and testing processes, enable supply chain traceability, assist in recipe and formulation management, ensure compliance with food regulations, and support product development and research. By leveraging computer-aided tools, food companies can monitor and mitigate risks, maintain product quality, track and record information throughout the supply chain, optimize production processes, and meet regulatory standards, ultimately delivering safe and high-quality food products to consumers [22, 61–65].

3.3 Techniques and tools in computer-aided quality control and quality assurance

3.3.1 Statistical process control (SPC) software

Statistical Process Control (SPC) software is a valuable tool in computer-aided quality control and quality assurance (CAQC and CAQA) that enables organizations to monitor and analyze process data for effective quality management. SPC software provides a range of features and functionalities that support data collection, analysis, visualization, and decision-making. Here is an overview of SPC software and its applications [6, 22, 66–69]:

- Data collection and analysis: SPC software collects real-time process data from various sources, stores it centrally, and applies statistical techniques for analysis, providing insights into process performance.

- Control charting: SPC software offers diverse control charts to visualize process variation, detect deviations from quality standards, and automatically generate alerts.

- Process capability analysis: SPC software calculates indices to measure process capability, identify areas for improvement, and ensure compliance with customer specifications.

- Trend analysis and predictive analytics: SPC software visualizes process data over time, identifies trends, patterns, and potential quality impacts, and uses predictive analytics for proactive quality management.

- Collaboration and reporting: SPC software facilitates stakeholder collaboration, centralized data access, and generates comprehensive reports for effective communication and decision-making.

- Integration with other systems: SPC software seamlessly integrates with MES, ERP, or QMS, enabling efficient data exchange, process monitoring, and organization-wide quality control.

3.3.2 Quality management software

Quality Management Software (QMS) is a vital tool in computer-aided quality control and quality assurance (CAQC and CAQA) that helps organizations streamline and automate their quality management processes. QMS software provides a centralized platform for managing quality-related activities, documentation, and data. Here is an overview of QMS and its applications [66, 70–75]:

- Document control: QMS software centralizes and controls quality-related documents, managing creation, revision, approval, and distribution while ensuring compliance and maintaining an audit trail.

- Non-conformance management: QMS software provides a workflow for recording, investigating, and resolving non-conformances, tracking details, assigning responsibilities, and monitoring CAPAs to improve efficiency.

- Change management: QMS software facilitates structured change evaluation, approval, and implementation, ensuring proper documentation, review, and communication to minimize risks to quality.

- Audit management: QMS software plans, executes, and tracks audits, creating checklists, scheduling, assigning auditors, and capturing findings to streamline the process and ensure compliance.

- Training and competence management: QMS software tracks and manages employee training records, certifications, and competencies, identifying needs, scheduling sessions, and monitoring progress.

- Performance measurement and reporting: QMS software defines KPIs, tracks real-time quality performance, generates reports and dashboards for visualization, and supports continuous improvement efforts.

3.3.3 Quality information systems

Quality Information Systems (QIS) are essential tools in computer-aided quality control and quality assurance (CAQC and CAQA) that enable organizations to manage and analyze quality-related data. QIS software provides a centralized platform for collecting, storing, and analyzing data to support decision-making and continuous improvement. Here is an overview of QIS and its applications [70–75]:

- Data collection and integration: QIS software collects and integrates quality data from various sources, eliminating data silos and ensuring accuracy and completeness.

- Data analysis and visualization: QIS software applies statistical techniques, analyzes quality data, and provides interactive dashboards and reports for informed decision-making.

- Supplier management: QIS software evaluates and monitors supplier quality, tracking certifications, audits, and non-conformances to ensure high-quality materials.

- Corrective and preventive actions (CAPA): QIS software manages CAPAs, recording, tracking, and resolving quality issues to prevent recurrence.

- Risk management: QIS software identifies, assesses, and mitigates quality-related risks, prioritizing improvement efforts.

- Compliance and regulatory requirements: QIS software ensures compliance with quality standards and regulations, supporting documentation and audit processes.

3.3.4 Computer-aided quality function deployment (QFD)

Computer-Aided Quality Function Deployment (QFD) is a powerful technique used in computer-aided quality control and quality assurance (CAQC and CAQA) to ensure that customer requirements are translated effectively into product or service design characteristics. QFD helps organizations align their offerings with customer expectations, prioritize design features, and enhance the overall quality of their products or services. Here is a more detailed exploration of Computer-Aided QFD and its applications [66, 72–74, 76, 77]:

- Customer requirements analysis: Computer-Aided QFD software systematically analyzes customer requirements, capturing feedback and conducting market research for accurate representation in product or service design.

- Cross-functional collaboration: Computer-Aided QFD promotes effective collaboration among departments, ensuring a comprehensive understanding of customer requirements and facilitating the development of design characteristics.

- Relationship matrix development: QFD software simplifies the creation of relationship matrices, visually representing the connections between customer requirements and design characteristics.

- Design trade-off analysis: Computer-Aided QFD enables thorough design trade-off analysis, assessing the impact of design decisions on customer requirements and optimizing for desired quality outcomes.

- Supplier integration: QFD software facilitates collaboration with suppliers, ensuring alignment with design requirements and selecting high-quality components.

- Continuous improvement: Computer-Aided QFD supports continuous improvement through monitoring performance, gathering feedback, and refining offerings based on customer requirements and market trends.

3.4 Benefits and drawbacks of computer-aided quality control and quality assurance

Computer-aided Quality Control (QC) and Quality Assurance (QA) offer significant benefits, including increased accuracy, efficiency, real-time monitoring, improved decision-making, compliance adherence, and enhanced collaboration. However, challenges include initial investment costs, complexity, data quality, customization, technical dependencies, and change management (**Figure 3**).

3.4.1 Benefits of computer-aided QC and QA

Computer-Aided Quality Control (QC) and Quality Assurance (QA) systems offer significant advantages to organizations, driving improvements in overall quality, efficiency, and customer satisfaction. These systems provide increased accuracy and consistency by automating processes and utilizing advanced algorithms, minimizing human errors and delivering reliable results. Efficiency and productivity are enhanced through streamlined data collection, analysis, and reporting, allowing organizations to allocate resources more efficiently and focus on higher-value activities, leading to improved operational efficiency. Real-time monitoring capabilities enable prompt detection of quality deviations, reducing waste and rework and facilitating timely interventions. Moreover, computer-aided QC and QA systems

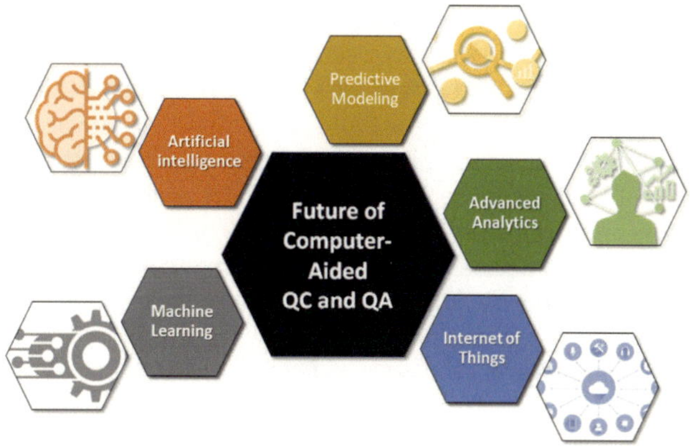

Figure 3.
Summary of benefits and challenges of computer-aided QC and QA [72].

enable advanced data analysis, supporting data-driven decision-making and extracting valuable insights from large datasets. They also assist organizations in meeting compliance requirements and foster collaboration and communication among stakeholders involved in quality control and assurance, ultimately driving a culture of quality throughout the organization [72, 78–81].

By implementing computer-aided QC and QA systems, organizations can unlock a range of benefits that positively impact their operations. These systems provide increased accuracy and consistency, enhancing the reliability of quality control and assurance activities. They also improve efficiency and productivity by automating tasks and allowing for more efficient resource allocation. Real-time monitoring capabilities enable prompt intervention to prevent defects and non-conformities, leading to more efficient processes. Advanced data analysis capabilities enable organizations to extract valuable insights and make informed decisions. Additionally, computer-aided QC and QA systems assist organizations in meeting compliance requirements and foster collaboration and communication among stakeholders, promoting a culture of quality. Embracing computer-aided QC and QA empowers organizations to enhance accuracy, efficiency, real-time monitoring, decision-making, compliance adherence, and collaboration, resulting in improved overall quality and increased customer satisfaction [72, 78–81].

3.4.2 Challenges and limitations of computer-aided QC and QA

Implementing computer-aided Quality Control (QC) and Quality Assurance (QA) systems provides significant advantages to organizations but comes with its share of challenges and limitations. One challenge is the initial investment and implementation costs, which can be a barrier for small and medium-sized enterprises (SMEs) with limited budgets. The complexity and learning curve associated with these systems also pose challenges, requiring specialized knowledge and training for effective utilization. Data quality and integration are crucial considerations, as inaccurate or incomplete data can lead to unreliable analysis. Striking a balance between customization and flexibility is important, ensuring the system aligns with organizational needs without compromising its adaptability. Technical dependencies and compatibility issues may arise when integrating with existing IT infrastructure, necessitating regular updates and maintenance [72, 78–81].

Another challenge lies in the cultural shift required for successful implementation. Resistance to change and lack of employee acceptance can hinder adoption, making change management strategies and fostering a culture of quality essential. Overcoming these challenges requires ongoing support and training for employees, as well as demonstrating the benefits of the system. By addressing these challenges, organizations can harness the benefits of computer-aided QC and QA systems, such as improved accuracy, efficiency, and decision-making. Awareness of these factors empowers organizations to make informed decisions and effectively overcome potential drawbacks, leading to successful implementation and enhanced overall quality [72, 78–81].

3.4.3 Overcoming the challenges of computer-aided QC and QA

Implementing computer-aided Quality Control (QC) and Quality Assurance (QA) systems may present challenges, but organizations can proactively address these obstacles to ensure successful implementation. Some of the strategies for overcoming the challenges associated with computer-aided QC and QA systems [72, 78–81]:

1. A comprehensive planning and assessment phase are crucial before implementing computer-aided QC and QA systems. Organizations should conduct a thorough assessment of their needs, existing processes, and available resources. This evaluation helps identify specific requirements, evaluate feasibility, and estimate costs accurately. By developing a comprehensive implementation plan, organizations can anticipate challenges and allocate resources effectively.

2. Selecting the right system and vendor is key to successful implementation. Organizations should carefully evaluate different software options, considering factors such as functionality, scalability, ease of use, compatibility, and vendor support. Engaging with vendors through demonstrations and pilot projects provides valuable insights into system capabilities and compatibility with existing infrastructure.

3. Providing adequate training and support is essential to overcome the learning curve associated with computer-aided QC and QA systems. Organizations should invest in comprehensive training programs that cover software usage, data interpretation, and statistical analysis techniques. Ongoing support and assistance should be available to address any questions or difficulties that employees may encounter during system implementation and usage.

4. Ensuring data quality and integration is critical for successful implementation. Organizations should establish data management protocols, including defining data standards, implementing data validation processes, and ensuring compatibility between different data sources and systems. Regular data audits and maintenance activities are necessary to ensure data accuracy and integrity.

5. Balancing customization and standardization are crucial when implementing computer-aided QC and QA systems. While customization allows alignment with specific organizational needs, excessive customization can introduce complexity and hinder system flexibility. Standardizing processes and workflows where feasible helps streamline implementation and ensures compatibility with future updates and upgrades.

6. Addressing organizational culture and change management is vital for successful implementation. Leaders should communicate the benefits of computer-aided QC and QA systems to employees and involve them in the decision-making process. Engaging employees through training, workshops, and feedback sessions fosters a sense of ownership and helps overcome resistance to change. Creating a supportive environment that encourages continuous learning and improvement is essential.

By implementing these strategies, organizations can overcome the challenges associated with computer-aided QC and QA systems and ensure successful implementation. Proactive planning, selecting the right system and vendor, providing adequate training and support, ensuring data quality and integration, balancing customization and standardization, and addressing change management and employee engagement contribute to a smooth implementation process and maximize the benefits of computer-aided QC and QA systems in quality management.

3.5 Implementation considerations for computer-aided quality control and quality assurance

This subsection highlights key considerations for selecting and integrating software and tools in quality management as referred by [46, 72, 78–85].

3.5.1 Selection and integration of software and tools

The selection and integration of software and tools are critical factors in the successful implementation of computer-aided Quality Control (QC) and Quality Assurance (QA) systems. To begin, organizations should conduct a comprehensive needs assessment and gather specific requirements before selecting any software or tools. This involves identifying key functionalities, considering unique processes and industry requirements, and involving relevant stakeholders in the decision-making process. By understanding their specific needs, organizations can narrow down the options and select software that aligns with their requirements.

Vendor evaluation and due diligence are vital steps in the selection process. Organizations should conduct research, request product demonstrations, and obtain references or customer testimonials. Assessing the vendor's reputation, experience, customer support capabilities, and future development plans is crucial. Engaging with the vendor through discussions and evaluations helps determine the suitability of the software for the organization's needs.

Compatibility and integration with existing systems and infrastructure are critical considerations. Organizations should evaluate the compatibility of the selected software and tools with current hardware, operating systems, databases, and other software applications. Assessing the ease of integration and the availability of APIs or data exchange capabilities ensures seamless data flow between different systems. However, scalability and future requirements should also be considered. Organizations need software that can accommodate their growth and evolving needs over time. The software should be flexible enough to adapt to changes in processes, handle increased data volumes, support additional users, and provide advanced analytics or reporting capabilities.

User-friendliness and training requirements are essential for effective adoption and user engagement. Organizations should evaluate the ease of use, intuitiveness of the user interface, and the learning curve associated with the software. Assessing the training and support provided by the vendor ensures that employees can quickly learn and utilize the software effectively.

Moreover, performing a cost analysis and evaluating the return on investment (ROI) associated with the software and tools is crucial. This includes considering upfront costs, licensing fees, maintenance fees, training costs, and any additional expenses. It is important to assess the potential benefits in terms of improved productivity, enhanced quality, reduced defects, and cost savings.

By considering these factors and making informed decisions during the selection and integration process, organizations can successfully implement computer-aided QC and QA systems. Thorough needs assessment, vendor evaluation, compatibility and integration assessment, scalability considerations, user-friendliness evaluation, cost analysis, and ROI assessment contribute to the selection of software and tools that best meet the organization's needs and drive effective quality management.

3.5.2 Training and skill development

Raining and skill development are vital for the successful implementation of computer-aided Quality Control (QC) and Quality Assurance (QA) systems. Organizations should design comprehensive training programs that cover system functionalities, data entry, report generation, data interpretation, and analytical capabilities. These programs can be delivered through workshops, online modules, tutorials, and hands-on practice sessions to ensure employees have a deep understanding of the software. Customized training materials, such as user manuals and reference guides, should be developed to provide step-by-step instructions and troubleshooting tips tailored to the specific software and tools being implemented. Additionally, adopting a train-the-trainer approach allows key individuals to receive extensive training and become internal trainers who can conduct sessions for other employees. This approach promotes consistent training and builds internal expertise, ensuring a shared understanding of the software's capabilities.

To foster continuous learning and skill development, organizations should provide ongoing training opportunities beyond the initial implementation phase. This can include refresher courses, advanced training sessions, webinars, and participation in industry conferences or workshops. By encouraging continuous learning, employees can enhance their expertise and stay up to date with the latest advancements in QC and QA practices. Cross-functional training is also essential to promote collaboration and a shared understanding across different departments and stakeholders involved in the implementation process. Training sessions that involve representatives from various departments facilitate knowledge exchange and foster a collaborative environment. It is crucial for organizations to establish mechanisms to monitor and evaluate the effectiveness of training programs, such as assessments, feedback surveys, and performance evaluations. Regular feedback from users helps identify areas for improvement and ensures that training programs meet the evolving needs of employees. By prioritizing training and skill development, organizations can maximize the adoption and utilization of computer-aided QC and QA systems, building employee expertise and driving successful implementation.

3.5.3 Data security and privacy

Protecting personal and sensitive information is a top priority in quality management, and organizations can implement effective measures to ensure its safeguarding.

One crucial step is to raise employee awareness through comprehensive training on data security and privacy best practices. This involves educating employees about the risks and consequences of data breaches, emphasizing the importance of protecting personal information, and fostering a culture of data security. Training sessions should cover topics such as password hygiene, recognizing phishing attempts, and reporting security incidents. Regular awareness campaigns further reinforce data security practices among employees, instilling a sense of responsibility and vigilance.

At the technical level, organizations should establish robust access controls and encryption techniques to limit unauthorized access to personal and sensitive information. Implementing strong authentication mechanisms and regularly reviewing user access privileges ensure appropriate levels of authorization. Conducting risk assessments and vulnerability scans helps identify and address security gaps proactively. Data backup and disaster recovery plans ensure the availability and integrity of information, while clear incident response protocols facilitate prompt and effective handling of data breaches.

By implementing these measures, organizations can significantly enhance the protection of personal and sensitive information in quality management. Employee awareness and training, technical security measures, risk assessments, data backup and recovery plans, and incident response protocols collectively contribute to a robust data security framework that safeguards information from unauthorized access and potential breaches.

3.6 Future trends in computer-aided quality control and quality assurance

Computer-aided Quality Control (QC) and Quality Assurance (QA) systems are being transformed by technologies such as Artificial Intelligence (AI), Machine Learning (ML), Internet of Things (IoT) integration, advanced analytics, and predictive modeling (**Figure 4**).

3.6.1 Artificial intelligence and machine learning

Artificial Intelligence (AI) and Machine Learning (ML) have emerged as transformative technologies in computer-aided Quality Control (QC) and Quality Assurance

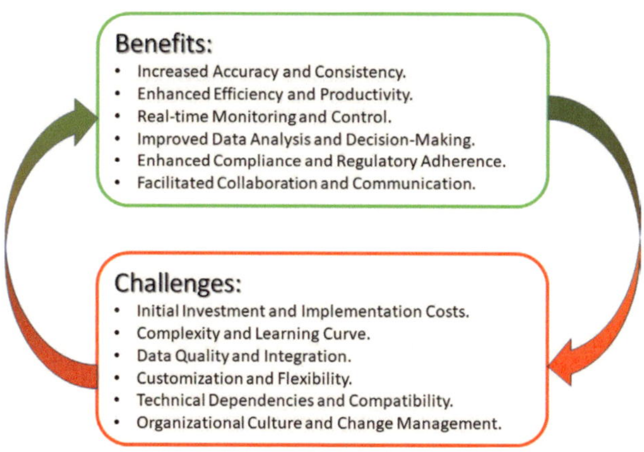

Figure 4.
Future trends in computer-aided QC and QA.

(QA) systems. These technologies excel in data analysis and pattern recognition, enabling organizations to gain valuable insights from large volumes of QC and QA data. By leveraging AI and ML algorithms, organizations can uncover hidden patterns and make informed, data-driven decisions to enhance their quality processes. Predictive analytics is another significant application, as it allows organizations to forecast potential quality issues and take proactive measures to prevent them. Automation is also revolutionized through AI and ML, enabling automatic defect detection and real-time monitoring of quality parameters, leading to increased accuracy and efficiency in quality control activities [86–90].

Furthermore, AI techniques like Natural Language Processing (NLP) enable the analysis of unstructured data sources such as customer feedback and product reviews. By extracting meaningful insights from these sources, organizations can understand customer sentiment, identify quality issues, and implement improvements based on customer feedback. Decision support systems powered by AI and ML algorithms provide real-time recommendations to QC and QA professionals, assisting them in identifying improvement opportunities, optimizing process parameters, and implementing corrective actions. The integration of diverse data sets from QC systems, manufacturing execution systems, supply chain data, and customer feedback further enhances the understanding of quality performance, enabling effective problem-solving and process optimization [86–90].

Looking ahead, the future of QC and QA holds tremendous potential with advancements in AI and ML. Real-time quality monitoring, predictive maintenance to prevent equipment failures, and adaptive and self-learning quality systems are among the anticipated developments. Integration with emerging technologies such as the Internet of Things (IoT) and Big Data analytics will further amplify the capabilities and impact of QC and QA systems. As AI and ML continue to evolve, QC and QA processes will become increasingly efficient, data-driven, and adaptable, paving the way for improved product quality and customer satisfaction [86–90].

3.6.2 Internet of things (IoT) integration

The integration of Internet of Things (IoT) technology with computer-aided Quality Control (QC) and Quality Assurance (QA) systems opens up new possibilities for improving quality management processes. This section explores the key aspects and future trends of IoT integration in QC and QA. IoT integration offers real-time monitoring and control by deploying sensors and actuators throughout the production process. Organizations can capture real-time data on quality parameters, enabling continuous monitoring of critical variables and immediate corrective actions. It also enables remote quality assurance, allowing centralized monitoring and assessment across multiple locations or supply chains. This facilitates benchmarking, performance tracking, and the implementation of standardized quality control procedures [91–95].

Another significant benefit of IoT integration is predictive maintenance. IoT sensors monitor equipment health and provide data for analysis, enabling organizations to predict failures, schedule maintenance, and prevent unexpected downtime. IoT devices also enhance traceability and quality control by capturing data on product origin, manufacturing conditions, storage parameters, and transportation conditions. This end-to-end visibility ensures compliance with quality standards and streamlines the recall process if needed. Moreover, the wealth of data generated by IoT integration can be leveraged for data-driven decision-making. Advanced analytics tools help identify trends, patterns, and correlations, empowering organizations to enhance quality control processes and optimize resource allocation [91–95].

Furthermore, IoT integration fosters collaboration and communication among stakeholders involved in quality management. Real-time data sharing, remote access to quality information, and collaborative platforms facilitate effective communication and coordination among quality professionals, production teams, suppliers, and customers. This streamlined collaboration enables rapid problem-solving, quick responses to quality issues, and continuous improvement efforts. Overall, the integration of IoT with QC and QA systems presents new opportunities to enhance quality management processes and drive improved quality performance [91–95].

3.6.3 Advanced analytics and predictive modeling

Advanced analytics and predictive modeling techniques have revolutionized computer-aided Quality Control (QC) and Quality Assurance (QA) systems, and this section delves into their key aspects and future trends. These techniques enable organizations to extract valuable insights from complex datasets, uncovering hidden patterns and correlations that significantly impact quality performance. By analyzing historical quality data, organizations gain a deeper understanding of quality-related factors, facilitating more informed decision-making processes. Predictive modeling plays a crucial role by leveraging historical data to develop statistical models that forecast future quality outcomes. This empowers organizations to proactively detect problems, allocate resources effectively, and optimize quality control processes. Furthermore, advanced analytics techniques facilitate root cause analysis, helping organizations identify the underlying causes of quality issues and implement targeted improvement efforts to prevent similar problems from recurring. Process optimization is also a significant benefit, as it allows organizations to identify key process parameters that influence product quality, leading to enhanced process stability, reduced variability, and improved overall quality performance [96–100].

Real-time quality monitoring is made possible through advanced analytics techniques, enabling organizations to continuously monitor quality parameters in real-time. By analyzing data streams from sensors, production lines, and quality control devices, organizations can promptly detect quality deviations, trigger automated alerts or corrective actions, and ensure consistent product quality throughout the production process. This real-time monitoring capability facilitates rapid response and minimizes the potential impact of quality issues. Additionally, prescriptive analytics takes predictive modeling to the next level by providing actionable recommendations for optimal actions to improve quality outcomes. Considering various constraints, goals, and performance indicators, prescriptive models offer valuable insights and guidance to quality professionals, helping them make informed decisions and achieve their quality objectives. Advanced analytics and predictive modeling have transformed the field of QC and QA, enabling organizations to harness data-driven insights, enhance process stability, and ultimately deliver higher quality products and services [96–100].

3.7 Summary of Section 3

The section provides an overview of computer-aided Quality Control (CAQC) and Quality Assurance (CAQA) technologies and their applications in various industries, including healthcare, manufacturing, service, construction, and food. It emphasizes the benefits of these technologies, including real-time monitoring, improved decision-making, and collaboration, while also highlighting the challenges in implementing them, such as initial costs, technical dependencies, customization, and change management. The document discusses popular techniques and tools in CAQC and CAQA and provides recommendations for selecting and integrating software and tools. It also discusses the future trends in computer-aided Quality Control and Quality Assurance.

4. Chapter conclusion

This chapter has illuminated the pivotal role of Quality Control (QC) and Quality Assurance (QA) in modern-day quality management. It began by emphasizing their significance in upholding product and service quality, aligning with customer needs, and fulfilling organizational goals. Quality Control, as the guardian of quality maintenance, works in tandem with Quality Assurance, ensuring strict adherence to predefined requirements.

In Section 1, we explored the fundamental importance of QC and QA, showcasing their power in defect prevention, early issue identification, and the journey of continuous improvement. Through real-world applications spanning diverse industries, we witnessed the tangible results of these practices in achieving consistent quality, operational efficiency, and regulatory compliance.

Section 2 delved deeper, unveiling a range of techniques underpinning Quality Control and Quality Assurance. Statistical Process Control (SPC), Six Sigma, Lean Manufacturing/Management, Failure Mode and Effects Analysis (FMEA), and Quality Audits and Inspections, each with its unique approach, converged on a common mission: ensuring quality consistency, waste reduction, elevated customer satisfaction, and ongoing enhancement.

In Section 3, we ventured into the domain of Computer-Aided Quality Control (CAQC) and Quality Assurance (CAQA) technologies. Their transformative influence, from real-time monitoring to informed decision-making and collaboration, was evident. Yet, we acknowledged the implementation challenges encompassing initial costs, technical complexities, customization needs, and change management. We concluded by offering practical advice for software and tool selection and glimpsed into future trends, reflecting the ever-evolving landscape of quality management.

In conclusion, the principles of Quality Control and Quality Assurance, enriched by technological advancements, serve as catalysts for elevating product and service quality, exceeding customer expectations, and attaining organizational triumph. This chapter serves as a guide, illuminating the path to quality excellence and heralding the transformative potential of QC, QA, and technology in shaping the future of business and industry.

Quality Tools, Technologies, and Techniques: Enhancing Product and Service Excellence
DOI: http://dx.doi.org/10.5772/intechopen.113994

Author details

Rawhi Alrae
United Arab Emirates University, United Arab Emirates

*Address all correspondence to: alrae.rawhi@gmail.com

References

[1] Juran JM, Gryna FM. Quality Planning and Analysis: From Product Development through Use. New York, USA: McGraw-Hill; 1993

[2] Oakland JS. Total Quality Management: Text with Cases. Oxford, United Kingdom: Butterworth-Heinemann; 2003

[3] Pyzdek T, Keller PA. The Six Sigma Handbook: The Complete Guide for Greenbelts, Blackbelts, and Managers at all Levels. New York, USA: McGraw-Hill Education; 2014

[4] Womack JP, Jones DT, Roos D. The Machine That Changed the World: The Story of Lean Production. New York, USA: Harper Perennial; 1991

[5] International Organization for Standardization (ISO). ISO 9001:2015–Quality Management Systems–Requirements. Geneva, Switzerland. 2015

[6] Montgomery DC. Introduction to Statistical Quality Control. Hoboken, New Jersey, USA: John Wiley & Sons; 2017

[7] Feigenbaum AV. Total Quality Control: Engineering and Management. New York, USA: McGraw-Hill; 1991

[8] Kanji G. Total Quality Management: The Second Generation. London, United Kingdom: Routledge; 1996

[9] Goetsch DL, Davis S. Quality Management for Organizational Excellence: Introduction to Total Quality. Upper Saddle River, New Jersey, USA: Pearson; 2014

[10] Kaplan RS, Norton DP. The Balanced Scorecard: Translating Strategy into Action. Boston, Massachusetts, USA: Harvard Business Press; 1996

[11] Westcott RE, Westcott PC. The Certified Six Sigma Green Belt Handbook. Milwaukee, Wisconsin, USA: ASQ Quality Press; 2015

[12] Hartman HG, DesJardins DC. Business Ethics: Decision Making for Personal Integrity and Social Responsibility. New York, USA: McGraw-Hill Education; 2017

[13] International Organization for Standardization. ISO 9000:2015–Quality management systems–Fundamentals and vocabulary. [Online]. Geneva, Switzerland. Available from: https://www.iso.org/standard/62085.html

[14] Dale BG, van der Wiele T. Total Quality Management and Operational Excellence: Text with Cases. London, United Kingdom: Routledge; 2018

[15] Oakland JS. Total Quality Management and Operational Excellence: Text with Cases. London, United Kingdom: Routledge; 2014

[16] Pyzdek T, Keller PA. The Six Sigma Handbook. New York, USA: McGraw-Hill Education; 2014

[17] Juran JM, Godfrey AB. Juran's Quality Handbook. New York, USA: McGraw Hill Professional; 2016

[18] George ML. Lean Six Sigma: Combining Six Sigma Quality with Lean Production Speed. New York, USA: McGraw-Hill Education; 2002

[19] Hillson D, Murray-Webster R. Understanding and managing risk attitude. Farnham, Surrey, United Kingdom: Gower Publishing; 2017

[20] Juran JM, Godfrey AB. Juran's Quality Handbook. New York, USA: McGraw-Hill Education; 1998

[21] Stamatis DH. Failure Mode and Effect Analysis: FMEA from Theory to Execution. Milwaukee, Wisconsin, USA: ASQ Quality Press; 2013

[22] ASQ. The Certified Quality Auditor Handbook. 4th ed. Milwaukee, Wisconsin, USA: ASQ Quality Press; 2012

[23] AIAG. Potential Failure Mode and Effects Analysis (FMEA). 4th ed. Southfield, Michigan, USA: Automotive Industry Action Group; 2019

[24] FDA. Hazard Analysis and Risk-Based Preventive Controls for Human Food: Guidance for Industry. Silver Spring, Maryland, USA: U.S. Food and Drug Administration; 2020

[25] American Society for Quality. Quality Management System. [Online]. Milwaukee, Wisconsin, USA. Available from: https://asq.org/quality-resources/quality-management-system

[26] International Organization for Standardization. ISO 9000 Family – Quality Management. [Online]. Geneva, Switzerland. Available from: https://www.iso.org/iso-9001-quality-management.html

[27] The Chartered Quality Institute. Quality Control and Quality Assurance. [Online]. London, United Kingdom. Available from: https://www.quality.org/article/quality-control-and-quality-assurance

[28] HSE. Risk Management – Quality Assurance. [Online]. London, United Kingdom. Available from: https://www.hse.gov.uk/risk/controlling-risks.htm

[29] Heizer J, Render B. Operations Management: Sustainability and Supply Chain Management. Upper Saddle River, New Jersey, USA: Pearson Education; 2016

[30] Antony J. Design of Experiments for Engineers and Scientists. Oxford, United Kingdom: Butterworth-Heinemann; 2014

[31] Pande PS, Neuman RP, Cavanagh RR. The Six Sigma Way: How GE, Motorola, and Other Top Companies Are Honing their Performance. New York, USA: McGraw-Hill Education; 2000

[32] Womack JP, Jones DT, Roos D. The Machine That Changed the World: The Story of Lean Production. New York, USA: Simon and Schuster; 1991

[33] Liker JK. The Toyota Way: 14 Management Principles from the World's Greatest Manufacturer. New York, USA: McGraw-Hill Education; 2004

[34] Shingo S. A Study of the Toyota Production System: From an Industrial Engineering Viewpoint. Boca Raton, Florida, USA: CRC Press; 1989

[35] Womack JP, Jones DT. Lean Thinking: Banish Waste and Create Wealth in your Corporation. New York, USA: Simon and Schuster; 2003

[36] Bicheno J, Holweg M. The Lean Toolbox: The Essential Guide to Lean Transformation. Totnes, Devon, United Kingdom: Picsie Books; 2019

[37] Liker JK, Meier D. The Toyota Way Fieldbook: A Practical Guide for Implementing Toyota's 4Ps. New York, USA: McGraw-Hill Education; 2007

[38] Smith DR, Reinertsen DG. Developing Products in half the Time: New Rules, New Tools. New York, USA: John Wiley & Sons; 1991

[39] Akhavan P, Rabbani M. Failure Mode and Effect Analysis: FMEA from Theory to Execution. Boca Raton, Florida, USA: Taylor & Francis; 2017

[40] Kubiak TM. The Certified Six Sigma Green Belt Handbook. Milwaukee, Wisconsin, USA: ASQ Quality Press; 2019

[41] Stamatis DH. The FMEA Pocket Handbook. Boca Raton, Florida, USA: CRC Press; 2017

[42] Harrington HJ. Total Improvement Management: The Next Generation in Performance Improvement. New York, USA: McGraw-Hill Education; 2018

[43] Grant EL, Leavenworth RS, Seward LE. Statistical Quality Control. New York, USA: McGraw-Hill Education; 2018

[44] Senapati S. Six Sigma: Methodology, Tools, and Implementation. Boca Raton, Florida, USA: CRC Press; 2014

[45] Goetsch DL, Davis SB. Quality Management for Organizational Excellence: Introduction to Total Quality. Upper Saddle River, New Jersey, USA: Pearson Education; 2014

[46] Hoyle D. ISO 9000 Quality Systems Handbook. London, United Kingdom: Routledge; 2017

[47] American Medical Association (AMA). Digital Health Implementation Playbook. Chicago, Illinois, USA; 2020

[48] World Health Organization (WHO). The Global Use of Medicines: Outlook through. Geneva, Switzerland: WHO; 2015;**2011**

[49] McNeill A, Mate KS. Globalization, climate change, and human health. New England Journal of Medicine. Boston, Massachusetts, USA. 2013;**368**(14):1335-1343

[50] Lee RT, Seo B. The impact of occupational stress on quality of care and resilience among nurses. Journal of Nursing Management. Chichester, West Sussex, United Kingdom. 2013;**21**(6):735-743

[51] Bitner MJ, Ostrom AL, Morgan FN. Service blueprinting: A practical technique for service innovation. California Management Review. Los Angeles, California, USA. 2008;**50**(3):66-94

[52] Fitzsimmons JA, Fitzsimmons MJ. Service Management: Operations, Strategy, and Information Technology. New York, USA: McGraw-Hill Education; 2013

[53] Zeithaml VA, Bitner MJ, Gremler DD. Services Marketing: Integrating Customer Focus across the Firm. New York, USA: McGraw-Hill Education; 2018

[54] Parasuraman A, Zeithaml VA, Berry LL. SERVQUAL: A multiple-item scale for measuring consumer perceptions of service quality. Journal of Retailing. New York, USA. 1988;**64**(1):12-40

[55] Zeithaml VA, Berry LL, Parasuraman A. The behavioral consequences of service quality. Journal of Marketing. Chicago, Illinois, USA. 1996;**60**(2):31-46

[56] Eastman C, Teicholz P, Sacks R, Liston K. BIM Handbook: A Guide to Building Information Modeling for Owners, Managers, Designers, Engineers, and Contractors. Hoboken, New Jersey, USA: Wiley; 2011

[57] Construction Industry Institute (CII). Best Practices Guide: Quality

Management in Construction Projects. Austin, Texas, USA: CII Publications; 2017

[58] National Institute of Standards and Technology (NIST). National BIM Program. [Online]. Gaithersburg, Maryland, USA; 2017. Available from: https://www.nist.gov/programs-projects/national-bim-program

[59] American Society of Civil Engineers (ASCE). Quality in the Constructed Project: A Guide for Owners, Designers, and Constructors. Reston, Virginia, USA: ASCE Publications; 2013

[60] Hartmann T, Messner JI. Strategic application of automated technologies in construction. Automation in Construction. Amsterdam, Netherlands. 2013;**33**:1-10

[61] Food and Agriculture Organization of the United Nations (FAO). Hazard Analysis and Critical Control Point (HACCP) System and Guidelines for its Application. Rome, Italy: FAO Publications; 2020

[62] U.S. Food and Drug Administration (FDA). Food Safety Modernization Act (FSMA). [Online]. Silver Spring, Maryland, USA; 2019. Available from: https://www.fda.gov/food/food-safety-modernization-act-fsma

[63] ICCI (International HACCP Alliance). HACCP Training Manual. College Station, Texas, USA: ICCI Publications; 2020

[64] Luning PA, Marcelis WJ, van Boekel MA, Veer PV. Scenario studies on future application of knowledge management in food supply chains: Identifying challenges for HACCP-based quality management. Trends in Food Science & Technology. Amsterdam, Netherlands. 2002;**13**(6):211-224

[65] Verghese J, Lewis H, Lockshin L, Spence J. Examining the feasibility of traceability in the Australian beef supply chain. British Food Journal. Bradford, United Kingdom. 2012;**114**(3):303-323

[66] Juran JM, Godfrey AB. Juran's Quality Handbook: The Complete Guide to Performance Excellence. New York, USA: McGraw-Hill Education; 2016

[67] Oakland JS. Statistical Process Control. London, United Kingdom: Routledge; 2014

[68] Chiang SY, Lin GH. Applying statistical process control to improve quality performance in a manufacturing company. Journal of Quality Assurance in Hospitality & Tourism. 2019;**20**(4):359-375

[69] Mitra A, Asif M. Predictive analytics for quality control in the automotive industry: A case study on SPC chart interpretation. International Journal of Quality & Reliability Management. 2018;**35**(1):97-112

[70] Dale B, Plunkett J. Quality Management in Construction Projects. London, United Kingdom: Routledge; 2016

[71] Bertsche B. Fundamentals of Quality Control and Improvement. Hoboken, New Jersey, USA: John Wiley & Sons; 2016

[72] Goetsch DL, Davis SB. Quality Management for Organizational Excellence. Upper Saddle River, New Jersey, USA: Pearson; 2014

[73] ISO 9000 family of standards: Quality management, International Organization for Standardization; 2021. [Online]. Geneva, Switzerland. Available from: https://www.iso.org/iso-9001-quality-management.html

[74] ASQ. The Quality Toolbox, Second Edition. Milwaukee, Wisconsin, USA: ASQ Quality Press; 2019

[75] ISO 13485:2016—Medical devices—Quality management systems—Requirements for regulatory purposes, International Organization for Standardization. Geneva, Switzerland. 2016

[76] Akao Y, Mazur GH. The leading edge in QFD: Past, present, and future. International Journal of Quality & Reliability Management. 2003;**20**(1):20-35

[77] Hauser JR, Clausing D. The house of quality. Harvard Business Review. 1988;**66**(3):63-73

[78] Pyzdek T, Keller PA. The Handbook for Quality Management, Third Edition. New York, USA: McGraw-Hill Education; 2019

[79] Azevedo SG, Carvalho MS. Using Quality Tools and Techniques: A Practical Guide. Boca Raton, Florida, USA: CRC Press; 2008

[80] Besterfield-Sacre M, Besterfield DH, Sacre CW, Besterfield GH. Quality Control, Eighth Edition. Upper Saddle River, New Jersey, USA: Pearson; 2016

[81] Harrington HJ, Voehl F. Organizational Excellence Handbook, Fourth Edition. Boca Raton, Florida, USA: CRC Press; 2012

[82] Fitzsimmons JA, Fitzsimmons MJ. Service Management: Operations, Strategy, Information Technology. New York, USA: McGraw-Hill Education; 2019

[83] Zeithaml VA, Bitner MJ, Gremler DD. Services Marketing: Integrating Customer Focus across the Firm. New York, USA: McGraw-Hill Education; 2019

[84] Verghese J, Lewis H, Lockshin L, Spence J. Examining the feasibility of traceability in the Australian beef supply chain. British Food Journal. Bradford, United Kingdom. 2015;**117**(3):981-997

[85] Mitra A, Asif M. Predictive analytics for quality control in the automotive industry: A case study on SPC chart interpretation. International Journal of Quality & Reliability Management. 2016;**33**(4):478-497

[86] Salvendy G, editor. Handbook of Human Factors and Ergonomics. Hoboken, New Jersey, USA: John Wiley & Sons; 2012

[87] Chen CLP, Zhang CY. Data-intensive applications, challenges, techniques and technologies: A survey on big data. Information Sciences. 2014;**275**:314-347

[88] Provost F, Fawcett T. Data Science for Business: What you Need to Know about Data Mining and Data-Analytic Thinking. Sebastopol, California, USA: O'Reilly Media; 2013

[89] LeCun Y, Bengio Y, Hinton G. Deep learning. Nature. 2015;**521**(7553):436-444

[90] Cortes C, Vapnik V. Support-vector networks. Machine Learning. 1995;**20**(3):273-297

[91] Zanella A, Bui N, Castellani A, Vangelista L, Zorzi M. Internet of things for smart cities. IEEE Internet of Things Journal. 2014;**1**(1):22-32

[92] Atzori L, Iera A, Morabito G. The internet of things: A survey. Computer Networks. 2010;**54**(15):2787-2805

[93] Gubbi J, Buyya R, Marusic S, Palaniswami M. Internet of things (IoT): A vision, architectural elements, and future directions.

Future Generation Computer Systems. 2013;**29**(7):1645-1660

[94] Bandyopadhyay D, Sen J. Internet of things: Applications and challenges in technology and standardization. Wireless Personal Communications. 2011;**58**(1):49-69

[95] Lee I, Lee K. The internet of things (IoT): Applications, investments, and challenges for enterprises. Business Horizons. 2015;**58**(4):431-440

[96] Witten IH, Frank E, Hall MA, Pal CJ. Data Mining: Practical Machine Learning Tools and Techniques. San Francisco, California, USA: Morgan Kaufmann; 2016

[97] Hand DJ, Mannila H, Smyth P. Principles of Data Mining. Cambridge, Massachusetts, USA: MIT Press; 2001

[98] Kim J. Data analytics methods and techniques. Journal of Open Innovation: Technology, Market, and Complexity. 2015;**1**(1):20

[99] Berry MJ, Linoff GS. Data Mining Techniques: For Marketing, Sales, and Customer Relationship Management. Hoboken, New Jersey, USA: John Wiley & Sons; 2011

[100] Montgomery DC. Introduction to Statistical Quality Control. Hoboken, New Jersey, USA: John Wiley & Sons; 2012

Chapter 2

Design of Experiments (DOE): Applications and Benefits in Quality Control and Assurance

Sheriff Lamidi, Rafiu Olalere, Adekunle Yekinni and Khairat Adesina

Abstract

This chapter explores the applications and benefits of Design of Experiments (DOE) in the context of quality control and quality assurance. DOE is a statistical methodology that enables researchers and practitioners to systematically investigate and optimize processes, identify critical factors affecting quality, and reduce variability and waste. This chapter begins by introducing the overview and definitions of DOE, covering topics such as the history of DOE, types of DOE, steps involved in conducting DOE, and key components of DOE. The specific applications of DOE in quality control and quality assurance were explored, highlighting their importance across various industries. It demonstrates how DOE can be effectively applied to optimize products and processes, reduce defects and variation, improve quality, implement Six Sigma, and validate and verify processes. It then delves into the specific applications of DOE in quality control and assurance, highlighting its significance in various industries and sectors. Furthermore, the book addresses challenges and considerations in implementing DOE in real-world scenarios, such as resource constraints, experimental constraints, and data analysis complexities. It provides basic information on software tools commonly used in DOE.

Keywords: design of experiments (DOE), quality, quality control, process variability, optimization

1. Introduction

Quality control and assurance are crucial aspects of any manufacturing or industrial process. Ensuring high-quality products and services is essential for customer satisfaction, brand reputation, and overall business success. Quality is a measure of the level of conformance of a product to design specifications or the ability of a product or service to satisfy user requirements. The duo of quality assurance and quality control helps deliver a defect-free product or service. Quality assurance focuses on preventing defects by ensuring the approaches, techniques, methods, and processes designed for the projects are implemented correctly. Quality control, on the other hand, focuses on identifying defects by ensuring that the approaches, techniques, methods, and processes designed in the project are followed correctly [1]. Quality assurance is process-oriented and a managerial tool, whereas quality control is product-oriented and a remedial tool, according to [1]. One powerful tool used in quality control and assurance is the design of experiments (DOE). According to engineers and technologists, they often make use of DOE methodologies for various applications ranging from the design of new products, improvement of design, maintenance, control and improvement of manufacturing processes, maintenance and repair of products, and several others [2–4]. This chapter aims to explore the applications and benefits of DOE in quality control and assurance. Design of experiments (DOE) is a statistical method for planning and conducting experiments. DOE is used to identify the factors that affect a process and to determine the optimal levels of those factors, as shown in **Figure 1**. DOE can be used to improve the quality of products and processes, reduce costs, and increase efficiency [5]. Businesses and manufacturing companies can use Design of Experiments (DOE) in a variety of ways to differentiate themselves from the competition by constantly redesigning their products or creating new products to establish a presence in other markets. First, DOE can be used to identify the factors that most affect the quality of a product. By understanding which factors are most important, businesses can focus their efforts on improving those factors. This can lead to a product that is more reliable, durable, and user-friendly than the competition's products. Second, DOE can be used to reduce the cost of manufacturing a product. By identifying the most efficient way to produce a product by optimizing manufacturing processes using DOE methodologies, businesses can improve the quality of their products and save money on labor, materials, and other costs [6]. This can lead to a lower price for the product, which can make it more competitive. Third, DOE can be used to develop new products that meet the needs of a specific market. By understanding the needs of the target market, businesses can develop products that are more likely to be successful. This can help businesses gain a foothold in new markets and increase their market share.

1.1 Objective of the chapter

The objective of this chapter is to provide an in-depth understanding of DOE and its applications in quality control and assurance. We will explore various

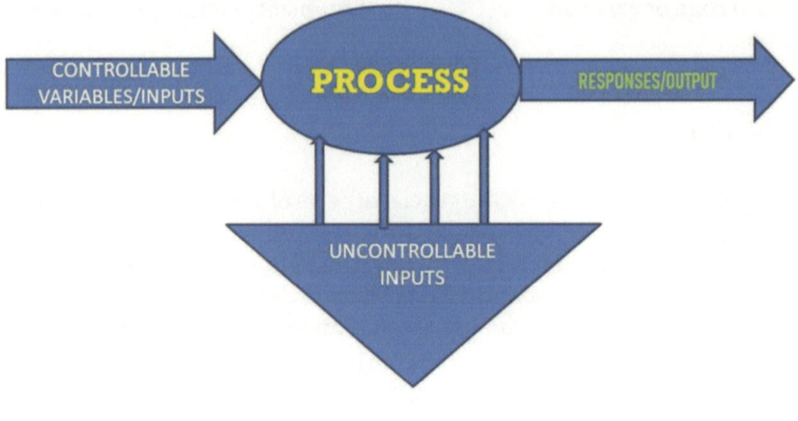

Figure 1.
Design of Experiment (DOE).

experimental designs, statistical techniques, and methodologies that are commonly used in DOE. Additionally, we will discuss its practical applications across various fields and the benefits and advantages that DOE offers in ensuring and improving quality standards.

1.2 An overview and definitions of design of experiments (DOE)

Definitions of Design of Experiments (DOE): DOE is a statistical methodology used to systematically plan, conduct, analyze, and interpret experiments to obtain valid and reliable results. It allows researchers to efficiently explore and identify the significant factors influencing a process or product's performance. DOE is an important statistical method used in controlling input factors or variables in order to ascertain the level of relationships with the output (responses) according to **Figure 1**, so as to ensure product or process quality. DOEs are usually carried out in five stages [7, 8] as shown in **Figure 2**. They are:

1.3 History of DOE

DOE has its roots in the work of Sir Ronald Fisher, who developed the basic principles of DOE in the early twentieth century. Fisher's work was initially applied to agricultural research, but it was soon adapted for use in other fields, including manufacturing, engineering, and medicine [9]. Since then, many scientists and statisticians have contributed to DOE development and its application in different fields [9–12].

1.4 Types of DOE

There are many different types of DOE, each with its own strengths and weaknesses [8, 13]. The best type of DOE to use will depend on the specific situation. Factors to consider include the number of factors, the number of levels for each factor, the desired level of confidence, and the time and budget constraints (**Figure 3**) [10, 14–16].

The most common types of DOE are:

1.4.1 Full factorial designs

These designs involve testing all possible combinations of factors. For example, if there are two factors with two levels each, there would be four possible combinations (2 × 2 = 4). Full factorial designs are the most comprehensive, but they can also be the most time-consuming and expensive.

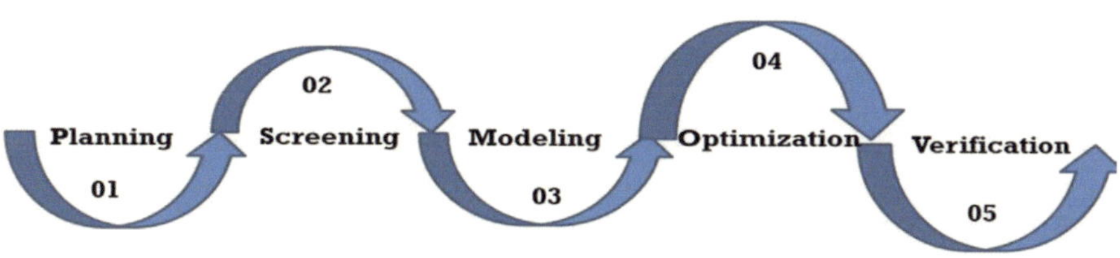

Figure 2.
Five stages of DOE.

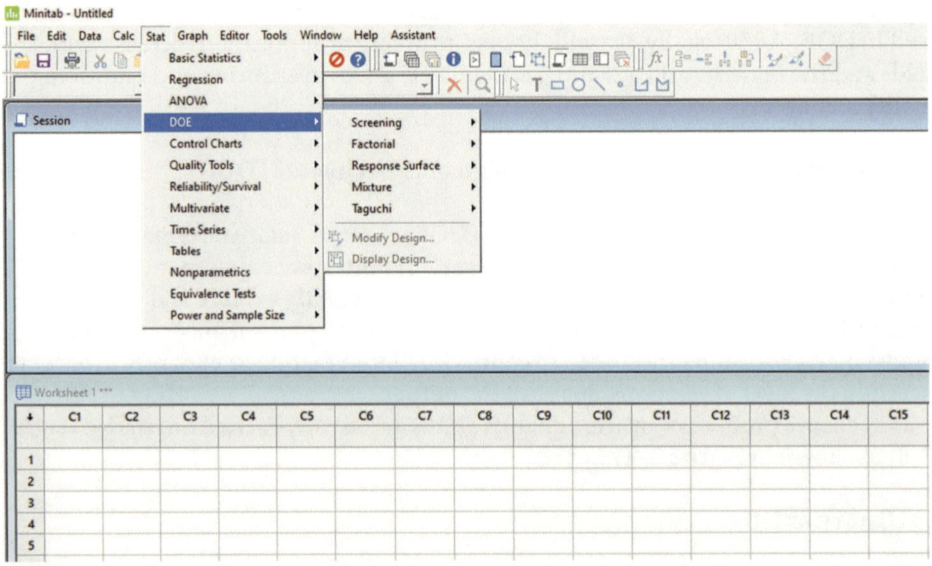

Figure 3.
Types of DOE.

1.4.2 Screening/fractional factorial designs

Fractional factorial experiments are a type of factorial experiment that uses fewer experimental runs than a full factorial design. These designs involve testing a subset of the possible combinations of factors. Fractional factorial designs are less comprehensive than full factorial designs, but they can save time and money.

1.4.3 Response surface methodology (RSM) designs

RSM is a type of DOE that is used to fit a mathematical model to the response variable. RSM can be used to identify the optimal levels of the factors and predict the response variable for new combinations of factors. These designs are used to study the relationship between a response variable and multiple factors.

1.4.4 Mixture designs

These designs are used to study the relationship between a response variable and multiple factors that are mixed together. Mixture designs are often used in the food and beverage industry to understand how the different ingredients in a product affect the taste, texture, and other properties of the product.

1.4.5 Taguchi designs

These designs are a type of fractional factorial design that is specifically designed for quality improvement. Taguchi designs are often used in manufacturing, where it is important to produce products that meet the required quality standards.

1.5 Steps involved in conducting DOE

In order to obtain good results from a DOE, the following 8 steps shown in **Figure 4** below are necessary:

- Define clear objectives
- Select Process variable
- A feasible experimental design must be selected.
- Execute the selected design.
- Ensure that the data are consistent with the experimental assumptions.
- Analyze and interpret the results.
- Results presentation, and application for decision making.
- Conclusions

In the application of the concept of DOE methodology for quality control and assurance, the following terminologies, otherwise known as components of DOE, are commonly used:

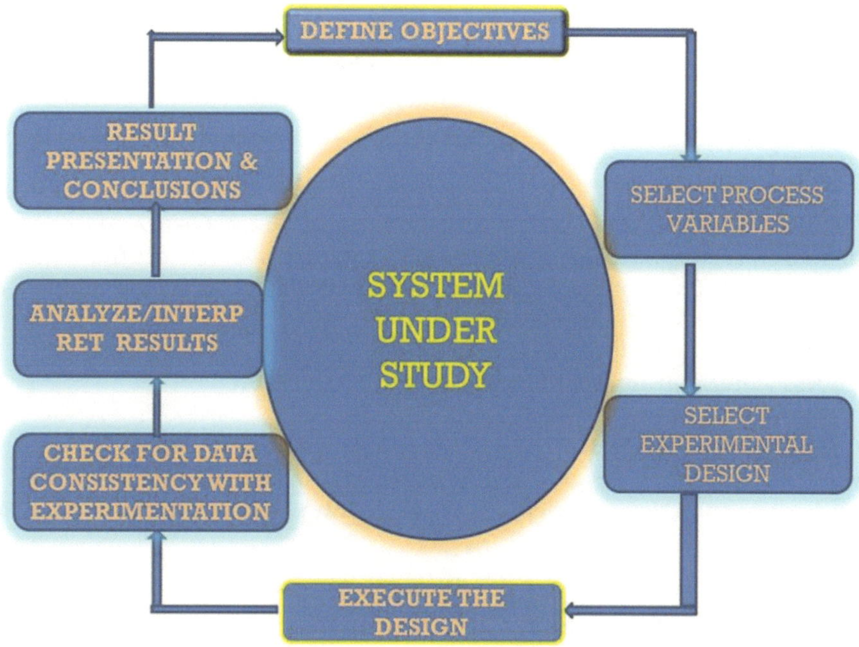

Figure 4.
Steps required in DOE.

2. Key concepts/components of DOE

a. Factors: Variables that may influence the outcome of an experiment.

b. Levels: The values at which factors are set during an experiment.

c. Response Variable: The outcome or output variable that is measured or observed.

d. Experimental Units: The entities or subjects on which the experiments are conducted.

e. Treatment: The combination of factor levels applied to an experimental unit.

f. Replication: The process of repeating the experiment to reduce variability and enhance reliability.

3. Applications of DOE in quality control and assurance

3.1 Product and process optimization

DOE enables the systematic exploration of various factors and their interactions to optimize product and process performance. By identifying the key factors and their optimal levels, manufacturers can improve quality, reduce costs, and enhance efficiency.

3.2 Defects and variation reduction DOE

Helps identify the root causes of defects and variations in a manufacturing process. By conducting experiments and analyzing the results, quality engineers can pinpoint the factors that contribute to defects and develop strategies to reduce or eliminate them.

3.3 Quality improvement and six sigma

DOE is an integral part of Six Sigma methodologies, which aim to achieve process excellence and reduce variation. By using DOE, organizations can identify critical process parameters, set optimal levels, and implement strategies to minimize defects and variations, thus improving overall quality.

3.4 Process validation and verification

DOE plays a crucial role in the validation and verification of manufacturing processes. By conducting designed experiments, organizations can gather data on process performance, determine critical process parameters, and establish robustness and reliability of their processes.

4. Key factors affecting quality optimization processes and waste reduction

Optimizing quality processes and reducing waste is a crucial aspect of many industries. While the specific factors influencing these processes can vary depending on the context, here are some key factors that commonly affect quality optimization and waste reduction.

- Process Design and Standardization: Well-designed processes with clear specifications and standard operating procedures (SOPs) play a vital role in optimizing quality and minimizing waste. Factors such as process layout, equipment selection, workflow efficiency, and error-proofing mechanisms can significantly impact the quality of output and waste generation [17, 18].

- Quality Control and Monitoring: Effective quality control measures, including robust inspection protocols, real-time monitoring systems, and statistical process control (SPC) techniques, help identify and rectify quality issues promptly. Monitoring critical process parameters and implementing quality control checks at various stages can minimize defects and waste [12, 19].

- Training and Skill Development: Well-trained and skilled personnel are essential for maintaining quality standards and reducing waste. Adequate training programs that emphasize quality awareness, technical skills, and problem-solving capabilities contribute to consistent quality optimization and waste reduction [20].

- Continuous Improvement and Lean Practices: Embracing continuous improvement methodologies, such as Lean Six Sigma, can drive quality optimization and waste reduction. Tools like value stream mapping, root cause analysis, and Kaizen events enable organizations to identify and eliminate process inefficiencies, defects, and non-value-added activities [21].

5. Software tools commonly used in DOE

There are several software tools available that can assist in the design, analysis, and implementation of Design of Experiments (DOE). These tools provide a user-friendly interface and various statistical capabilities to simplify the DOE process. The choice of software tool depends on the specific requirements of the experiment, the level of complexity involved, and the user's familiarity with the software. Here are some commonly used software tools for DOE:

- Minitab: Minitab is a popular statistical software package widely used for DOE. It offers a comprehensive set of DOE tools, including factorial designs, response surface methods, and mixture designs. Minitab provides easy-to-use graphical and statistical analysis features, making it suitable for both beginners and experienced users.

- JMP: JMP is a powerful statistical software developed by SAS. It offers a range of DOE techniques, such as factorial designs, response surface methods, and mixture designs. JMP provides an interactive interface with drag-and-drop capabilities for designing experiments, analyzing data, and visualizing results.

- Design-Expert: Design-Expert is a specialized software tool specifically designed for DOE. It offers a wide range of experimental design options, including factorial designs, response surface methods, mixture designs, and Taguchi designs.

Design-Expert provides advanced graphical and statistical analysis features to facilitate the optimization of processes and product formulations.

- R: R is a popular open-source programming language for statistical computing and graphics. It has a rich collection of packages that support DOE, such as the 'DOE' package and 'rsm' package. R provides extensive flexibility and customization options for designing experiments, analyzing data, and performing advanced statistical modeling.

- Excel: Microsoft Excel, though not specifically designed for DOE, can be used for simple experimental designs and analysis. It offers basic statistical functions, charts, and data analysis tools that can be utilized for conducting DOE experiments and analyzing results.

6. Some examples of specific fields where DOE found its practical application

6.1 Manufacturing industry

DOE can be used to improve the quality of products, to reduce costs, and to increase efficiency. For example;

- DOE can be used to optimize the process of manufacturing a part, identify the root cause of a quality problem, or reduce the variability of a process, which is a measure of quality. It can be used to identify the causes of defects in a product or to find ways to reduce the time it takes to manufacture a product. DOE can be adopted in the manufacturing industry by an industry that desires to manufacture a machine part (from Al-Si alloy material) with minimum surface roughness by combining three controllable variables (cutting speed, feed rate, and depth of cut). Due to the combinations of many variables, the Design of Experiments (DOE) could be used to carry out the study of the effect of the three machining variables (cutting speed, feed rate, and depth of cut) on the surface roughness (Ra) of Al-Si alloy [22, 23].

- In the automotive industry, DOE is used to improve the fuel efficiency of cars and trucks. By identifying the factors that most affect fuel efficiency, businesses can design cars and trucks that use less fuel. This can lead to lower emissions and a reduced cost of ownership for the customer.

- In the food industry, DOE is used to improve the taste and texture of food products. By understanding how different factors affect the taste and texture of food, businesses can develop products that are more appealing to consumers. This can lead to increased sales and a stronger brand reputation.

- In the pharmaceutical industry, DOE is used to develop new drugs that are more effective and less harmful than existing drugs. By understanding how different factors affect the effectiveness and toxicity of drugs, businesses can develop drugs that are more likely to be approved by the agency in charge. This can lead to increased profits and to a better quality of life for patients.

6.2 Engineering

DOE is used in engineering to improve the design and performance of products and processes. According to [10] in their research presented 77 cases of quality improvement through practical applications of DOE in Engineering. For example;

- DOE can be used to optimize the design of a car engine or to improve the yield of a chemical reaction [10]. DOE can be used to improve the design of products and processes, reduce costs, and increase efficiency. Furthermore, DOE can be used to optimize the design of a bridge, identify the root cause of a failure, or reduce the weight of a product.

- DOE can be used to optimize the design of composite materials for specific structural applications. Factors such as fiber type, fiber volume fraction, resin content, and curing parameters can be varied systematically to achieve desired mechanical properties such as strength, stiffness, and impact resistance [24].

- DOE is used to optimize process parameters to improve yield and quality in semiconductor engineering. Factors such as temperature, pressure, etching time, and gas flow rates can be varied systematically to identify the optimal settings that result in minimal defects and enhanced performance [25].

6.3 Medicine

DOE is used in medicine for diverse applications. DOE found its applications in medical research and practice, including drug formulation, medical device manufacturing, and radiation therapy. Most of these applications widely demonstrate how DOE can be used to systematically investigate and optimize factors influencing key outcomes in medical applications. Some of the key examples are;

- DOE can be used to optimize drug formulations to improve the bioavailability of pharmaceutical products. Factors such as excipient composition, drug concentration, and manufacturing parameters can be systematically varied to identify the optimal combination that maximizes drug absorption and efficacy [26].

- DOE can be used to optimize the dosage of a drug, to identify the side effects of a treatment, or to reduce the risk of a disease. DOE is used in medicine to improve the effectiveness of treatments and to reduce the side effects of drugs. For example, DOE can be used to identify the best dose of a drug to treat a particular condition or to find ways to reduce the toxicity of a drug.

- DOE can be utilized to optimize treatment protocols in radiation therapy for cancer patients. Factors such as radiation dose, treatment duration, and beam angles can be systematically varied to identify the optimal combination that maximizes tumor control while minimizing side effects on healthy tissues [24, 27].

6.4 Agriculture

Agriculture: DOE is used in agriculture to improve crop yields and to reduce the use of pesticides and fertilizers. DOE can be used to determine the best combination of fertilizer and irrigation rates to maximize crop yields, optimization of plant growth conditions in controlled environments and many more. For example;

- DOE can be used to evaluate the effects of fertilizer formulations on crop yield. In the work of [17], DOE was used to investigate the impact of different fertilizer formulations on crop yield. The study involved varying factors such as nitrogen, phosphorus, and potassium concentrations in the fertilizer mix. By systematically designing and conducting experiments, they were able to determine the optimal combination of nutrients that maximized crop yield while minimizing the amount of fertilizer required.

- DOE can be used to assess the impact of irrigation techniques on water use efficiency. The effect of different irrigation techniques on water use efficiency in crop production was evaluated by [28]. Various factors, such as irrigation frequency, irrigation duration, and water application rate, were manipulated and studied. The experiments allowed the researchers to identify the optimal combination of irrigation practices that resulted in improved water use efficiency without compromising crop yield [28].

- Optimization of Plant growth conditions in controlled environments was conducted by [29] using DOE. The growth conditions of the plant was optimize in controlled environments such as greenhouses or growth chambers. Factors such as light intensity, temperature, humidity, and CO_2 levels were systematically varied to determine the optimal combination that promoted plant growth, development, and yield.

6.5 Marketing

DOE is used in marketing to improve the effectiveness of advertising campaigns and to increase sales. DOE is a powerful tool that can be used to improve the performance of products, processes, and organizations. It is a valuable tool for anyone who wants to improve their results. Here are some specific examples of how DOE has been used in practice:

- DOE can be used to optimize product packaging design elements such as color, shape, size, and labeling to understand their impact on consumer perception and purchase behavior. By systematically varying these factors, marketers can identify the optimal packaging design that maximizes consumer appeal and product sales [30].

- DOE can be employed to evaluate different pricing strategies and their impact on consumer behavior, purchase intent, and profitability. Factors such as price levels, discount offers, and promotional strategies can be systematically varied to determine the optimal pricing strategy that maximizes sales and profitability [31, 32].

- DOE can be utilized to test and optimize various elements of advertisements, such as visual design, headline, copywriting, and call-to-action. By systematically varying these factors and measuring consumer responses, marketers can identify the optimal combination that maximizes advertisement effectiveness and consumer engagement [33].

7. Benefits of DOE in quality control and assurance

- Efficient Resource Utilization: DOE allows organizations to allocate their resources efficiently by identifying the most influential factors. By focusing on these factors, companies can optimize their processes and achieve significant improvements in quality without unnecessary expenditure.

- Cost Reduction: By systematically exploring process factors and their interactions, DOE helps identify cost-effective solutions. By reducing defects, eliminating waste, and optimizing process parameters, organizations can save costs associated with rework, scrap, and material consumption.

- Improved quality: DOE can be used to improve the quality of products and processes by identifying and reducing the variability of the process.

- Reduced costs: DOE can be used to reduce costs by identifying the most efficient way to produce a product or by reducing the amount of waste.

- Increased efficiency: DOE can be used to increase efficiency by identifying the root cause of problems and by improving the design of products and processes.

- Enhanced Decision Making: DOE provides a structured approach to experimentation, resulting in reliable and statistically valid data. This enables informed decision making based on evidence rather than intuition or guesswork. By using DOE, organizations can make data-driven decisions to improve quality and minimize risks.

- Faster Time to Market: DOE facilitates the identification of critical process parameters and optimal levels, leading to faster process optimization. By reducing the time required for experimentation and process development, organizations can accelerate product development cycles and bring products to market more quickly.

8. Challenges and considerations in implementing DOE in real-world scenarios

- Resource Constraints: Limited resources such as time, budget, and availability of equipment or materials can pose challenges in implementing DOE. Conducting experiments may require significant time and financial investments. It is essential to carefully plan and allocate resources to ensure the feasibility and success of DOE studies [19, 34].

- Experimental Constraints: Some experiments may face practical constraints due to factors such as safety regulations, ethical considerations, or limitations in the process or system under investigation. Researchers must identify and address these constraints to design experiments that are feasible and align with regulatory requirements [21, 35].

- Data Analysis Complexities: Analyzing experimental data and interpreting the results can be challenging, particularly when dealing with complex designs or large datasets. Specialized statistical knowledge may be required to properly analyze and draw meaningful conclusions from the data obtained through DOE. Consideration should be given to the appropriate statistical methods and software tools for analyzing the experimental results [35, 36].

- Planning for Interactions and Confounding: Identifying and addressing potential interactions among factors and confounding effects can be complex in DOE. Interactions and confounding can affect the interpretation of experimental results and lead to incorrect conclusions. Careful consideration and appropriate experimental design strategies, such as fractional factorial designs, can help mitigate these challenges [16, 37, 38].

9. Conclusion

In this chapter, the authors focus on the applications and benefits of DOE in quality control and assurance. DOE is a very important statistical methodology that enables both scientists, engineers, researchers, and various other professionals to design, develop, and optimize high-quality products and services. DOE has many applications in various fields. Application of DOE ensures high-quality products and services, increases customer satisfaction, efficient resource utilization, cost reduction, enhanced decision making, improves brand reputation, and ensures overall business success. Having presented in this chapter the numerous applications of DOE in various fields, it can be concluded that DOE is a powerful research tool and methodology for quality control and assurance.

Acknowledgements

The opportunity to have this chapter review included in the main book, Quality Control and Quality Assurance - Techniques and Applications, is greatly appreciated by the authors. We also commend the rigorous peer review process.

Author details

Sheriff Lamidi*, Rafiu Olalere, Adekunle Yekinni and Khairat Adesina
Lagos State University of Science and Technology (Formerly Lagos State Polytechnic), Lagos, Nigeria

*Address all correspondence to: lamidi.s@lasustech.edu.ng

References

[1] Kiran Kumar Panigrahi. 2023. Available from: https://www.tutorialspoint.com/differences-between-quality-assurance-and-quality-control

[2] Tanco M, Viles E, Ilzarbe L, Álvarez MJ. Manufacturing industries need Design of Experiments (DoE). In: Proceedings of the World Congress on Engineering. London, UK: Newswood Limited; 2007

[3] Smith J, Johnson A. Application of design of experiments (DOE) methodologies in engineering and technology. Journal of Engineering and Technology Applications. 2020;**15**(2):45-62

[4] Taguchi G. System of Experimental Design. New York: Wiley; 1987

[5] Dahlgaard JJ, Dahlgaard-Park SM, Boesgaard KM. Enablers of waste management in Danish companies: Training, commitment and motivation. Journal of Cleaner Production. 2007;**15**(18):1765-1775

[6] Gutiérres PH, De la Vara SR. Análisis y diseño de experimentos. Journal of Chemical Information and Modeling. 2012. DOI: 10.1017/CBO9781107415324.004

[7] Wu CFJ. Design of Experiments: Theory and Practice. Hoboken, NJ: Wiley; 2006

[8] McCool J. Using the Weibull Distribution: Reliability Modelling, and Interference. New Jersey: John Wily and Sons; 2012

[9] Madadian Bozorg N, Leclercq M, Lescot T, Bazin M, Gaudreault N, Dikpati A, et al. Design of experiment and machine learning inform on the 3D printing of hydrogels for biomedical applications. Biomaterials Advances. 2023;**153**:213533. DOI: 10.1016/j.bioadv.2023.213533

[10] Ilzarbe L, Álvarez MJ, Viles E, Tanco M. Practical applications of design of experiments in the field of engineering: A bibliographical review. Quality and Reliability Engineering. 2008;**24**(4):417-428. DOI: 10.1002/gre.909

[11] Assia C. Designer's Guide to Lab Practice. 1st ed. London: Routledge; 2023

[12] Ebadi M, Mozdianfard MR, Aliabadi M. Employing full factorial design and response surface methodology for optimizing direct contact membrane distillation operational conditions in desalinating the rejected stream of a reverse OSMOSIS unit at Esfahan refinery. Water supply Journal. 2018;**19**(2):492-501. DOI: 10.2166/ws.2018.094

[13] Myers RH, Montgomery DC. Response Surface Methodology: Process and Product Optimization Using Designed Experiments. United Kingdom: John Wiley & Sons; 2016

[14] Michael Sadowski. 2022. Available from: https://www.synthace.com/blog/types-of-doe-design-a-users-guide

[15] Terry WA. Introduction to Experimental Methods. 1st ed. Boca Raton: Taylor and Francis Group; 2023

[16] Khavekar RS, Hari VS. Analyzing the need for a comparative study of Shainin DoE and traditional DoE tools for deploying six sigma in Indian manufacturing companies. IOP Conference Series: Materials Science and Engineering. 2018;**376**:012121

[17] Zhang Y, Yu Z, Wang J, Li Y, He X. Optimization of fertilizer formulations for sustainable agriculture using design

of experiments and response surface methodology. Journal of Plant Nutrition. 2018;**41**(6):713-725

[18] Yang CH, Huang JT. Process optimization and layout design for electronics manufacturing service industry. Mathematical Problems in Engineering. 2020;**2020**:1-13

[19] Montgomery DC. Design and Analysis of Experiments. 8th ed. Hoboken, NJ: Wiley; 2013

[20] Antony J, Banuelas R. Key ingredients for the effective implementation of six sigma program. Measuring Business Excellence. 2002;**6**(4):20-27

[21] Anderson MJ, Whitcomb PJ, Montgomery DC. Design of experiments in quality engineering. In: Encyclopedia of Statistics in Quality and Reliability. 2nd ed. New York City, USA: John Wiley & Sons; 2017. pp. 1-15

[22] Kasali Aderinmoye A, Sheriff BL. Effect of dry cutting system on surface finish in turning operation of Al-Si alloy. Journal of Multidisciplinary Engineering Science and Technology. 2021;**8**:10

[23] Lamidi S, Olaleye N, Bankole Y, Obalola A, Aribike E, Adigun I. Applications of response surface methodology (RSM) in product design, development, and process optimization. In: Response Surface Methodology - Research Advances and Applications. London, UK: IntechOpen; 2023. DOI: 10.5772/intechopen.106763

[24] Wang Y, Lu H, Zhu Q, Tang Y. Design optimization of composite materials using DOE for stiffness. Polymer Testing. 2019;**80**:106119

[25] Lee Y, Kim C, Park S. Optimization of the etching process parameters in plasma etching of Al_2O_3 using a design of experiment. Journal of the Korean Physical Society. 2020;**77**(3):219-224

[26] El-Helaly SN, Mostafa SA, Hussein AK. Optimization of theophylline oral disintegrating tablets formulation using design of experiments for enhanced bioavailability. Drug Development and Industrial Pharmacy. 2020;**46**(1):33-43

[27] Vicente V, Zerón H, Martínez Á, Palomo A, Garayoa J. Optimization of treatment protocols in radiation therapy using design of experiments: A review. International Journal of Radiation Oncology *Biology* Physics. 2017;**99**(4):839-846

[28] Rashid A, Aslam M, Azizullah A, Shah M. Enhancing water use efficiency in wheat crop using different irrigation techniques. Environmental Monitoring and Assessment. 2019;**191**(1):9

[29] Körner O, Challa H. Plant phenotyping using DOE and machine learning algorithms in controlled environments. In: Proceedings of the 10th International Conference on Computer Vision Theory and Applications (VISAPP). Vol. 1. Lisbon Portugal: SciTePress; 2017. pp. 547-554

[30] Hoegg J, Hofmann J, Meckel K. Packaging aesthetics and consumer preferences: A cross-cultural study using experimental auctions and choice tasks. Journal of Marketing. 2017;**81**(1):86-101

[31] Ariely D, Wertenbroch K. Procrastination, deadlines, and performance: Self-control by precommitment. Psychological Science. 2002;**13**(3):219-224

[32] Box GEP, Hunter WG, Hunter JS. Statistics for Experimenters: Design, Innovation, and Discovery. New York, USA: Wiley; 2005

[33] Liu X, Dong Y. Designing optimal advertising messages for different cultures: An experimental study. International Journal of Advertising. 2018;**37**(4):588-610

[34] Montgomery DC. Design and Analysis of Experiments. Arizona, USA: John Wiley & Sons; 2017

[35] Hicks CR, Turner KV. Fundamental Concepts in the Design of Experiments. United Kingdom: Oxford University Press; 1999

[36] Roberto F, Alberto M, Luca P, Luigi S. Design of experiments and machine learning with application to industrial experiments. Stat Papers. 2023;**64**:1251-1274. DOI: 10.1007/s00362-023-01437-w

[37] Box GEP, Hunter WG, Hunter JS. Statistics for Experimenters: Design, Innovation, and Discovery. 2nd ed. Hoboken, NJ: Wiley; 2005

[38] Grant EL, Leavenworth RS. Statistical Quality Control. New York City, USA: McGraw-Hill Education; 2013

Chapter 3

Shewhart Control Chart: Long-Term Data Analysis Tool with High Development Capacity

Vladimir Shper, Elena Khunuzidi, Svetlana Sheremetyeva and Vladimir Smelov

Useful application of control charts starts and ends with education

Kaoru Ishikawa

Abstract

This chapter suggests some of the ways in which we can enrich our understanding of the theory of variability when we extend our attention to a gap between the real problems any practitioner may encounter and the traditional theory of control charts stated in textbooks, guides, standards, etc. The benefits are about more than just covering additional ground, for this expanded focus also provides insights into how many real problems are being ignored, many new types of charts turn out to be excessively difficult for engineers, many tacit assumptions that traditional theory is based on stay not being understood by practitioners. We are going to consider the impact of different types of process instability, data homogeneity, nonnormality, and nonrandomness on the right application of Shewhart control charts. We also propose the recommendations to practitioners on how to avoid the above-mentioned problems and improve data-based decision-making.

Keywords: exploratory data analysis, shewhart control chart, process stability, assignable causes of variation, nonhomogeneity, nonrandomness, capability indices

1. Introduction

On 16 May 2024, the World will be celebrating the 100th anniversary of the Shewhart Control Chart (ShCC) – the most essential tool of process stability analysis used successfully in practically all areas of human activity. This 100-year-old tool of exploratory data analysis is described in many old and new books (see, e.g., [1–11], to name a few), and there are international standards devoted to control charting [12] as well as numerous sites on the Internet. On the other hand, most quality professionals, managers and engineers, CEOs, and even not a small amount of statisticians either are not familiar with this tool of data analysis or have just very superficial understanding of it [13]. Almost 30 years ago, R. Hoyer and W. Ellis claimed this thought in their paper ([14], *63*)[1] in the following form: "... our experience indicated that a sizable majority of quality professionals are not knowledgeable about basic issues of statistics and Statistical Process Control (SPC). Our instructional activities in a broad range of academic, industrial, and service delivery environments have convinced us that there are many individuals who are "doing SPC" without understanding what it is about". Then they made a conclusion (*ibid*): "Although it is disappointing that the technical content of the quality improvement discipline has progressed so little during the past 65 years, that is probably not the most significant problem facing process control initiatives during the next decade. Instead, there is every reason to be concerned that many quality professionals are directing continuous process improvement activities without a sound understanding of the basic issues". We are sure that this situation has not changed notably since those times [13], but this chapter is devoted just to "the technical content of the quality improvement discipline". It is noteworthy that such prominent experts in Statistical Process Control (SPC) area as Lloyd Nelson and William Woodall argued emphatically [15] against the assertion of Hoyer and Ellis that "the technology of the quality science has been intellectually dormant for the past 65 years" ([16], *73*). The authors of [14] agreed with the experts that many interesting issues have been going on in the research journals devoted to quality improvement. "But, unfortunately, it appears to be primarily academics... and there are many real-world settings in which these results are only marginally useful" ([15], *93*). That is just what our paper is focused on. We will discuss several traditional assumptions that cover a minuscule part of reality and lead to many gaps between real process behavior and corresponding math models. Moreover, there are many practical questions that are not being discussed in the current literature at all, and have not been ever discussed in the past. In Section 2, we will give a brief survey of the problems that seem to be the most important to us. Then, Section 3 is devoted to the ambiguity in the notion of assignable causes of variation – one of the basic ideas of SPC. In Section 4, we will present our considerations on the data nonrandomness. Then, in Section 5, we will discuss the issues of highly skewed processes. Section 6 contains some examples revealing the limitation of current theory of ShCCs. Finally, in Section 7, we will share our concerns about the interaction between SPC and metrology, SPC and management. Our proposals for further research are given in conclusion.

2. The most important problems of Shewhart control chart application

It was outlined in [13] that "There is no friendship between business and the theory of variation – incomprehension is going on and on" [13]. This is not just our viewpoint. We have already mentioned above similar statements of Hoyer and Ellis in [14–16]. A well-known expert in SPC W. Woodall in 2000 published a survey, "Controversies and Contradictions in Statistical Process Control" [17]. One of the main problems discussed there was the relationship between hypothesis testing and control charting. The main Woodall's conclusion on this issue is that it is a simplification to consider control charting as something equivalent to hypothesis testing ([17], *343*). This approach can be a serious obstacle to the right application of ShCCs in Phase I of process analysis. We agree. Moreover, this widely spread view of equivalency between these two entities can prevent the right use of control charts both in

[1] Here and everywhere below the figure printed in Italic after comma indicates the page in the reference.

Phase I and in Phase II. What is even more important is that very few practitioners know about this problem and hardly ever ponder on it.

D. Steinberg wrote in 2016 a survey of the state-of-the-art in industrial statistics, where he mentioned the following problems in SPC: multivariate data, profile data, and data from phasor measurement units [18]. He expressed his concern that too many research papers rely on unrealistic assumptions that do not exist in practice.

In 2017, W. Woodall wrote a follow-up to his 2000 survey called "Bridging the gap between theory and practice in basic statistical process monitoring" [19]. This time, he revisited some of the same problems he had talked about before, and discussed some new ones. Among the old problems, there was again the relation between statistical theory and practice. At the end of that paper, Woodall made several useful suggestions that could have improved the quality of statistical papers in the area of SPC and the quality of related studies. Simultaneously, he made a proposal that we think is completely unacceptable. According to Woodall, the use of the moving range chart should be ceased. We beg to differ from this suggestion and will explain our viewpoint below.

In 2022 W. Woodall published a new paper on SPC issues titled "Recent Critiques of Statistical Process Monitoring Approaches". In the introductory section, he wrote: "Hundreds of flawed papers on statistical process monitoring (SPM) methods have appeared in the literature over the past five to ten years. The presence of so many flawed methods, and so many incorrect theories, reflects badly on the SPM research field. Critiques of the various misguided approaches have been published in the last two years in an effort to stem this tide" [20]. Let us look at the flawed methods enlisted by Woodall: Use of Inadvisable Weighted Averages, Use of Auxiliary Information, Rules Equivalent to Runs Rules, Neutrosophic Methods, Mixing Various Charts, The Generally Weighted Moving Average Chart, Misuses of the EWMA Statistic, Repetitive Sampling Methods, using the coefficient of variation, the multivariate coefficient of variation, and various capability indices, etc.

It is noteworthy that overwhelming majority of practitioners uses only simple ShCCs because many new types of charts (e.g., CUSUM, EWMA, changepoint, etc.) turn out to be too difficult for engineers, operators, and workers.

We are sure that there are at least two root causes of such a sad situation. One, mentioned above, was described in Hoyer, Ellis paper [14–16]. Another one is more fundamental. In the report written in 1996, G. Box noted [21]: "An important issue in the 1930s was whether statistics was to be treated as a branch of Science or Mathematics. Unfortunately, to my mind, the latter view has been adopted in the United States and in many other countries. Statistics has for some time been categorized as one of Mathematical Sciences, and this view has dominated "university teaching, research, the awarding of advanced degrees, promotion, tenure of faculty, the distribution of grants by funding agencies and the characteristics of statistical journals". Judging by above-mentioned papers nothing has changed since 1996. All flawed techniques enlisted by Woodall in [20] are math's exercises or "Statistical Gymnastics" as caustically noted Ch. Quesenberry in [22]. Shewhart's close friend and associate W. Edwards Deming, ending the foreword to the 1939 Shewhart's book, wrote: "Another half-century may pass before the full spectrum of Dr. Shewhart's contributions has been revealed in liberal education, science, and industry" ([23], *ii*). It seems like another half-century may pass before all who are trying to use control charts efficiently have understood the main ideas of Shewhart and Deming.

D. Steinberg, in his paper [18], cited well-known statistician B. Gunter, who wrote in 2008 panel discussion in *Technometrics* on the future of industrial statistics: "I fear that

Technometrics has evolved from primarily making connections to the real, hard, and complex questions of scientific practice to primarily producing artificial formulations of those questions suitable for compact "solution" by mathematical characterization. To understand what is useful and not merely wrong in industrial statistical practice, we need to pay much more attention to the messy details that make up reality". Then Steinberg did not agree with Gunter's statement that most academic papers "has become completely cut off from real problems". But he agreed "that many of the most challenging and exciting problems arising today are not getting space in our journals and that we need better theory to guide us in attacking such problems" ([18], *52*).

Let us sum up the main idea of all papers cited above: too many statistical works went far away from real practice and do not help practitioners in solving their real problems. This is a direct contradiction to Shewhart-Deming approach and to the basic idea of Shewhart control chart, which "stands out as the only one that actually examines the data for the internal consistency which is a prerequisite for any extrapolation into the future. Thus, unlike all "tests" and "interval estimates" of statistical inference Shewhart's process behavior charts are tools for Analytic Studies. Rather than mathematical modeling, or estimation, Shewhart's charts are concerned with taking appropriate actions in the future based upon an analysis of the data from the past. Out of all the statistical procedures available today, they alone were designed for the inductive inferences of the real world" ([9], *19*). We see this tendency to disregard reality for the world of math models and also ignore the problems of simple control charts in favor of more and more complex designs. Many books and standards that are being widely used by practitioners all over the world teach the theory of control charts based on very unrealistic assumptions about real processes and their behavior (see [1–5, 7, 12], to name a few). In the following discussion, we will examine more thoroughly some issues, such as the various types of assignable causes of variations, the examples of unanswered questions in the theory of control charts, and other related topics.

3. What is an assignable cause of variation, and how it changes a process?

According to Shewhart ([24], *14*), "...in the majority of cases there are unknown causes of variability in the quality of a product which do not belong to a constant system...these causes were called *Assignable*". He, further, explains that an assignable cause of variation is one which can be found without excessive waste of time and money. What is mostly important for us, Shewhart outlines the principal impossibility to establish a criterion of revealing an assignable cause a priori either by formal or by mathematical method.

A famous statistician and quality guru, Dr. Deming wrote in his foreword to Shewhart's book ([23], *ii*): "The great contribution of control charts is to separate variation into two sources: (1) the system itself ('chance causes', Dr. Shewhart called them), the responsibility of management; and (2) assignable causes, called by Deming 'special causes', specific to some ephemeral event ..." A process is called statistically controllable or stable or predictable if all assignable causes of variation are removed.

On the other hand, Wheeler and Chambers [6] defined assignable causes of variation as follows: "Uncontrolled variation that is characterized by a pattern of variation that changes over time". In a paper written later, Wheeler [25] pointed out that chance or common causes differ from assignable causes due to their impact on a process. So, they are not principally contradictory.

Woodall [17] provided the following definition: "'Common cause' variation is considered to be due to the inherent nature of the process and cannot be altered without changing the process itself. 'Assignable (or special) causes' of variation are unusual shocks or other disruptions to the process, the causes of which can and should be removed".

Finally, Montgomery [10] stated that: "...common causes are sources of variability that are embedded in the system or the process itself, while assignable causes usually arise from an external source."

Thus, there are slightly different views on whether assignable causes of variation are a result of intervention into the system from the outside or not; however, there is a full agreement that they are "some ephemeral events that can usually be discovered ...and removed". A tool for distinguishing assignable causes of variation from common causes is the control chart, coined by W. Shewhart in 1924. Up to 2010 there were more than 4000 research papers published on this topic [26]. Intensive analysis of books and main reviews in this area showed [1–10, 25, 27, 28] that practically all papers on the ShCCs considered very simple model of chart's behavior. Almost all researchers studied the statistical properties of simple ShCC charts when some assignable cause of variation changed either the mean or the standard deviation or both of the underlying distribution, which stayed of unchanged type (and almost always was normal).

Alternatively, if one looks through the explanation of assignable causes of variation in the very popular SPC Manual [29], used in auto industry for many years, she/he will see the picture (page 30 in [29]), which clearly shows that an assignable cause of variation can lead to an arbitrary change of the distribution function type[2]. Just this was the main idea of the work [30]. The authors studied the case'when after a special cause of variation emerged, the underlying normal distribution function (DF) transformed into either uniform or log-normal distribution. It was found that the probabilities of detecting a shift in the mean changed radically from the case of normal DF (see Figures 4–7 in Ref. [30]). Indeed, as soon as one accepts the opportunity of DF type to change after the impact of assignable cause of variation, a lot of different possibilities emerges, and only one of them has been investigated and described in the literature. It was proposed in Ref. [30] to introduce two types of assignable causes of variation: not changing the underlying DF and changing it. It is worth stressing that this idea can be generalized, as ShCCs do not need any assumptions about DF type. So, a more general proposal could be an introduction of two types of assignable causes of variation: not changing the system where the process is going on, and changing that system. Though more than 10 years have passed since the paper [30] was published, this idea has not been either supported or refuted by statistical community. Here, we would like to revisit this notion from a different angle. Let us look at **Figure 1**, taken from our work [31]. One can see in **Figure 1**, the ShCC with two red circles and two green ovals on it. Red circles relate to the points falling beyond chart's limits, and ovals show the points where the process mean jumped. Obviously, both situations emerged due to some special causes of variations. But is there any difference between these two cases: one when the assignable cause was evanescent, and the system has not changed, and second when the assignable cause has changed the system? As far as we know, such a question has never been discussed in SPC literature. Does it deserve to be discussed? We are sure it does because in the first case the search

[2] It is noteworthy that using an appropriate ShCC to analyze stability of key processes is a mandatory requirement in the automotive standard ISO/TS 16949.

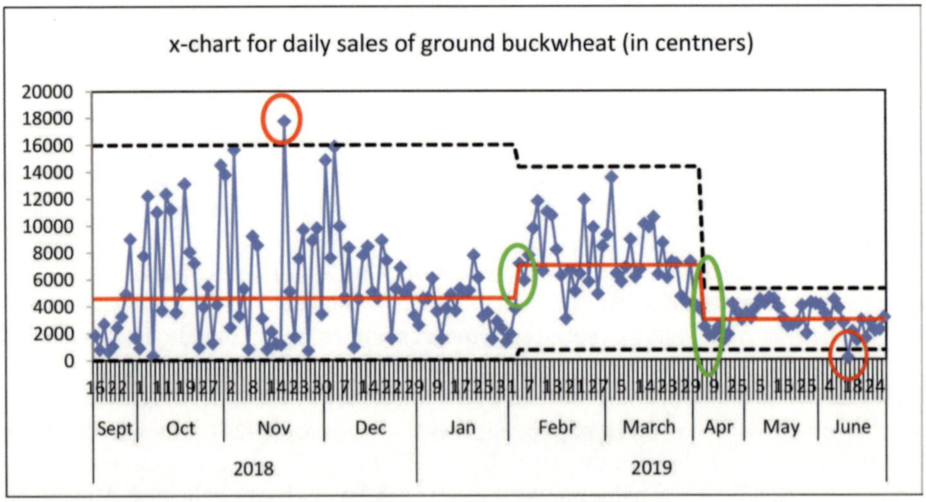

Figure 1.
Daily sales of a distribution network. Here 1 centner = 100 kilograms.

for the root cause of interference in the process has to be made by the process team (engineers, operators, linear managers, etc.), and in the second case this search is an act of top management – only CEOs are responsible for the system as a whole. This forces us to come back to the idea of different types of assignable causes of variations. Having generalized and slightly modified the definitions from [30, 31] we suggest the following version:

Definition 1: An assignable cause of variation of type I (*Intrinsic*) does not change the system within which a process works (e.g., does not change the type of the underlying DF). As a result, it is quite natural to consider such a type of assignable causes as belonging to the system (though this is not a necessary condition).

Definition 2: An assignable cause of variation of type X (eXtrinsic) changes the system within which a process works (e.g., changes the parameters, or type or both of the underlying DF). As a result, it is quite natural to consider such a type of assignable cause as, most probably, not belonging to the system (though this is also not necessary).

If the statistical community agrees with our suggestions, then the difference between dissimilar types of assignable causes of variation will help practitioners to grasp who, in the first place, has to interfere in the process. This is highly important knowledge to improve the process with success. The instability due to assignable causes of type I requires searching for a root cause within the system. The instability due to assignable causes of type X requires searching for a root cause outside the system.

4. The importance of time order of process values

As stressed by Shper and Adler [32], the problem of data nonrandomness has been underestimated in recent years, though it was of primary importance to Shewhart. More than once, W. Shewhart returned to this issue, explaining the key role of the order of points for understanding whether a process is stable or not. On page 12 of [23], he clarifies that all attempts to determine some DF that could thoroughly describe a state of statistical control are useless and senseless. Some statisticians considered the normal law to be such DF, but these hopes turned out to be refuted completely. Further on page 27, W. Shewhart continues: "...the significance of observed order is independent of the frequency distribution..." and "... are primitive". Shewhart's conclusion about the importance of the point order was firmly supported by such outstanding gurus as W. Deming (see Deming's Foreword in [23]) and G. Box [33].

Let us consider a revealing example of an erroneous conclusion caused by the neglecting of the role of the order of points in real processes. A well-known expert in SPC, W. Woodall has long since his survey of 2000 [17], stood up for the elimination of the moving range chart from an arsenal of SPC tools. His arguments were based on the results of random data simulation, which follows from the paper of Rigdon et al. [34], which Woodall referred to in [17]. However, the moving range automatically takes into account the order of points due to the structure of successive differences. That is why it contains much more important information about the process than the standard deviation (SD). The SD completely ignores the succession of process points. Therefore, using the average of the moving range of two (AMR) to estimate the ShCC limits allows anyone to take into account the order of points within a process. If we eliminate the moving range chart from using, we automatically neglect many patterns within a process.

Another famous SPC expert, L. Nelson, also recommended avoiding using the moving ranges because of problems with their interpretation [34]. But, in fact, he considered the moving ranges necessary to calculate the limits of the x-chart because they are better than SDs. The reason is clear: the moving range measures variations from point to point irrespective of their level, which can vary due to trends, oscillations, patterns, etc. Now, one may ask: how often are any patterns present in real processes, and if they influence the outcomes or not?

Without any doubt patterns are present at all real processes, but sometimes their influence may be small, and, consequently, ignored. The famous Box adage about models is working, of course. But, in practice, the level of pattern influence is rarely known beforehand. Shewhart foresaw this many years ago: "... a sequence is called random if it is known to have been produced by a random operation, but is assumed to be nonrandom if occurring in experience and not known to have been given by a random operation" ([23], *17*). This means that if an observed sequence is or is not random, it can be verified only in the future and not by any ingenious math. It follows immediately from this Shewhart remark that there is no and cannot even exist any universal indicator of process nonrandomness. So, what should we do in such a situation? We need to have a variety of dissimilar metrics/indices/rules revealing different types of nonrandomness. There are a number of such indices that are well-known and widely used. The so-called additional rules for ShCC interpretation are the first coming to mind. These rules are described in practically all books, guides, standards, etc., of SPC (see, e.g., §5.7 in [9]). It is noteworthy that each such rule can reveal only one single case of nonrandomness. In other words, the standard set of rules covers a minuscule part of potential opportunities. Except for these rules, there is a run test on randomness [35] – a useful rule based on the number of series in data. Again, it reveals only one type of nonrandomness – a relationship between a number of points lying above and below a chart central line.

A new test on data randomness was proposed in [32]. It is the ratio of AMR to SD. As noted above, the value of AMR is very sensitive to any patterns of nonrandomness, so this ratio deviates from its standard value as soon as data have some kind of

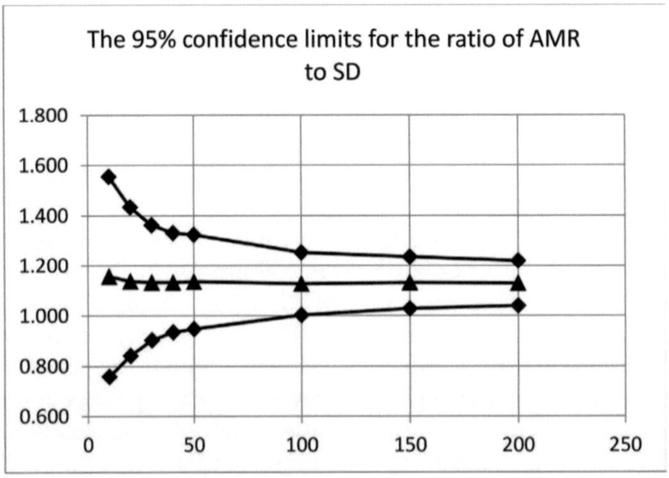

Figure 2.
The limits of AMR/SD ratio for k = 10, 20, 30, 40, 50, 100, 150, and 200.

nonrandomness[3]. What is important about this index is that it covers not a single case of nonrandomness but, rather, an unknown set of possibilities. In **Figure 2** (taken from [32]), one can see the dependence of confidence limits for AMR/SD on the number of points (k) under investigation. If the value of AMR/SD lies above or below the lines on that picture, then one may state that with 95% confidence, the data are not random. The exact values of probabilities are given in [32].

There was the question raised in Shper and Adler [32]: why do current studies of ShCCs ignore the order of points? Our version is as follows. This ignorance is caused by the lack of understanding of the difference between analytic and enumerative studies. As Deming explained many times (see, e.g., [37]), an enumerative study deals with units taken from some population that have a definite value of mean, SD, DF, etc. An analytic study deals with a process that is going on and on; it does not have a definite mean or DF because it changes permanently. Many statistical methods such as confidence intervals and hypothesis testing can be applied to enumerative studies but are inapplicable to analytical ones. Deming's conclusion in Ref. [37] turned out to be quite unambiguous: "Statistical theory for analytic problems has not yet been developed" [37].

5. The impact of non-normality on the control limits of ShCCs

It is well-known that the traditional theory of ShCCs is based on (1) assumption of data randomness and (2) assumption of data being distributed normally. We discussed the first assumption in the previous paragraph. Below, we will consider the second.

It is assumed here that the reader is somewhat familiar with the construction and use of ShCCs. According to a generally accepted view, each ShCC has limits, separating the zone of system variability from the area of the assignable cause habitation. These limits can be calculated with very simple formulas based on the three-sigma

[3] The reverse ratio of SD to (AMR/$d2$) was called a stability index in [36] ($d2$ is one of the constants used to construct ShCC). We think that this name is incorrect, but this is a topic of another paper.

rule suggested by Shewhart [24]. Typical formula for the control limits (CL) of many popular control charts for variables looks like [38]:

$$\text{CL} = \text{Average} \pm \text{Scaling Factor} \bullet \text{Some Measure of Dispersion} \quad (1)$$

Scaling factors in (1) are frequently named as control chart constants and are usually denoted by different capital letters with indices, for example, A_2, D_3, D_4, and E_2 (to name a few most widely used). They, in turn, depend on the so-called bias correction factors d_2, d_3, and d_4 ([9], *416*):

$$A_2 = \frac{3}{d_2\sqrt{n}} \quad (2)$$

$$D_3 = 1 - 3\frac{d_3}{d_2} \quad (3)$$

$$D_4 = 1 + 3\frac{d_3}{d_2} \quad (4)$$

$$E_2 = \frac{3}{d_2} \quad (5)$$

$$E_5 = \frac{3}{d_4} \quad (6)$$

The factors d_2, d_3, and d_4 (and, therefore, chart coefficients) are considered constant in the SPC literature, mostly widely used by practitioners (e.g., [6, 8–10]). Though, statisticians working in the field have known for a long time that non-normality has a great impact on bias correction factors. Why is this? On the one hand, this opinion is based on the works of many outstanding statisticians of the first half of the last century (see references and other details in [39], and, e.g., §7.3 in the excellent book [40]). On the other hand, the bias correction factors do change insignificantly in many cases but not in all possible ones. This issue was carefully studied in our work [39], where such skewed distributions as exponential, log-normal, Weibull, Burr and Pareto were simulated, and the values of d_2, d_3, and d_4 were estimated. The main results of that paper are as follows.

In conclusion to his landmark work of 1967 [41], Irwing Burr wrote: "... we can use the ordinary normal curve control charts constants unless the population is markedly non-normal. When it is, the tables provide guidance on what constants to use." Unfortunately, Burr did not point out what the words "markedly non-normal population" meant operationally. Moreover, there has been no discussion on this issue up to now. So, we proposed in [39] that a twofold increase in the probability of falling beyond the control limits be considered as the condition of significant deviation. Then, after simulation, the results shown in **Tables 1–3** were obtained. In **Table 1**, the parameters of investigated DFs and their notation are presented. β_1 and β_2 are the squared skewness and traditional kurtosis, respectively. The means of d_2, d_3, d_4, and

#	DF	Parameters of DF	β_1	β_2	Notation
1	Normal	Mean = 0 SD = 1	0.00	3.00	Gau
2	Uniform	Mean = 0 SD = 1	0.00	1.80	Uni
3	Exponential	λ = 0.1; 0.2; 0.01	4.00	9.00	Exp
4	Log-normal	Mean = 0 SD = 0.5	3.06	8.90	Lgau1
5		Mean = 0 SD = 0.6	5.11	13.27	Lgau2
6		Mean = 0 SD = 0.7	8.34	20.79	Lgau3
7		Mean = 0 SD = 1.0	38.25	113.94	Lgau4
8	Weibull	Shape parameter = 0.8	7.92	15.7	Wei1
9		Shape parameter = 1.5	1.15	4.40	Wei2
10		Shape parameter = 0.7	12.24	23.54	Wei3
11	Logistic	Mean = 0 SD = 1	0.00	4.20	Lgi
12	Burr's	c = 2, k = 4 from Burr's table I	2.04	7.36	B5
13	Pareto	Shape parameter = 20	5.48	9.13	Prt

Table 1.
Notations and parameters of DFs studied in [39].

	d_2	d_3	d_4	E_2	E_5	D_4	δE_2, %**	δE_5,%	δD_4, %
Gau*	1.129	0.851	0.955	2.66	3.14	3.26	−0.10	−0.12	−0.21
	1.128	*0.853*	*0.954*	*2.660*	*3.145*	*3.268*			
Uni	1.156	0.816	1.017	2.60	2.95	3.12	−2.44	−6.21	−4.60
Exp	1.003	0.994	0.699	2.99	4.29	3.97	12.44	36.47	21.58
Wei1	0.927	1.068	0.567	3.24	5.29	4.46	21.66	68.24	36.36
Wei2	1.091	0.899	0.871	2.75	3.44	3.47	3.37	9.52	6.24
Wei3	0.866	1.112	0.476	3.46	6.30	4.85	30.23	100.40	48.48
Lgau1	1.041	0.957	0.782	2.88	3.84	3.76	8.34	21.98	14.99
Lgau2	1.003	0.995	0.718	2.99	4.18	3.98	12.44	32.85	21.67
Lgau3	0.965	1.032	0.655	3.11	4.58	4.21	16.87	45.63	28.77
Lgau4	0.827	1.144	0.456	3.63	6.58	5.15	36.37	109.19	57.59
Lgi	1.104	0.884	0.902	2.72	3.33	3.40	2.16	5.75	4.11
B5	1.067	0.926	0.835	2.81	3.59	3.60	5.70	14.24	10.27
Prt	0.978	1.021	0.658	3.07	4.56	4.13	15.32	44.97	26.44

**This row gives the generally accepted values of bias correction factors.*
***δE_2, % denotes the relative deviation of E_2 from its standard value 2.66. Similar – δE_5,% and δD_4, %.*

Table 2.
Results for x-mR chart.

the corresponding values of A_2, E_2, *and* E_5, calculated by using Eqs. (2)–(6), as well as their relative deviations from the standard values for the x-mR chart, are given in **Table 2**. **Table 3** provides similar information for $\overline{X}$ -R chart with subgroup size n = 2 and 3. The DFs having a relative probability increase of less than twofold are excluded from **Table 3**. Almost all figures in **Table 2** show the values that relate to DFs having

	d_2	d_3	d_4	A_2	D_4	δA_2,%**	δD_4, %
Gau*	1.128 *1.128*	0.852 *0.8525*	0.953 *0.954*	1.881 *1.880*	3.266 *3.268*	0.03	−0.06
				$n = 2$			
Exp	1.000	0.998	0.692	2.121	3.994	12.84	22.22
Wei1	0.921	1.074	0.560	2.303	4.498	22.51	37.65
Wei3	0.865	1.120	0.475	2.452	4.884	30.45	49.46
Lgau1	1.149	0.995	0.904	1.846	3.598	−1.80	10.10
Lgau2	1.005	0.994	0.718	2.111	3.967	12.27	21.39
Lgau3	0.956	1.041	0.642	2.219	4.267	18.03	30.56
Lgau4	0.818	1.155	0.447	2.593	5.236	37.94	60.22
Prt	0.978	1.027	0,657	2.169	4.150	15.37	27.00
				$n = 3$			
Gau*	1.690 *1.693*	0.890 *0.8884*	1.584 *1.588*	1.025 *1.023*	2.578 *2.575*	0.18	0.12
Exp	1.504	1.115	1.233	1.152	3.226	12.65	25.26
Wei1	1.385	1.229	1.036	1.251	3.662	22.25	42.22
Wei3	1.298	1.301	0.899	1.334	4.007	30.44	55.61
Lgau1	1.557	1.062	1.327	1.112	3.046	8.74	18.30
Lgau2	1.503	1.125	1.231	1.152	3.246	12.65	26.04
Lgau3	1.434	1.191	1.120	1.208	3.492	18.07	35.60
Lgau4	1.215	1.365	0.807	1.429	4.381	39.70	70.14
Prt	1.463	1.159	1.164	1.184	3.377	15.73	31.13

**This row gives the generally accepted values of bias correction factors.*

Table 3.
Results for $\bar{X}$ -R chart.

more than a twofold increase in corresponding probabilities. It was recommended in [39] that for all cases when one encountered "markedly non-normal" data, he/she should use the algorithm to construct the ShCC proposed in [39] and the corrected values of the chart's constants given there.

6. What happens when the mean shift is transient?

This issue was studied in a paper published in 2021 [42]. We investigated the impact of transient shifts on the operational characteristics of ShCCs. Traditionally, all guides and standards tell the readers about the efficiency of the X-bar chart to detect the shift of the process mean. This efficiency notably surpasses the efficiency of x-chart, and the superiority grows when the sample size n increases. However, this is true but only for the so-called sustained shift. When the shift becomes transient, the situation may change radically, which most practitioners simply do not know because this issue is rarely discussed in the literature. In [42], we discussed an elementary model shown in **Figure 3**.

In order to compare the efficiency of different types of ShCCs one needs to calculate the so-called power function (PF) for each type of chart. In Ref. [42] there was

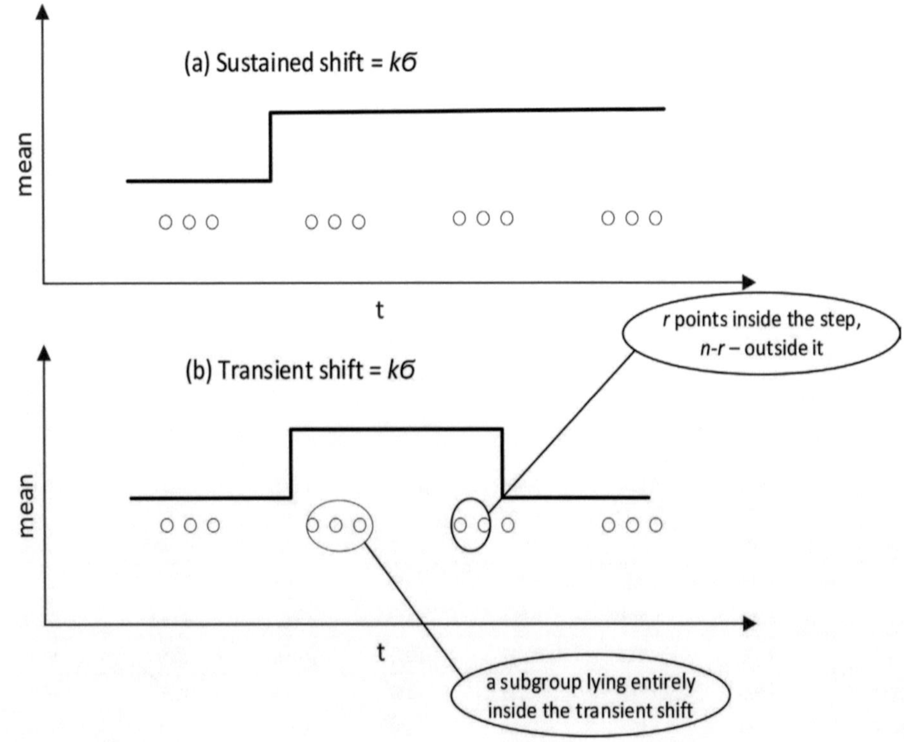

Figure 3.
(a) – sustained shift, and (b) – transient shift, n – sample size, here n = 3 The number of subgroups lying entirely inside the step we denoted as m. Here m = 1.

considered the case when the transient shift starts between subgroups and ends after m-shifted subgroups within the (m + 1)th. In other words, we assumed that the X-bar chart with subgroups of size n was being built and an assignable cause of variation emerged between subgroups and lasted so that it covered m subgroups totally and r points within the last (m + 1)th. In this case, it was obtained for the average of shifted subgroups:

$$\overline{X} = \frac{\left[nm(\mu_0 + k\sigma) + r(\mu_0 + k\sigma) + (n-r)\mu_0\right]}{n(m+1)} = \mu_0 + \frac{m}{m+1}k\sigma + \frac{rk\sigma}{n(m+1)} \tag{7}$$

Clearly, the bias of the average will be approaching the sustained value when $m \to \infty$, and the impact of the transient shift will be maximal for m = 0. In this case the average is equal to

$$\overline{X} = \frac{\left[r(\mu_0 + k\sigma) + (n-r)\mu_0\right]}{n} = \mu_0 + \frac{rk\sigma}{n} \tag{8}$$

Obviously, the shift of the average will be decreasing and therefore the probability of detecting it will be decreasing as well. The value of *SD* is assumed to be constant (at least at the first approximation) and equal to $\sigma / \sqrt{n}$. Thus, one can obtain:

$$PF = 1 - \Phi\left(3 - \frac{kr}{\sqrt{n}}\right) \tag{9}$$

Using of traditional assumptions of data normality and of being *i.i.d.* (identically independently distributed) was implied. The results of calculations on the base of (9) for different values of *n* and *r* are shown in **Figure 3** of [42]. One example of those pictures is given in **Figure 4**.

As one could expect, when *r* is equal to *n* the PF coincides with the traditional one given in many books (see, for example, Figure 10.1 in [9]).

As soon as not all the subgroup's points fall into the changed process, the PF starts to decrease, and for some value of *r* – in [42], we called this value boundary, r_b – the PF becomes less than the corresponding PF for $n = 1$ (the PF of the chart for the individual values). It is quite easy to find out the value of r_b for which the PF of X-bar chart becomes less than the PF of the chart for individuals:

$$r_b \leq \sqrt{n} \tag{10}$$

This means that the traditional conclusion about the superiority of X-bar chart over an X-chart does not work, at least when the duration of the process jump is shorter than the time to gather a subgroup. It all depends on the amount of points falling into the changed part of the subgroup. This example displays that even the simplest ShCCs are not as easy in practice as they may seem. And even for the values of $r > r_b$, the resulting values of probability to reveal the signal about the shift may be essentially less than what is traditionally mentioned in SPC textbooks. For example, for an X-bar chart with $n = 10$, the probability of detecting the mean's shift of one sigma falls down from 0.564 when the shift is sustained to 0.078 when only five points are within the changed process, and five points belong to the unchanged process.

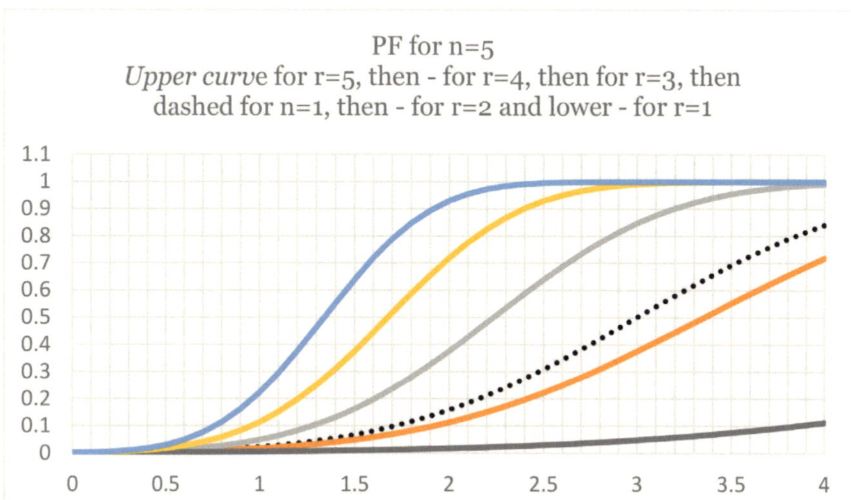

Figure 4.
An example of PF for sample of five with different values of r.

7. ShCCs and 2 M's: metrology and management

In the final section of this chapter, we would like to briefly discuss two areas of people's activity where ShCCs must be used most often and where they are not practically being used at all. These fields are metrology and management.

Metrology's primary concern is variation of measurement, so it seems very logical that all metrology-responsible people have to know the basics of SPC and have to be able to construct and interpret ShCCs. Unfortunately, reality looks quite the opposite [43, 44].

Management's primary concern is variation between people and its impact on the organization. Again, the knowledge of ShCCs looks like it leaves no alternative for managers. Once more, the dark side of reality has beaten us to it [45]. Dr. Deming wrote the honest truth in the foreword to [23]: "The fact is that some of the greatest contributions from control charts lie in areas that are only partly explored so far, such as applications to supervision, management, and systems of measurements, including the standardization and use of instruments, test panels, and standard samples of chemicals and compounds" ([23], *ii*). A detailed discussion of these issues would require writing another chapter that would most likely be much longer than this one. So we decided to confine ourselves to these ultrashort comments in order to put these topics to the next discussion.

8. Conclusions

We live in a World full of complexity, human irrationality, variability, and, consequently, unpredictability [45]. One of the main features of this world is the ubiquitous presence of variations. Variations differ from each other, so the question arises if one needs to react to them or leave them unnoticed. The ability to answer this question is not something innate to the human being. People should be taught, especially taking into account that knowledge about variability and how to cope against it emerged only after Shewhart's works of 1931 and 1939 [23, 24]. The tool invented by him – control chart – is the only way to understand if the process is controlled/stable/predictable or not. Without this knowledge, there are no chances to improve any system. Ironically, ShCCs turned out to be very simple technically and rather difficult to use profitably. The reason – contextual knowledge is vital for a genuine understanding of what the charts are trying to tell us. ShCC is a communication tool between a system where a process is going and the process owner. Its construction cannot be totally algorithmized [46]. In other words, everybody in any organization should be taught the art of constructing and interpreting ShCCs. In fact, the reality looks completely different. We started this chapter with the quote of Hoyer, Ellis about quality professionals who do not know the basics of SPC[4], and here we will continue citing their phrase "And why should they? Our review of a very large number of SPC textbooks reveals page after page of "cookbook" discussions of practically everything under the sun—with very little discussion on the foundation of SPC ([14], *63*)". Unfortunately, the basics of SPC is a set of rules and recipes which are essentially more narrow than wants of practice. We tried to widen this set in this chapter. To this end, we discussed some uncertainties in the interpretation of assignable causes of variation, and proposed to

[4] We are sure that engineers, managers, CEOs, teachers, physicians, etc. should be added to quality professionals.

differentiate between assignable causes that change the system, and those, that do not influence it. Then, we considered the impact of data order on ShCCs and suggested using the average moving range as an indicator of nonrandomness. The next part of the chapter was devoted to the influence of data asymmetry on the limits of ShCC. Recent results for some asymmetrical DFs were delivered. Further, we discussed a simple case of transient jump of process mean and demonstrated that some traditional and established rules should be essentially changed. At last, at the very end of the chapter, we formulated two important issues that should be discussed carefully in the near future. They relate to the link between ShCCs and 2 M's: management and metrology.

Conflict of interest

The authors declare no conflict of interest.

Author details

Vladimir Shper*, Elena Khunuzidi, Svetlana Sheremetyeva
and Vladimir Smelov
NUST "MISiS", Moscow, Russia

*Address all correspondence to: vlad.shper@gmail.com

References

[1] Schindowski E, Schürz O. Statistische Qualitätskontrolle. Berlin: Veb Verlag Technik; 1974

[2] Murdoch J. Control Charts. London: The Macmillan Press, Ltd; 1979

[3] Grant EL, Leavenworth RS. Statistical Quality Control. 5th ed. NY: McGraw-Hill; 1980

[4] Kume H. Statistical methods for quality improvement. The Association for Overseas Technical Scholarship (AOTS). 1985. p. 304

[5] Duncan AJ. Quality Control and Industrial Statistics. 5th ed. Homewood, IL: Irwin; 1986

[6] Wheeler DJ, Chambers DS. Understanding Statistical Process Control. 2nd ed. Knoxville: SPC Press, Inc.; 1992

[7] Rinne H, Mittag H-J. Statistische Methoden der Qualitätssicherung. Fernuniversität-Gesamthochschule-in-Hagen. Deutschland: Fachbereich Wirtschaftswissenschaft; 1993

[8] Alwan LC. Statistical process analysis. Irwin/McGrow-Hill Series in Operations and Decision Sciences. 2000. p. 752

[9] Wheeler D. Advanced Topics in Statistical Process Control. 2nd ed. Knoxville: SPC Press; 2004

[10] Montgomery DC. Introduction to Statistical Quality Control. 6th ed. Jefferson City: John Wiley & Sons; 2009

[11] Balestracci D. Data Sanity: A Quantum Leap to Unprecedented Results. USA: Medical Group Management Association; 2009

[12] ISO 7870-2:2013. Control Charts - Part 2: Shewhart Control Charts. ISO. 2013

[13] Sheremetyeva S, Shper V. Business and variation: Friendship or misunderstanding. Standards and Quality. 2022;**2**:92-97. DOI: 10.35400/0038-9692-2022-2-72-21 (in Russian)

[14] Hoyer RW, Ellis WC. A graphical exploration of SPC. Quality Progress. 1996;**6**(Part 2):57-63

[15] Hoyer RW, Ellis WC. Another look at “a graphical exploration of SPC”. Quality Progress. 1996;**11**:85-93

[16] Hoyer RW, Ellis WC. A graphical exploration of SPC. Quality Progress. 1996;**5**(Part 1):65-73

[17] Woodall W. Controversies and contradictions in statistical process control. Journal of Quality Technology. 2000;**32**:341-350. DOI: 10.1080/00224065.2000.11980013 [Accessed: February 20, 2023]

[18] Steinberg D. Industrial statistics: The challenges and the research. Quality Engineering. 2016;**28**(1):45-59. DOI: 10.1080/08982112.2015.1100453 Accessed: February 22, 2023

[19] Woodall W. Bridging the gap between theory and practice in basic statistical process monitoring. Quality Engineering. 2017;**29**(1):2-15. DOI: 10.1080/08982112.2016.1210449 Accessed: February 20, 2023

[20] Woodall W. Preprint. Available from: https://www.researchgate.net/publication/366090712_Recent_Critiques_of_Statistical_Process_Monitoring_Approaches_12722 [Accessed: February 21, 2023]

[21] Box G. Scientific Statistics, Teaching, Learning, and the Computer. CQPI Report, No. 146; 1996

[22] Quesenberry C. Statistical gymnastics. Quality Progress. 1998;**9**:77-79

[23] Shewhart W. Statistical Methods from the Viewpoint of Quality Control. N.Y.: Dover Publications, Inc.; 1939/1986

[24] Shewhart W. Economic Control of Quality of Manufactured Product. Milwaukee: ASQ Quality Press; 1931/1980

[25] Shewhart WD. Deming, and six sigma. In: W. Edwards Deming 2007 Fall Conference, Manuscript No. 187. Available from: http://www.spcpress.com/pdf/DJW187.pdf [Accessed: June 18, 2023]

[26] Adler Y, Maksimova O, Shper V. Shewhart Control Charts in Russia and Abroad: Brief Review of the State-of-the Art (Statistical Aspects). 2011. Available from: http://ria-stk.ru/upload/image/stq/2011/N8/082011-1.pdf (*in Russian*). [Accessed: June 18, 2023]

[27] Jensen WA, Jones-Farmer LA, Champ CW, Woodall WH. Effects of parameter estimation on control chart properties: A literature review. Journal of Quality Technology. 2006;**38**:349-364

[28] Chakraborti S, Human SW, Graham MA. Phase I statistical process control charts: An overview and some results. Quality Engineering. 2009;**21**:52-62

[29] Analysis MS, editor. Reference material. In: Chrysler Group LLC, Ford Motor Company. 4th ed. USA: General Motors Corporation; 2010

[30] Adler Y, Shper V, Maksimova O. Assignable causes of variation and statistical models: Another approach to an old topic. Quality and Reliability Engineering International. 2011;**27**(5):623-628. DOI: 10.1002/qre.1207

[31] Shper V, Sheremetyeva S, Smelov V, Hunuzidi E. Shewhart control charts – An irreplaceable tool of explanatory data Analysis with underestimated potential. International Journal for Quality Research. 2023;**18**(2) (In press)

[32] Shper V, Adler Y. The importance of time order with Shewhart control charts. Quality and Reliability Engineering International. 2017;**33**(6):1169-1177. DOI: 10.1002/qre.2185

[33] Box G, Narasimhan S. Rethinking statistics for quality control. Quality Engineering. 2010;**22**(2):60-72

[34] Rigdon SE, Cruthis EN, Champ CW. Design strategies for individuals and moving range control charts. Journal of Quality Technology. 1994;**26**(4):274-287

[35] Kenett RS, Zacks S. Modern Industrial Statistics: With Applications in R, MINITAB and JMP. London: John Wiley & Sons; 2014. p. 9

[36] Nelson LS. Control charts for individual measurements. Journal of Quality Technology. 1982;**14**(3):172-173

[37] Deming W. In: Orsini J, editor. The Essential Deming. Leadership Principles from the Father of Quality. NY: McGrow-Hill; 2013

[38] Wheeler DJ. Are you Sure we Don't Need Normally Distributed Data? 2010. Available from: https://www.qualitydigest.com/inside/six-sigma-column/are-you-sure-we-don-t-need-normally-distributed-data-110110.html [Accessed: August 2, 2021]

[39] Shper V, Sheremetyeva S. The impact of non-normality on the control limits of Shewhart's charts. Tyazheloe Mashinostroenie. 2022;**1-2**:16-29

[40] David HA. Order Statistics. John Wiley & Sons, Inc.; 1970

[41] Burr IW. The effect of non-normality on constants of $\overline{X}$ and R charts. Industrial Quality Control. 1967;**563**:566-569

[42] Shper V, Gracheva A. Simple Shewhart control charts: Are they really so simple? International Journal of Industrial and Operations Research. 2021;**4**:010. DOI: 10.35840/2633-8947/6510

[43] Adler Y, Shper V. Statistical thinking and metrology: problems and decisions. In: XX IMEKO World Congress. Busan, Republic of Korea: Metrology for Green Growth; September 9-14, 2012

[44] Adler P, Felichkina, Shper. The role of distribution functions in metrology. Journal of Physics: Conference Series;**1420**:012036

[45] Shper V. Solving 21st century challenges. Quality Progress. 2022;**9**:36-41

[46] Adler Y. Problems that cannot be solved algorithmically and artificial intelligence. Economics and Management: Problems and Decisions. 2018;**7/77**(5):17-24. (in Russian)

Chapter 4

Identification of Defects Causes: Ishikawa Diagram and 5 Whys in Theoretical and Practical Terms

Barbara Ciecińska

Abstract

The chapter briefly presents the issue of quality and its management in production. In the relevant part of the chapter, the Ishikawa diagram, its variations, and the possibility of extending its application to new areas will be discussed. Similarly, the 5 Whys method is presented. The application of the Ishikawa diagram and how to use the 5 WHYS method are presented with practical examples, enabling them to be transferred to the operation of real enterprises. These examples concern machining, laser processing, and gluing in the context of defective products and processes. The discussion includes case studies of what defects may appear in this type of (or similar, in the field of mechanical engineering) processes, a discussion of their importance in production. Ways to improve machining and assembly operations in the presented context are shown.

Keywords: Ishikawa diagram, 5 Whys, machining, laser processing, bonding

1. Introduction

Production management is currently a difficult but also interesting and dynamically developing field. In production plants, with a constantly changing range of manufactured products, the challenge is to maintain a certain acceptable level of production, especially in relation to product quality. In production, the method of carrying out the production process may change; modern technologies and different devices and materials are implemented. The range of produced products can be very different at different times of the year. An example of a particular challenge may be the concept of Industry 4.0, which talks about mass-scale production of individualized, unitary products. In addition, the manufacturing process is significantly affected by the environment in which legal and normative guidelines change; various concepts of production management appear, such as Total Productive Maintenance regarding the maintenance of the machine park in efficiency, Cleaner Production expressing the need to reduce the negative impact of production on the natural environment, and implementation of a systemic approach based on ISO 9001 (quality management), 14001 (environmental management) or 45001 (occupational safety management) standards. The classic production model draws attention to the input elements, such as materials, energy, and knowledge, and to the output elements—products, services, and profit, which are processed in a specific way conditioned by technology. All these elements are somehow "fastened together" by the clamp of the management philosophy. In addition to the abovementioned concepts, decisions are also influenced by social and cultural conditions, customs, financial resources, and many other factors [1, 2].

For these reasons, quality can be perceived in different ways. In one case, it will be identified with the characteristics of the product, in another—with a specific manufacturer. In still other situations, quality may be defined as specific operating conditions, functional features, and features of the work environment, in relation to the natural environment, esthetic values, and others [3].

Despite the constant changes in the reasons for production decisions, the tools and methods of quality management known for a long time remain constant and effective. An example is the Ishikawa diagram and the 5 Whys method, which make it possible to identify the causes, places, and problems with insufficient quality, not only of products but also of defects in the broader sense. The identification process can be supported by other techniques, for instance, a kinship diagram or teamwork (like brainstorming), which trigger creativity and some freedom in defining problems and organize corrective or improvement actions [4].

2. Quality and quality management

Crosby defined quality as the fulfillment of requirements and Juran defined it as usability or applicability. Tse Li used to say that quality is "excellence, perfection of workmanship". The EN ISO 9001:2015 standard says that quality is the extent to which a set of inherent properties meet requirements [5, 6].

Quality and the concepts related to it have evolved from antiquity to the present. Groundbreaking changes in the attitude to the quality of products and services took place thanks to the American scientists William Deming and Joseph Juran. Their work was used in Japan in 1950–1955, which, combined with Japanese culture and lifestyle, led to the concept of Total Quality Management (TQM). The concept of TQM refers to all the goals of the enterprise and is often called a management philosophy that is applied in activities related to all undertakings and processes of the organization. The main role in these processes is played by people: management, which has a leadership role, and employees who make up a team that creates quality. In this philosophical approach, quality is seen as a process of continuous improvement that applies to the entire organization, customers, and suppliers. Feedback is maintained with the client in mutual communication, and the client's expectations and needs are taken into account in the work on quality. Suppliers are also an important element in the company's value creation chain, as their inappropriate products or services may be the beginning of the company's problems. The ISO 9001 standard is the basis for a systemic approach in which customer needs are defined by contract and quality management focuses on [7–9]:

- quality planning—in order to determine the requirements and goals to be pursued, for example, in the context of a product, process, or other production areas.

- production control—through the use of tools and methods of conduct, implementation of operational activities necessary to observe the process, and prevention of the occurrence of defects and noncompliant products.

- quality assurance—by planning and repeating activities carried out systematically in order to achieve customer satisfaction.

- improving quality—that is, taking actions that increase the effectiveness and efficiency of processes and activities performed.

With regard to the concept of quality, kaizen talks about the need to improve in small steps. Kaizen is synonymous with rational and skillful time management, communication, creative teamwork, as well as cost reduction by avoiding defects. Kaizen is based on foundations such as [10]:

- standardization of work,

- organization of work,

- elimination of waste.

Kaizen postulates are closely related to other concepts of production management, such as Lean Manufacturing, by striving to achieve maximum productivity, efficiency, and products without errors. This goal can be achieved, among others, by the use of a variety of quality management techniques and tools, both older and newly developed, combined with modern production management [11, 12]. **Table 1** shows the quality management tools. The first column lists the so-called old ones, and the second column lists the new ones.

3. Ishikawa diagram and 5 Whys

The method of analyzing the causes of various defects and irregularities, called the fishbone diagram or Ishikawa diagram, was developed by Kaoru Ishikawa in 1943. This diagram makes it possible to link the causes and effects of various technological, organizational, and economic problems in a universal way. In Ishikawa quality problems, the diagram is often used where the chain of cause and effect can be extended. The fishbone diagram makes it possible to see previously invisible or unconscious relationships and dependencies. The diagram comes in different versions, which

"Old" quality management tools	"New" quality management tools
Control sheets	Affinity diagram
Histogram	Dependency diagram
Correlation diagram	Tree diagram
Control cards	Array diagram
Cause and effect diagram	Table data analysis
Pareto-Lorenz analysis	Decision planning diagram
Block diagram	Network diagram

Table 1.
Quality management tools [11, 13].

differ in the type of analyzed areas or their number. Typical lists of areas for analysis are given in **Tables 2–6**; however, each time, the team can establish its own set [5, 14, 15].

Analyzed areas	Problem
Man	Is the loss of quality due to human error?
Methods	What methods of work or production were used?
Machinery	Maybe the error is generated by the machines malfunctioning?
Materials	Could the problem be in the material that the product is made of?
Management	Do problems arise from a lack of supervision, planning, and work organization; poor motivation and leadership; lack of control; and lack of conceptual, technical, and social skills?

Table 2.
5M Ishikawa diagram—basic.

Analyzed areas	Problem
5M	Basic 5M
Environment	Is the production environment, working condition, and other people a problem?

Table 3.
5M+E Ishikawa diagram.

Analyzed areas	Problem
5M	Basic 5M
Measurement	Are the measurements taken correctly?
Mother Nature	Can nature/weather influence the effects of production?
Maintenance	Is the technical condition of machines and their maintenance important for quality?

Table 4.
5M+3 M Ishikawa diagram.

Analyzed areas	Problem
Product	What kind of products will we sell?
Price	What should be the price for the products we will sell?
Promotion	Are there any promotional campaigns? How many times a year? Are they effective?
Place	Is the location from which the company operates favorable?
Process	How are the affairs of the company handled?
People	Are the employees of the company and other companies cooperating with us competent?
Procedures	Does what we do allow the company to function well?
Policies	Does the company have any rules by which it operates? Do they bring results?

Table 5.
8P Ishikawa diagram—for marketing.

Analyzed areas	Problem
Surroundings	Could the problems be related to the environment or location of the company?
Suppliers	Do suppliers deliver ordered goods of the required quality?
System	Can faulty devices, software, and processes be affected?
Skills	Are the team's skills sufficient to achieve the desired quality?

Table 6.
4S Ishikawa diagram for services.

Table 2 shows the simplest, basic version of the five areas of analysis in the Ishikawa diagram. For each of the areas listed in the first column, a question was proposed to facilitate the general formulation of the problem (shown in the second column).

Table 3 lists the basic areas as 5M and an additional frequently used area, Environment.

Yet another version of the areas in the diagram is shown in **Table 4**. It is called 5M+3M because other areas, occurring in production conditions such as measurement and maintenance of machines in working order, as well as the impact of changing natural conditions, are added to the basic five areas.

A different set of areas for analysis are shown in **Table 5**. This is a variation of the Ishikawa diagram for situations where the product is already produced and we are not looking for reasons for the lack of quality of this product. It is intended for departments such as sales, promotion and, customer relations.

The last example shown in **Table 6**, on the other hand, is a suggestion for service providers who may be struggling to maintain delivery quality, logistics, and other non-production processes.

However, the principle of creating a fishbone diagram is one. In its basic form, the diagram resembles the skeleton of a fish, where the bones converge at the spine. Principal bones correspond to the causes in a given area of analysis (1 principal axis = 1 area). Major bones can have many minor bones that illustrate the possible causes of a problem with a given effect. The line of the spine is the line of the formulated problem (effect of causes) (**Figure 1**) [4, 16].

The construction of the diagram corresponds to the original assumption of its creator. According to Ishikawa, effects, or problems, always have a cause, and in most cases, there are a number of active factors that caused them. The purpose of using the diagram is to discover, collect, and classify the causes of the problem in terms of the importance of the impact exerted.

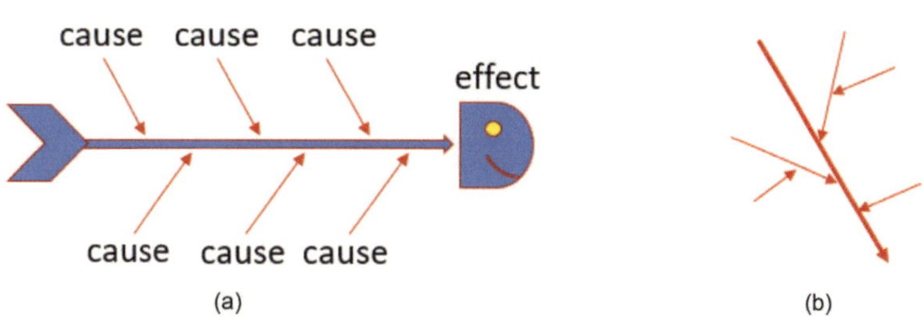

Figure 1.
Ishikawa diagram: (a) general concept, and (b) main and auxiliary bones.

Initially, the diagram as a tool for managing the production process was used to manage quality. It has now evolved and is used in other fields such as safety management to identify the causes of accidents or hazards, or in environmental management to identify the causes of negative environmental impact (e.g., waste generation, occurrence of industrial failures).

The cause and effect analysis should be carried out in an orderly manner:

Step 1—Creating a detailed description of the problem: What is wrong? When and where does the problem occur? How is it going?

Step 2—Creating an analysis group, that is, a team of employees who will use the "brainstorming" technique to generate as many potential causes of problems as possible. A leader should be chosen in the group, for instance, a department manager, who will coordinate the work of the team. The team should also include other competent people who come into contact with the problem due to their daily duties. Team work should take place in a calm, friendly, and creativity-enhancing atmosphere, assuming that there are no bad ideas at this stage, and you do not immediately look for a solution to the problem.

Step 3—Selecting and entering the main categories of causes into the diagram and matching the emerging proposals of their sources to them.

Step 4—Selecting the most probable causes of the analyzed problems and discussing the possibilities of remedying the situation.

The basic technique for determining the root causes of defects and process disruptions is the 5 Whys. It is a preventative technique, as opposed to the philosophy of taking action based on actual defects found. 5 Whys, originally written by Sakichi Toyoda and then developed by Taiichi Ohno, was based on the perception of a child's curiosity about the world. It is the child who, reaching the information, repeatedly asks "Why is it the way it is?" Hence, in quality management, 5 Whys is called "a child's play method for solving adult problems". 5 Whys is most effective in situations concerning work organization, employee relations, cooperation between departments, disputes, and where human error may be the cause [17, 18].

The 5 Whys technique requires the following [19]:

- choose a problem for analysis,
- organize a meeting of all people affected by the problem,
- ask at least 5 "Why?" questions,
- develop corrective actions,
- assign responsibility,
- supervise the implementation.

The diagram of the analysis is shown in **Figure 2**.

A variation of 5 Whys, called 5W2H, can be used to define the problem. Then, the questions are answered [20]:

1. What?—What exactly is the problem?

2. Who?—Who reported the problem?

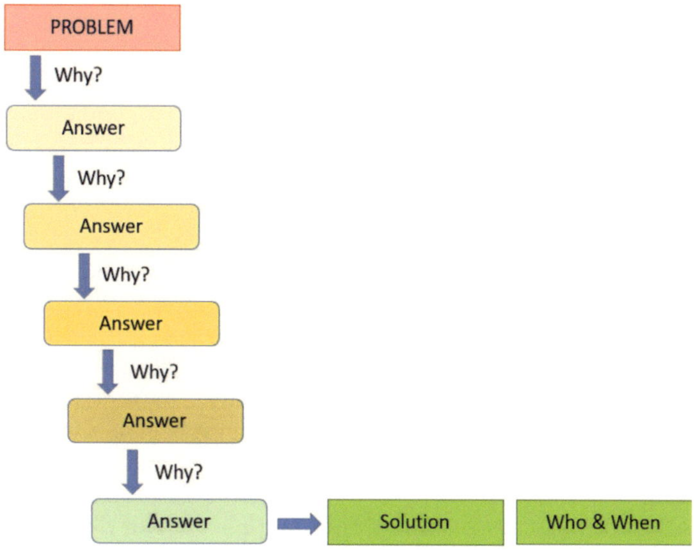

Figure 2.
5 Whys diagram.

3. When?—When did the problem start?

4. Where?—Where did he perform?

5. Why?—Why is this a problem?

6. How?—What is the effect? What are the consequences of this problem?

7. How many/How much?—What is the scale of the problem? (preferably when it is given in measurable values, e.g., % and pcs).

Then, you get full information about the nature and effects of the problem. The spreadsheet shown in **Table 7** can be used to define the problem.

No	Question	Answer	Verification	Action	Deadline
1.	What ....?				
2.	Who ....?				
3.	When ....?				
4.	Where ....?				
5.	Why ....?				
6.	How ....?				
7.	How much ....?				

Table 7.
Problem identification and solution—5W2H sheet.

4. Ishikawa diagram and 5 Whys—case studies

In order to demonstrate the usefulness of the Ishikawa diagram and the 5 Whys method in improving production, three different examples of their application are shown. Chapter 4.1 points out the execution difficulties in machining, where technological knowledge, knowledge of the phenomena occurring during cutting, and the ability to counteract the negative effects of the cutting force component play a key role. Different ways of carrying out the analysis of 5 Whys are shown, depending on how to define the problem to ensure correctly implemented production.

Chapter 4.2 shows the possibility of modifying the Ishikawa diagram in connection with the use of a new technological device, when a laser appears instead of a metal-cutting machine tool. Instead of the typical Machine and Management areas, the Laser and PC & Software areas were proposed. An example of a situation when the typical five questions do not lead to a solution to a problem is also presented. Then, the classic method framework should be abandoned and additional questions should be asked.

Chapter 4.3, in turn, shows an example of the application of the analyzed quality tools to solve waste problems using the example of a bonding operation. The problem to be solved is not a defective product, but a defective treatment of the environment and human health. Because of the possibility of different approaches to the responsibility for total waste and hazardous waste, two ways of carrying out the 5 Whys analysis are shown.

4.1 Cutting flat bars made of polypropylene

Machining of modern construction materials on CNC stations seems to be uncomplicated. However, cutting 5 mm thick flat bars made of polypropylene turned out to be a problem due to the length of the fixed blank and the behavior of the element during processing. Loss of stiffness during machining, deflection of the element under the influence of cutting force, inadequate fixation of the workpiece, and general incorrect selection of machining conditions are important technological problems (**Figure 3**).

The simplified 5W2H sheet (**Table 8**) was used to identify the problem.

In order to identify the causes of problems, the Ishikawa 5M+E diagram was developed (**Figure 4**).

a) b) c)

Figure 3.
Station with the problem of cutting flat bars: (a) CNC milling machine HAAS VF-1, (b) problem with cutting, and (c) problem of sticking chips.

No.	Question	Answer
1.	What is the problem?	The flat bar cannot be cut, and when appropriate stiffening is applied, the material on one side has a visible defect in the form of glued chips
2.	Who reported the problem?	Machine operator
3.	When did it happen?	At the start of production
4.	Where did it happen?	HAAS VF-1 CNC milling center
5.	Why is this a problem?	Because either the machining cannot be done or the part is not made to specification
6.	How is this happening?	The problem occurs after clamping the workpiece in a vice and trying to cut the material
7.	How many products failed to perform?	Not one so far

Table 8.
Simplified 5W2H sheet.

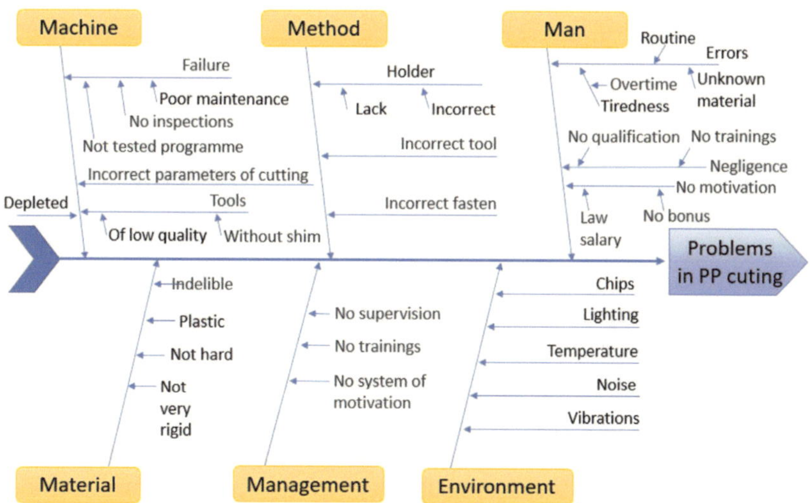

Figure 4.
Ishikawa diagram for polypropylene cutting.

Typical areas of analysis have been included in the diagram, and they have shown primarily human problems. In both Machine and Man areas, faults caused by a lack of knowledge, interest, and care and negligence were indicated. In the Method area, deficiencies in the tooling or faulty setup are indicated, but this also has its roots in human nature. The revealed difficulties should lead to rethinking whether flaccid elements should be cut at all in the presented way, for what reason this chosen position was considered justified. If, by decision of the management, CNC HAAS machining is maintained, then remedial steps will be necessary due to clear errors in the manufacturing technology. Due to the most important error, which is incorrect processing conditions, that is, incorrect mounting and loss of rigidity, the method of making the cut was first considered in 5 Whys, because so far, all products had been damaged (**Figure 5**). Only then were the defects verified—glued chips on the object (**Figure 6**).

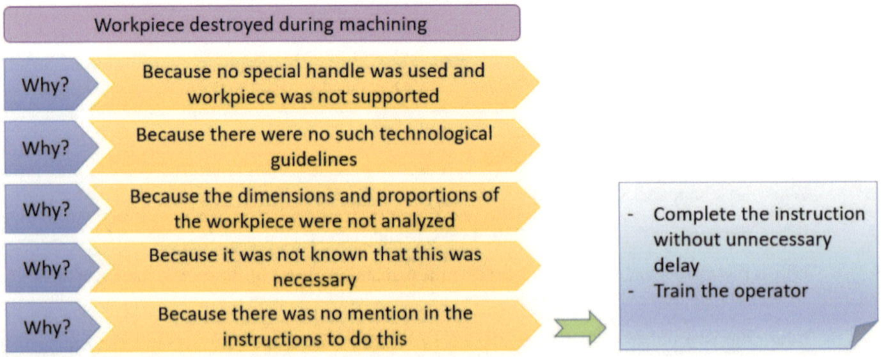

Figure 5.
5 Whys for destroyed workpieces.

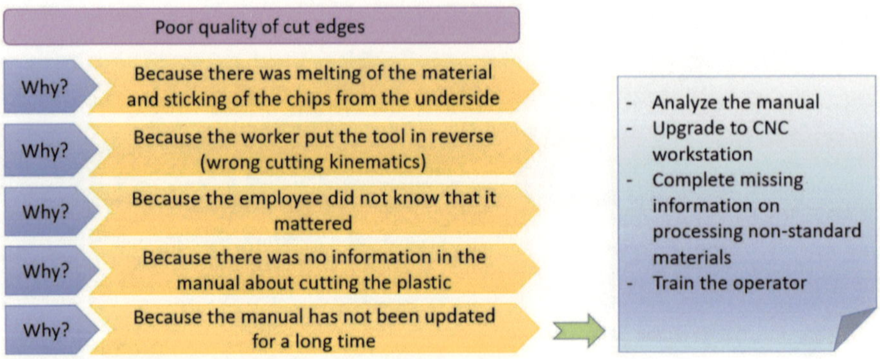

Figure 6.
5 Whys for poor quality of cut edges of workpiece.

After analyzing 5 Whys, the team came to the conclusion that the key issue is the substantive knowledge of employees. Deficiencies in this area should be made up as soon as possible through training. The second urgent task is to read and update the workstation manual, because the conditions for the implementation of the technological operation have changed dramatically along with the changed machine tool.

4.2 Laser cutting of small diameter holes

Lasers are now being used more and more commonly for various technological operations. Due to the still quite high cost of purchasing high-end devices, it is important in production to use your equipment in many different applications. The analyzed station with a fiber laser for marking and engraving was used to cut small holes in thin sheets. This method turned out to be effective, but various problems arose during the start of production (**Figure** 7). When developing the technological guide, it was found that with the given software, it is impossible to drill holes (you cannot apply a laser pulse multiple times in one place), but you can cut them. Cutting can be carried out according to different strategies. By setting the beam trajectory in a circle, burns are obtained at the end of cutting, depending on the beam parameters. On the other hand, the spiral trajectory allows for the gradual removal of material, but with an inaccurate definition of the spiral shape, a clear "teeth" was obtained.

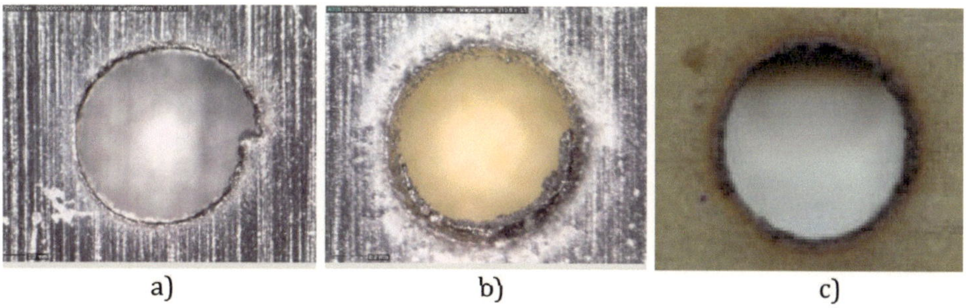

Figure 7.
Examples of poor quality during laser cutting holes: (a) wrong "spiral" cutting strategy, (b) edge irregularities and ovality, (c) edge burns.

On the other hand, during cutting with a simultaneous circle and spiral, a deviation of ovality appeared due to the deformation of the hole under the influence of heat. Therefore, machining with clearly defined, optimal parameters, saved in the program controlling this laser model, plays a key role. The plant wants to eliminate all mistakes and defects due to the need for flexible production under the Industry 4.0 concept.

Due to the fact that the station was new in the plant, a general Ishikawa diagram was prepared, modifying it to include the PC & Software area, due to the often signaled difficulties and impact on the course of activities at the station (**Figure 8**).

First, the laser bone was drawn in detail (**Figure 9**); attention was paid to the degree of detail in the construction of the device and the variety of possible problems. In this way, it was realized that knowledge about the device is necessary to maintain its efficiency. In addition, the way the laser works specifically affects the choice of method. Many parameters are adjustable, and they, ultimately affect the machining effect.

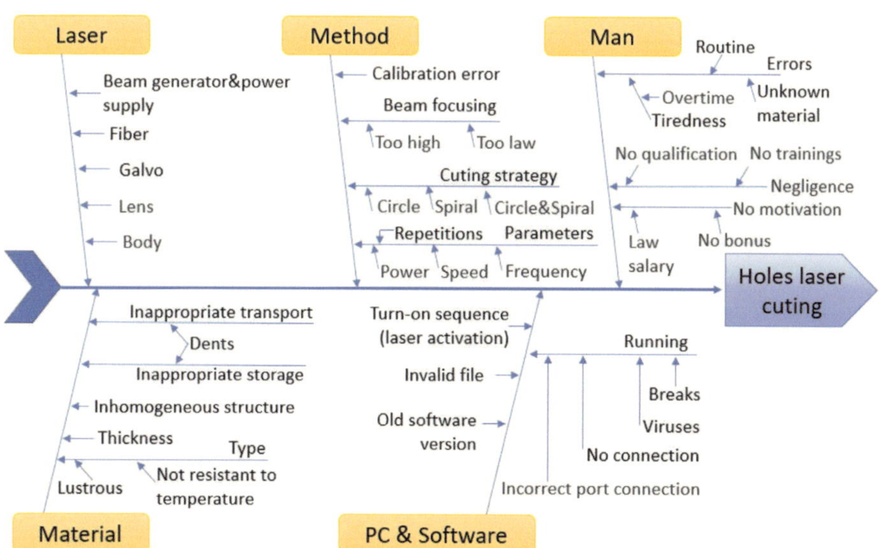

Figure 8.
Ishikawa diagram for hole laser cutting.

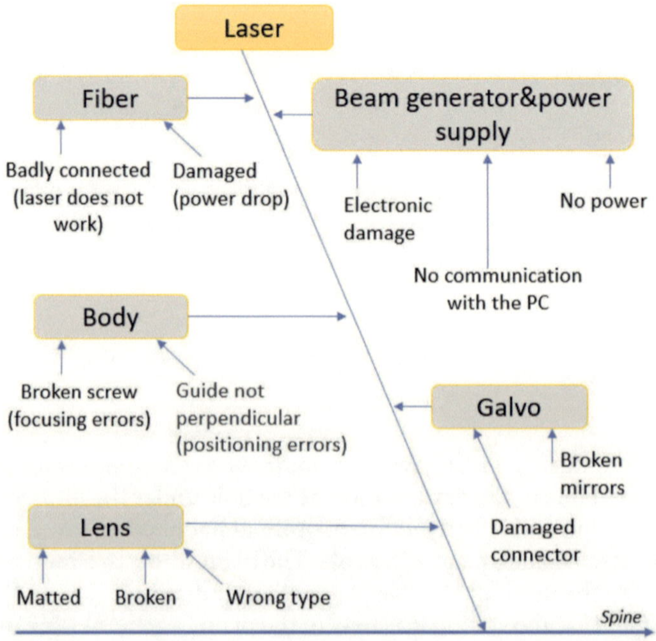

Figure 9.
Different problems in laser bone.

From the point of view of laser hole cutting, the most important issue is to maintain the optimal parameters of the laser beam. After a detailed analysis of the problem, it was determined that the employee has only one right program to run, but the thickness of the material changes. Thus, the location of the beam focus changes, which is crucial for the effective impact of the laser. An example of solving this problem is shown in **Figure 10**.

The above example is also a confirmation that sometimes rigid adherence to five questions may not give a satisfactory result; in the analyzed situation, one more question, the sixth question, was needed.

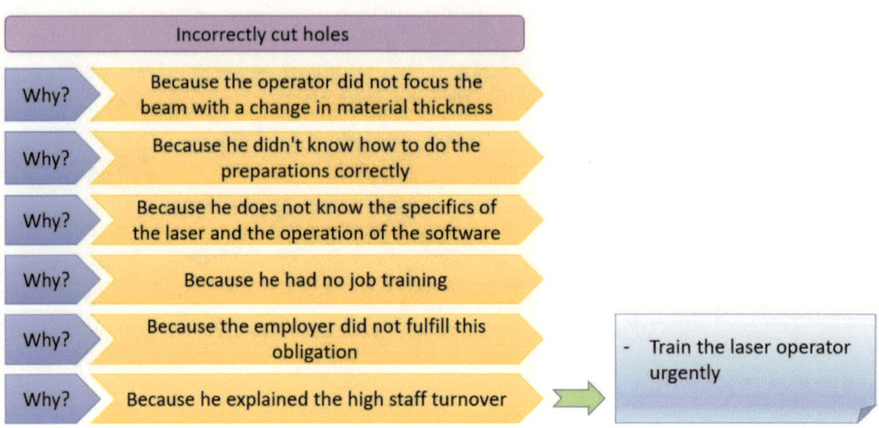

Figure 10.
5 Whys for holes made with defects.

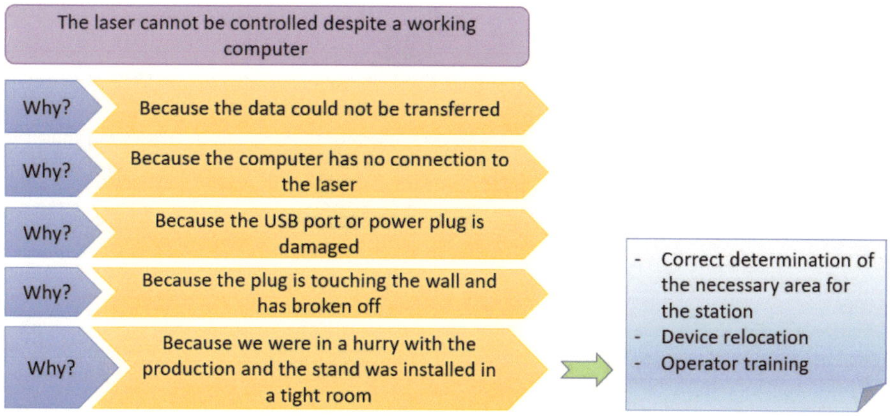

Figure 11.
5 Whys of the problem in PC & software area.

In turn, the analysis of the PC & Software area made the team think about why the laser often fails to start, despite a working computer and the right program. The 5 Whys results for this problem are shown in **Figure 11**.

Analysis of the Ishikawa diagram and the results of 5 Whys made it possible to conclude that the overriding issue is the efficiency of the device. All employees associated with this position are required to maintain the laser in this state. The operator has the greatest contact with the laser. His knowledge and general condition, indicated by the Man area, are important in the context of determining technological parameters. In addition to him, an important role is played by service technicians, electronics engineers, and programmers as people interfering with the construction and operation of the device. People can also have an impact during transport (and thus suppliers), when damage to components may occur, during assembly and organization of the station. Fortunately, the problem that was finally identified was trivial and was quickly eliminated.

4.3 Bonding of materials

In the gluing operations, various substances are used, necessary for the proper preparation of the surface of the materials before gluing. This diversity results from the type of bonded material, surface condition, selected adhesive mixtures, and gluing technology (surface preparation baths, rinsing, method of preparing the mixture, method of applying the adhesive, and method of curing) (**Figure 12**).

The problem of waste can be perceived in different ways in a plant. The Ishikawa diagram drawn for this problem made it possible to identify a number of reasons why a lot of waste from various materials and packaging contaminated with various substances remains after gluing operations (**Figure 13**).

Reflecting on the problem, attention was first drawn to the considerable amount of total waste that comes from the gluing process. 5 Whys for this thought process are shown in **Figure 14**.

On the other hand, by emphasizing the fact that gluing waste is hazardous waste, and additionally paying attention to its accumulation at workstations, 5 Whys may have a different course and result in other management decisions (**Figure 15**).

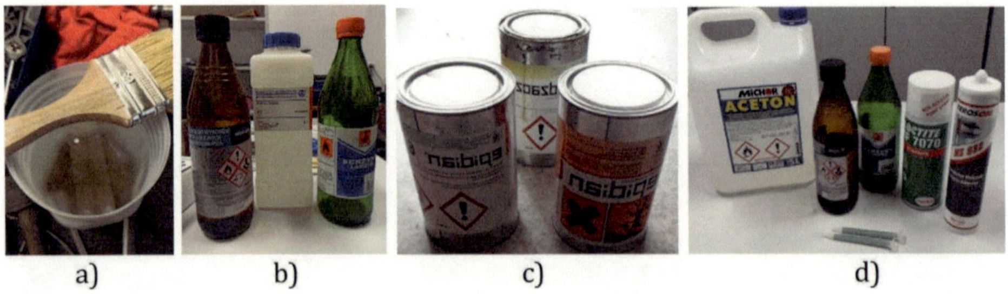

Figure 12.
The problem of a significant amount of waste after gluing: (a) used mixing containers and brushes for applying adhesives; (b) plastic and glass packaging from degreasers and rinsing liquids; (c) metal packaging for adhesives and hardeners; and (d) plastic, glass, and pressurized packaging with various applicators for washing, preparing, and activating the surface before gluing.

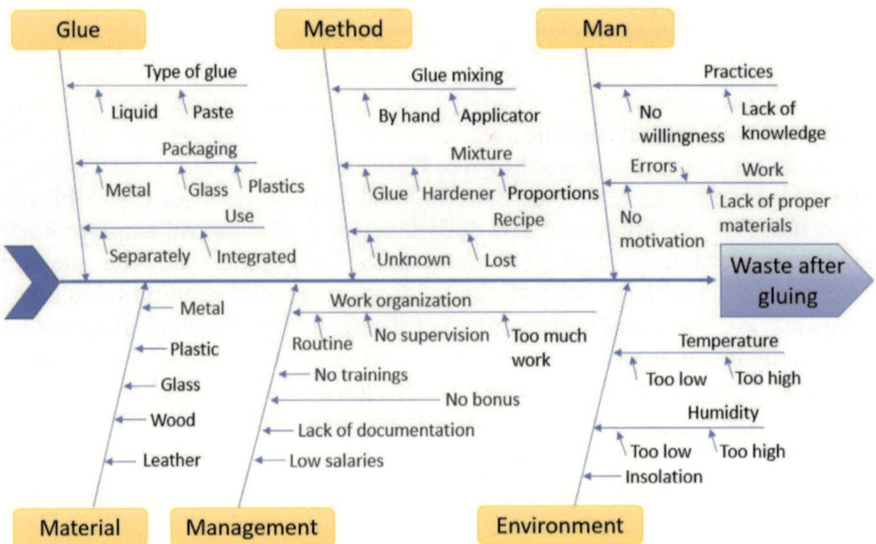

Figure 13.
Ishikawa diagram for gluing.

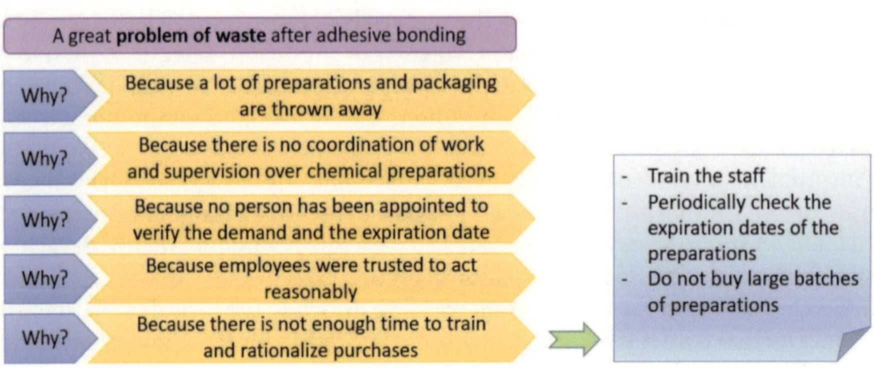

Figure 14.
The problem of the amount of waste after gluing.

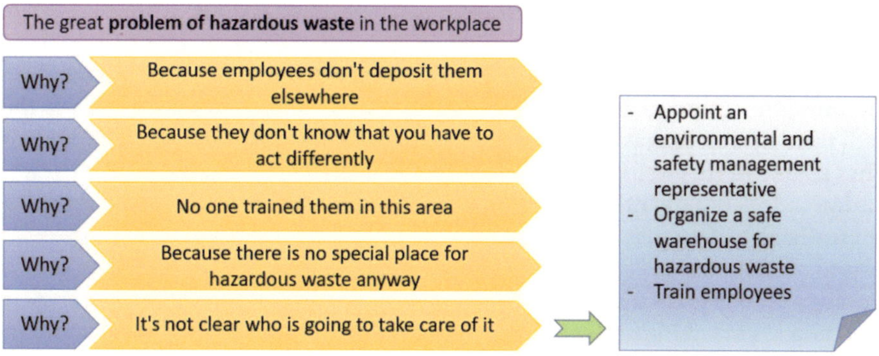

Figure 15.
The problem of hazardous waste after gluing at workstations.

5. Conclusion

The Ishikawa diagram is an effective tool for discovering the sources of problems in enterprises. It allows to classify and establish a hierarchy of a set of various statements, causes of faults, in a situation where this set is chaotic. However, it gives the opportunity to look at the production or activity of the company from the point of view of many areas that are often not suspected of affecting the broadly understood quality. When using the diagram, however, you may encounter difficulties:

- it is necessary to work in a group, so little efficiency is possible if the team leader is weak,
- for the effectiveness of the diagram, it is necessary to properly classify the causes and assign them to the appropriate areas,
- the analysis takes place in typical areas, and it is possible that the real cause of the problem is overlooked.

Difficulties can also arise when determining root causes using the 5 Whys:

- the symptom is perceived as the cause,
- the method seems trivially simple, and you might want to point out the reason right away, but in reality, it is not,
- hasty pointing of the guilty,
- lack of sincerity in the team, hiding some activities, conflicts, and fear of further employment,
- limited to 5 questions, even if the root cause has not been identified,
- conducting the analysis by one person, not a team.

The examples shown illustrate different production situations. A similar approach could be taken in many areas, for example, in the area of Man, by identifying the causes of errors inherent in the human mentality and mental and physical condition. Sometimes, the Ishikawa diagram was modified to include other issues, such as those related to computer operation and data flow. You can also encounter a situation when 5 Whys leads to a surprising discovery of the negligence of the employer, or draws attention to a seemingly insignificant problem that—as it turned out in point 4.3—can affect the health of employees.

Based on the examples, it can be concluded that the knowledge of manufacturing technology always plays a key role. However, in the production process, the subject is man, and his behavior determines the course of this process. Finally, it should be emphasized that the production process and its improvement should start with the management. The lack of awareness of the management and the lack of management skills can have a negative impact and manifest in the lack of training, motivation, and incentives for professional development of employees. Thus, it is difficult to talk about quality management of products and processes.

Conflict of interest

The author declares no conflict of interest.

Author details

Barbara Ciecińska
Rzeszow University of Technology, Rzeszow, Poland

*Address all correspondence to: barbara.ciecinska@prz.edu.pl

References

[1] Hoyle D. Quality Management Essentials. San Diego, CA: Elsevier; 2007

[2] Kaymak H. The relationship between total quality management practices and their effects on firm performance. Journal of Operations Management. 2003;**21**(4):405-435

[3] Chądzyńska M, Klimecka-Tatar D. Use the quality management tool which is the Ishikawa diagram on the example of a small leather business. Archives of Engineering Knowledge. 2017;**2**(1):20-22

[4] Voehl F. 5 whys. In: The Innovation Tools Handbook. 1st ed. Vol. 2. New York, NY: Productivity Press; 2016. pp. 1-10

[5] Rose KH. Project Quality Management: Why, What and How. Boca Raton, FL: J. Ross Publishing; 2005

[6] Juran JM. Juran on Leadership For Quality: An Executive Handbook. New York, NY: Simon and Schuster Inc.; 2003

[7] Ryan TP. Statistical Methods for Quality Improvement. Hoboken, NJ: John Wiley & Sons; 2011

[8] Mizuno S. Management for Quality Improvement: The 7 New QC Tools. Portland, OR: CRC Press; 2020

[9] Priede J. Implementation of Quality management system ISO 9001 in the world and its strategic necessity. Procedia Social and Behavioral Sciences. 2012;**58**:1466-1475

[10] Kan SH. Metrics and Models in Software Quality Engineering. 2nd ed. Boston, MA: Addison-Wesley Professional; 2003

[11] Dahlgaard JJ, Kristensen K, Kanji GK. Fundamentals of Total Quality Management. Process Analysis and Improvement. Routledge: Taylor & Francis Group; 2008

[12] Oakland JS, Oakland RJ, Turner MA. Total Quality Management and Operational Excellence. Text with Cases. Routledge: Taulor & Francis Group; 2021

[13] Midor K, Žarnovský J. Innovative use of Quality management methods for product improvement. Management Systems in Production Engineering. 2016;**24**(4):264-267

[14] Ilie G, Ciocoiu CN. Application of fishbone diagram to determine the risk of an event with multiple causes. Management Research and Practice. 2010;**2**(2):1-20

[15] Luca L. A new model of Ishikawa diagram for quality assessment. In: IOP Conference Series: Materials Science and Engineering. 20th Innovative Manufacturing Engineering and Energy Conference (IManEE 2016), Kozani, Greece. Vol. 161. 23-25 Sep 2016. pp. 1-6

[16] Ghatorga KS, Sharma R, Singh G. Application of root cause analysis to increase material removal rate for productivity improvement: A case study of the press manufacturing industry. Materials Today: Proceedings. 2020;**26**(Part 2):1780-1783

[17] Card AJ. The problem with '5 whys'. British Medical Journal of Quality & Safety. 2017;**26**:671-677

[18] Luca L, Pasare M, Stancioiu A. Study to determine a new model of the Ishikawa diagram for Quality improvement. Fiability & Durability. 2017;**1**:249-254

[19] Myszewski JM. On improvement story by 5 whys. The TQM Journal. 2013;**25**(4):371-383

[20] Benjamin SJ, Marathamuthu MS, Murugaiah U. The use of 5-WHYs technique to eliminate OEE's speed loss in a manufacturing firm. Journal of Quality in Maintenance Engineering. 2015;**21**(4):419-435

Chapter 5

Quality Determinants of Client Satisfaction: A Case Study of ACE – Australian Consulting Engineers, Sydney, Australia

Elham Said Hasham and Anthony Said Hasham

Abstract

The construction industry is one of Australia's fastest growing industries and its success is a result of a firm's client satisfaction with focus on product determinants such as price and quality. Ensuring quality at every phase is a must and building rapport with the client will go a long way. To capitalise on the growing demand for Engineering Consulting Firms (ECFs), we should "redefine the bottom line by allowing client satisfaction, high-quality standards, and profits to be the top priorities". Consequently, the emphasis should be on improving employee skills through various training provisions. Clients seek consistency and thus expect that all services should be similar in respect to quality and the ability to meet their needs. This calls for empowerment and comfortable work conditions to motivate employees and give them incentive to deliver quality and excellent output. The methodology utilised is triangulation-a combination of both quantitative and qualitative research. The case study-ACE was established in 1995 and has operations throughout Australia, the Philippines, Europe, U.A.E., K.S.A., and Lebanon. ACE is affiliated with key agencies and support organisations in the engineering industry with ISO certifications in Safety and Quality Management. The objective of this study is significant as it sheds light on employee motivation and client satisfaction as imperative determinants of the success of an organisation.

Keywords: organisational behaviour, leadership, satisfaction, motivation, quality service

1. Introduction

In today's market where competition is fierce, firms conduct research [1] to discover client preferences. Organisations keep close watch on the priorities of the clients and monitor their satisfaction levels so that they can improve their services because there is a direct relationship between satisfaction and loyalty. Accordingly, employees of Engineering Consulting Firms (ECFs) need the required resources, work conditions and professionalism to enable them to establish rapport and demonstrate positive attitude with clients.

Indeed, the client has a perceived expectations and this is a major factor that affects satisfaction [2]. Employees should be able to attend to clients' needs and resolve problems efficiently and obviously motivated employees will produce better results in this respect in comparison to discontented employees. Engineering consultants deal with a variety of clients in their profession. Therefore, it is important that engineering consultants understand the determinants of client satisfaction to improve their likelihood of project success in the short-term and continual business opportunities in the long-term.

Research [3] on the determinants of client satisfaction typically focuses on product determinants such as price and quality. However, research investigating the human resource (HR) determinants of engineering consultants and their impact on client satisfaction is limited. By understanding engineering consultancy from an HR perspective, decisions can be made more effectively to improve the likelihood of project success. The most important resources of an organisation are its employees and thus their contribution to the company's image is very crucial in terms of behaviour and actions.

According to [4], it can be said that the factors involved constitute an appropriate style of leadership, motivating the middle and lower-level employees and developing a warm relationship. Such factors are crucial for direction and are considered as the ability of the driver to acquire performance on the part of subordinates on individual basis in an organisation. Employees should perform significant roles in earning clients' trust. Companies are facing not only local but also international competition, thus they must invest in implementing more sophisticated technology and developing new products to ensure approval. Accordingly, companies need to hire professional consultants and encourage separate departments to brainstorm strategies to deal with competition.

In turn, to overcome the challenges that a firm may face in this twenty-first Century, Human Resources Management "HRM" is a very crucial variable. It helps in making the firm more effective and efficient in its day-to-day activities and facilitates the acquisition of talented individuals through training and organisational development [5]. Thus, the conditions for an ECF to achieve client satisfaction are clear, open communication and professionalism of service. It is essential to understand the needs of clients which can be followed up by a Clients Care Manager "CCM". Ensuring quality at every phase is a must and building rapport with the client will go a long way. All these exercises will result in client satisfaction. When the clients are satisfied, this will result in repeated business. Hence, new clients will be attracted due to 'word of mouth' and this will save marketing costs and ultimately, the firm's reputation will be enhanced. In sum, there is a direct relationship between employee motivation and client satisfaction; a firm that takes care of its employees will not have to worry about client satisfaction [2].

1.1 Importance of research

Most consulting firms depend on client satisfaction to achieve competitive advantage [6]. This is because with positive client satisfaction, consulting firms can enjoy increased loyalty, a reduced marketing effort through word of mouth and positive reviews. This also creates greater networking opportunities to grow the client base and builds a more attractive environment for recruiting the best talent [7]. In the consulting industry, clients have trouble perceiving a service advantage due to the intangible nature of the products [8]. An interesting challenge that engineering consulting firms face when trying to deliver client satisfaction is the double-edged sword of their pricing fees. Clients tend to prefer lower fees; however, lower fees provide the ECF with a smaller budget to perform their contract. With a smaller budget, it can be difficult to ensure the quality requirements that the client would prefer [9].

To capitalise on the growing demand for engineering consultants, ECFs must "redefine the bottom line by allowing client satisfaction, high-quality standards, and profits to be the top priorities" [10]. The choice of organisations available to the employees and the choice of products available to the clients have increased which has made it important for the organisations to adopt strategies that ensure retention for the organisation. It has also become important for organisations to maintain an employee friendly atmosphere. Competitive advantage is based on training, recruitment, effectiveness, and efficiency that will eventually impact the financial stability of the organisation. According to a study [11], employees that tend to be dissatisfied with their jobs are unable to provide quality service and become demotivated. Thus, it is essential for organisations to identify the performance loopholes and provide adequate training opportunities to ensure employee development. Empirical evidence [11] suggests that organisations that offer opportunities of personal development tend to have higher employee satisfaction.

Moreover, strategies for training programmes and employee engagement activities can help an organisation to develop sustainable employee satisfaction. One study [12] shows that clients base their perceptions on employee intercommunication and the quality of services provided. An employee can build a relationship of trust and loyalty with the clients when they are satisfied with their jobs and are well aligned with the mission and goals of the organisation [12]. In addition, another study [13], confirms the direct association between employee motivation and client satisfaction. Accordingly, the findings of [13], suggest that it is crucial for the organisation to realise the value of talent retention to attain a stable financial position for the organisation.

In short, an organisation must display continuous commitment to employee welfare. However, the findings of a study [14], suggest that it becomes difficult for an organisation to attain desired levels of employee satisfaction because of the diverse nature of employees. As thus, it becomes difficult for employers to understand the needs and motivational factors amongst the employees [14]. Consequently, employers need to be aware of and analyse the behaviour of employees to adopt strategies that increase motivational levels. In accordance, [15] recommend that companies wishing to enhance the level of client satisfaction need to focus on building holistic models that include the concepts of employee satisfaction.

Consequently, this study will evaluate the impact of employee motivation on client satisfaction in Engineering Consultancy Firms (ECFs). The research findings of the study will assist in taking adequate steps for achieving desired levels of employee motivation and client satisfaction, and as a result, in improving the reputation of the organisation in the business environment. Thus, the findings of this study can prove to be helpful for employers and top management. The findings of the study will also increase the knowledge base of the present literature and will help researchers validate future results for the other organisations in the business environment. Because employee motivation and client satisfaction are important determinants of organisational success, this study will determine the objectives of increased productivity, employee motivation and client satisfaction.

1.2 Research questions

Initially, this research aims to fill the research/knowledge gaps identified in the presented literature review. By conducting a study based on these gaps, ECFs will be able to understand the link and how it can benefit them. More precisely, this research will attempt to identify and test the key factors that lead to client satisfaction (CS) in the context of ECFs. From the various literature reviews that were undertaken, these five human resource factors were identified and will be investigated hereafter: experience and knowledge, motivation, proximity, professionalism, and care factor. In addition, this research will investigate the causal relationship between client satisfaction, client retention, and "word of mouth", as conventional relationship marketing consequences.

Moreover, the analysis in the study is conducted with an aim to find relevant and precise answers to the research questions. It should be noted that these questions are intended to increase knowledge on the subject of interest and hence, they will remain within the scope of the study as they have the potential to both increase the complexity and the difficulty of the statistical analysis. According to a study conducted by [16], research questions are considered as catalysts of a research study because they define the focus of the research and depict what should be the result of the research study.

The objectives of this research are:

- To investigate the literature pertaining to the instigators of client satisfaction, especially human resource determinants.
- To investigate if client satisfaction increases the likelihood of client retention and improved organisational reputation through "word of mouth" for ECFs.
- To provide a set of recommendations for engineering consulting firms (ECFs) to improve their client satisfaction and retention.

Furthermore, to fulfil the aim of the study, the following research questions have been framed:

1. What theories can be applied to Small or Medium Engineering Consulting Firms to achieve better performance results?
2. What is the influence of the Human Resource Management on Client Satisfaction?
3. What are the relationships between Client Satisfaction, Client Retention, "Word of Mouth" and reputation?
4. What benefits can Engineering Consulting Firms (ECFs) achieve from understanding the HR determinants of Client Satisfaction?

1.3 The case study: ACE an engineering consulting firm

Dr. Anthony S. Hasham is the founder and Managing Director of Australian Consulting Engineers Pty Ltd. "ACE". He is a Certified Practicing Engineer

(C.P.Eng.-Aust.), a Fellow member of the Institute of Engineers (F.I.E.-Aust.), a Member of the Order of Engineers and Architects Lebanon (OEA) and is found on both the National Engineers Register (N.E.R.-Aust.) and Register of Practicing Engineers Queensland (R.E.P.Q.). Furthermore, Dr. Hasham has sat on numerous boards with the most current positions being President of the Maronite Catholic Society and Vice President of the Australian Lebanese Chamber of Commerce.

In 1993 he joined the board of directors for Unaveave Consulting Pty Ltd., and in 1995 established Australian Consulting Engineers, initially providing consultancy services in the field of Structural, Civil, and Stormwater Engineering. ACE has since expanded operations throughout Australia and internationally offering consultancy services in the fields of Structural; Post-tension; Remedial; Civil/Stormwater; Temporary Works and Building Services (Mechanical, Electrical, Hydraulics and Fire Services) engineering, with over 200 staff Australia wide and internationally, spread throughout Australia, the Philippines, Europe, U.A.E, K.S.A., and Lebanon.

Furthermore, ACE is affiliated with key agencies and support organisations in the engineering industry. In addition to being a corporate member of Consult Australia and member of Engineers Australia, ACE is certified by the NSW Government Work Health and Safety Management Systems. ACE is a fully certified ISO9001:2015 Quality Management Systems Company across all disciplines. Being involved with over 15,000 projects during his 30 years in engineering consulting, Dr. Anthony Hasham has a vast experience in customer service, with a large customer base from which to extract data for the purpose of this research.

2. Literature review

Employee motivation and client satisfaction are two vital concerns of any organisation. Their contribution to the company's image is crucial in determining the views that clients keep in mind whilst thinking about the organisation's behaviour and actions. Loyalty to the organisations' clients incurs capital investments, which should include employees' motivation to achieve the company's non-monetary and economic objectives. Organisations are increasing their efforts to ensure that harmony exists between employee motivation and client's satisfaction to increase the chances of success.

In Ref. [17], proposed that ultimately, client satisfaction is based on previous studies and some theorised "cognitive framework of satisfaction-related concepts" through various literature. Consumer satisfaction is believed to influence attitude, change and purchase intention. In Ref. [17] further suggests that performance-specific expectation and expectancy disconfirmation play a key role in satisfaction decisions into a framework of satisfaction-related concepts. In turn, satisfaction increases as the performance/expectation ratio increases. This view was based mainly on the results of a seminal laboratory study by [18].

According to [19], client satisfaction is a significant indicator of client retention, and it also indicates their purchase intentions. The level of client satisfaction provides the company with a real picture about the degree to which products and services are satisfactory to the end consumer. Likewise, client satisfaction has a strong link to organisational success. Consulting firms that can ensure a positive client experience can gain a strong competitive advantage over their rivals. Client satisfaction is a measure of a consulting firm's success [10] and similarly enhances organisational reputation. Therefore, ECFs that can ensure positive client satisfaction will experience a wealth of benefits. In sum, client satisfaction is imperative for it provides a foundation to generate revenue and to gain profits at all levels.

The Institution of Professional Engineers New Zealand guided practicing engineers on the importance of client relationships in their handbook, "Developing and Maintaining Client Relationships" (2005). The handbook outlines the significance of client relationships and explains that the clients experience usually determines feedback and plays an integral role in either making or breaking the engineer's reputation. As such, client experience can be two-fold. Positive client satisfaction can bring a wealth of benefits to both the organisation and the individual. On the other hand, negative client satisfaction can severely hamper the ability for the organisation and the individual to grow and succeed.

For these reasons, [20] stresses that client retention is crucial for a company. Retaining a client is more important than acquiring a new one. Clients' satisfaction is thus considered as a basis for client retention as our goal is to meet their needs and requirements. When clients are happy, there are limited chances that such clients will shift to the rival for the same or comparable products. Thus, the chances of client retention are high, and this helps the company save money in acquiring new clients. Furthermore, organisations focus on client satisfaction strategies because without satisfied clients there is no business [21].

Besides, client satisfaction is the comparison between clients' expectations and their actual perspective of the service once it has been delivered [22, 23]. Without a doubt, a satisfied client is more likely to return for repeated business and create a positive word of mouth that may improve an organisation's reputation [24]. With the engineering consulting industry becoming more competitive, it is now easier for dissatisfied clients to compare services and do business with a direct competitor. As well as this, dissatisfied clients are more likely to tell other people about their experience and consequently, create a negative image [21].

Indeed, word of mouth (WOM) marketing is a tactic used to generate natural discussions about and recommendations for a product or company. Essentially, getting people to talk about your brand, product, or business is the goal of this type of marketing. Everyone wants people talking about their products because those conversations are what creates a buzz and drive business. Word-of-mouth marketing (WOM marketing, also called word of mouth advertising) differs from naturally occurring word of mouth, in that it is actively influenced or encouraged by organisations. Whilst it is difficult to truly control WOM, research [25] has shown that there are three generic avenues to 'manage' WOM for the purpose of WOM marketing: (1) a strong WOM foundation (sufficient levels of satisfaction, trust and commitment); (2) indirect WOM marketing management which implies that managers only have a moderate amount of control (controversial advertising, teaser campaigns, customer membership clubs); and (3) direct WOM marketing management, which has higher levels of control (paid WOM 'agents', "friend get friend" schemes).

Evidently, satisfied clients become a potential source of advancement for a company. When clients are satisfied with the products and services of the company, then individuals provide a positive word of mouth [26]. Overall, this helps the company to spread awareness to a wider population without incurring high costs of advertisement. Similar viewpoints [27] are pertinent to the positive implications of client satisfaction on the goodwill of a company as this provides an enhancement of reputation in the marketplace. Accordingly, this research aims to ascertain whether there is an association between the empowerment of managers and the delivery of quality service. There are several factors that influence the motivation of managers, such as perceived fairness of the promotion system, employee rewards and recognition, interpersonal relationships, and work security. Against this backdrop, the current study focuses on the active role played by managers in the provision of quality service that facilitates satisfaction on the part of clients. In other words, it identifies six key factors that play a pivotal role in the motivation of managers: sense of purpose to work, challenges at work, leadership in the workplace, financial gain, career progression, and the work environment.

It should be noted that the impact of trust has been investigated [28] along with its implications on future relationships and choice of future purchase. High organisational performance is then a prediction of managers' commitment to organisational goals, which, in turn, can be reflected in their attitudes and actions. The objectives of an organisation are influenced by the level of employee satisfaction, support, respect, and working conditions. It is to the benefit of an organisation to ensure the social and psychological well-being of their employees to ultimately ensure greater productivity.

In his famous Hierarchy of Needs, Abraham Maslow intensely talked about the second need-the sense of belonging. This stresses that managers with a sense of belonging are more, resolute, engaged, and motivated when executing assigned tasks. They are known to have greater confidence, interest, enjoyment, and enthusiasm when engaged in organisational activities. Managers not only work harder to provide better services, but also step outside their comfort zone to deliver quality services that are in alignment with client needs. A study [29] demonstrated that employees with a sense of belonging exhibit greater dedication in delivering services with improved quality. The argument is grounded on the theory of equity in social justice, which preaches that transactions can potentially pave the way for high-quality relationships [30]. When managers are accorded social gifts that are beyond their expectations, they tend to develop a sense of belonging, responsibility, and satisfaction, which, in turn, makes them more focused on addressing the needs and concerns of their clients.

Another pillar of human resource management is training and development and there is overwhelming evidence in favour of its effectiveness in team building [31]. In Ref. [32] concurs with this viewpoint and stresses that training programmes must be targeted at evidence-based strategies. Some clients are hard to deal with and they may become abusive, impatient, and angry. Effective client service involves the maximisation of aid that can be offered to clients whilst tolerating their behaviour. In this regard, training and development can help guide employees on the need to show respect to their clients by avoiding destructive remarks, recognising the regular clients, and not imposing negative attitudes [33].

In addition to this, the inability to cultivate relationships with clients often accompanies the failure to understand different personalities and communication styles, but this is a problem that can be addressed through effective training. Training and development provide a framework of understanding clients, which then makes it easier for members of an organisation to understand the needs and the problems of clients. Thereafter, the appropriate strategies can be developed to resolve these problems whilst minimising stress [34].

Above all, Abraham Zaleznik wrote the first landmark scholarly article that encapsulated the distinction between leadership and management. In this regard, [35] asserted that whilst both managers and leaders are needed in an organisation, they each make different contributions. For instance, with respect to goal setting, leaders articulate a vision and create the future; on the other hand, managers execute the plan and improve the present. Managers are branded with the tendency to exercise authority, promote stability, and focus on getting things done. In contrast, leaders pay close attention to the adoption of changes, new approaches of boosting productivity; they focus on the beliefs of people to boost their commitment [35, 36]. Certain scholars [37] contend that the two are diametrically opposite as they speculate that a good manager cannot be a good leader and vice versa.

To continue the motivational process for employees, it is crucial that the organisation provides them with the appropriate opportunities that will lead to growth. This nurturing process should exist not only for their career but also for personal development that directs individuals for future career progression [38]. Consequently, it is the responsibility of the leader or the supervisor in the organisation to develop career advancement paths and fair promotion policies for all the employees in the organisation.

In turn, the organisation has an obligation to establish various development and training programmes that will ensure the personal development of the employee whilst being a part of the organisation. It becomes essential for the organisation to understand and acknowledge that investing in training and development programmes will undoubtedly steer the organisation to enhance productivity and motivation and hence, effective contribution towards the goals of the organisation [39]. This managerial concept is known as Investment in People (IiP) [36]. Without a doubt, the lack of an adequate career path and promotion policy in the organisation can negatively impact the motivation level of the employees.

Subsequently, client satisfaction and employee rapport—'a close and harmonious relationship in which people and/or groups concerned understand each other's feelings or ideas and communicate well', have a positive link to client loyalty [40]. A commitment in relationship management [41] can be created through interactions, reciprocity, and trust. By investing the time and effort into affective commitment, it "creates psychological bonds that encourage clients to stay in that relationship and sets an expectation of reciprocation" [41]. Apart from this, [42] found three elements that constitute the quality of employee relationships-conflict handling, commitment, and trust. In Ref. [43] also consider commitment and trust as the elements of relationship quality. Within the context of the consumer market, the involvement or interaction of clients does not differentiate amongst the person offering the service and the company itself.

Last, but not least, we will now focus on the employee care factor. Highly engaged employees that care about their work and their clients are a key link to client satisfaction, performance outcomes and loyalty. When employees know that they are cared for, they become more engaged, dedicated, committed, and take more initiative in their work [44]. Furthermore, employees with the highest levels of commitment usually show higher performance by about 20% [45]. Some organisations define employee engagement as "going above and beyond" what is expected of them in their work roles [44]. Briefly, employee engagement is often defined as an "emotional and intellectual commitment" to the organisation [44]. It is a state where employees can value, enjoy, and believe in what they are working for. Employees that are not engaged will often waste their time and effort on tasks that do not matter and are often dissatisfied and do not complete tasks with full commitment. So, when employees feel more engaged in their work, they are more likely to offer better quality service, thus improving client satisfaction and promoting client loyalty [45].

The methodology adopted in this paper will be further discussed in detail in Section 4.

3. Hypotheses

This study used mixed methods research to analyse the causality and correlation between clients' gratitude and managers' motivation. Triangulation, both quantitative and qualitative investigation was utilised. Qualitative analysis was used to explore the market and justify the set of hypotheses that were devised. Qualitative research methods often use small sample sizes and involve techniques such as in-depth interviews, open-ended surveys, focus groups, and uninterrupted observations [46]. On the other hand, quantitative analysis is described as *"being scientific in nature"* due to its more rigorous approach in its use of *"scientific methods for data collection and analysis"* [47]. In turn, quantitative methods are used to verify data and models, and due to their large sample sizes and ease of replicability, the results can be confidently extrapolated to the general population [47]. Since the theme of this research "The Human Resource "HR" Determinants of Client Satisfaction from a Client Based Perspective", the researchers chose not only quantitative but also qualitative methodology to analyse and verify the effects.

Evidently, a study conducted by [12] shows that clients tend to base their perceptions of an organisation on the way in which employees communicate with them as well as on the quality of services provided. In other words, an employee can build a relationship of trust and loyalty with clients when they are satisfied with their jobs and feel they are well aligned with the mission and goals of the organisation. As such, for organisations to ensure the highest success rate, it is essential for them to develop positive work environments. Research [48] recommended that if a company wants its employees to be fully dedicated, they should provide them with learning and growth opportunities. In addition, it found that providing effective learning or growth opportunities and benefits to employees can significantly increase employee dedication. In turn, the dedicated employees will strive to help the organisation to meet its strategic objectives. Moreover, this research proved that employee pro-social behaviour and dedication are positively related to the improvement in the perceived transformational leadership execution. For these reasons, when employees feel that managers are implementing a high-level transformational leadership approach, they display dedication and pro-social behaviour [42]. Refer to Figure below for the Final Conceptual Model.

4. Methodology

This section will highlight the methods, processes and activities that were performed in this research. Background knowledge for the analysis will also be provided to explain what software contributed to the data analysis and as a result, answered the questions presented in part one and justified the hypotheses presented in part three. The most important research question for ECFs is to investigate the main human resource determinants on client satisfaction and the consequences that will lead to client retention and free marketing in the form of word-of-mouth.

For one thing, the introductory questionnaire included items pertaining to the general background of the firms. The various other questionnaires were each specific to the variables that formulated both the questions and hypotheses. For instance, the variables covered client satisfaction, needs, preference; employee experience, knowledge, motivation, professionalism; and the overall care factor, retention, and word-of-mouth. The Final Conceptual Model used is illustrated in **Figure 1** below. Ultimately, the process of preparation, codification, distribution, collection, and analysis of data was initiated. The results were used to establish the proposed model and were tested using statistical software such as, the Statistic Package for Social Science Statistics 24: IBM-SPSS Statistics 24 and IBM-SPSS-Amos 26. The primary use of *SPSS* is for data and variable organisation and management and provides a valuable tool to manage the data that was imported into *SPSS* from a *Microsoft Excel file* after the results were extracted from the survey package Jot-form.

For clarification, the introductory questions requested information from the clients to validate the respondent ensuring that they belong to the target audience. This information was also used to group each respondent by certain demographics and controls, such as age, position, experience with engineering consultants, value of project, and type of engineering consultant used. Anonymity of the respondents was assured and so questions were designed carefully as not to identify participants; the engineering firms that were part of the sample are prominent and successful members of Australia's elite. To elaborate, the seven-point Likert scale was employed so that the survey was able to meet the requirements of validity and reliability that were measured by Cronbach's Alpha.

At this time, there was much debate as to how large the sample should be and in accordance with the context of the study, numerous sample sizes were considered. A consensus amongst researchers is that a sample size of approximately 200 is usually suitable [49, 50]. There is also evidence that a meaningful analysis can be conducted with at least 100 samples [51]. Nonetheless, ACE decided to opt a random selection of 1000 that had been their clients for the last 3 years. The online survey on jot-forms was distributed via email to the above-mentioned clients. Of the 1000 emails, 273 bounced back, indicating that it was the wrong email or that the person no longer worked at this firm or address. Along with this, of the remaining 727 emails that were delivered, only 256 clients responded indicating a response rate of approximately 35%.

Once the data had been screened, the dataset was ready for the exploration of the quality of the measurement scales using an Exploratory Factor Analysis (EFA).

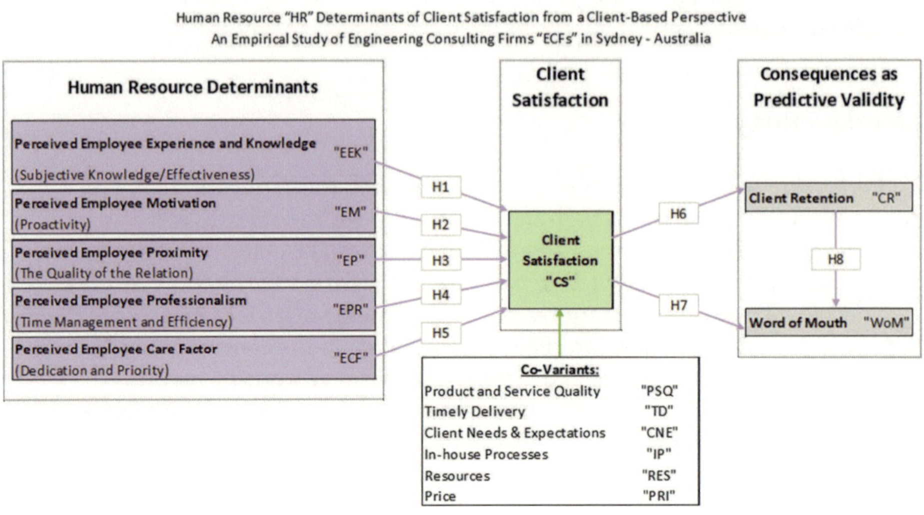

Figure 1.
Final Conceptual Model used in the analysis.

EFA is used for the exploration of the underlying structure amongst initial variables (or items) included in the analysis. It helps determine if theses initial variables are factorizable, and if the computed factors are interpretable. The *Exploratory Factor Analysis (EFA)* is a statistical approach that groups variables with strong correlations together to give a factor structure [52]. An important aspect of the EFA is that the analysis is not dependant on these constructs [52]. Rather, the EFA will identify the number of latent factors and the variables that strongly group with each latent factor based on the correlations of the observed variables.

Moreover, Structural Equation Modelling (SEM), is a technique used primarily in social and behavioural sciences and can be described as "*a combination of factor analysis and regression or path analysis*" [53]. A major advantage of SEM is its ability to measure theoretical constructs by using *latent variables*. Latent variables are factors that cannot be directly observed by themselves but can be inferred mathematically from the use of a number (usually 3–6) of observed variables [54]. In contrast, as the name suggests, an *observed variable* is a factor that can be directly measured by itself. In the case of this research, the latent variables will represent the HR determinants whilst the observed variables will be the survey question responses which can be directly measured and allows researchers to assess their theories using a testable model [55].

Overall, there are two types of variables: latent variables (unobserved variables) and measured variables (observed variables). In social science research, many variables cannot be directly measured such as the behavioural dimensions and so are placed as latent variables in the model. The reason why the behavioural dimensions are latent variables instead of measured variables is that these concepts are complex and multifaceted—a single measure will not adequately cover the whole dimension. For example, there are many variables that will be needed to cover the entirety of employee professionalism such as their appearance, manners, and ethics. Measured variables on the other hand, are variables where there is a measurement for that data-set. For instance, in the Client Retention questionnaire, one statement was: "In the future, I will use the ACE consultant for most of my engineering consultancy needs" and participants gave a score between one and seven giving an exact measurement for their retention level.

Likewise, the reliability of the measurement scales will be tested using Cronbach's alpha which is widely used to measure reliability in social sciences [56]. Cronbach's alpha measures the connection and consistency for a set of variables that are placed together as a factor. As a rule, the Cronbach's alpha value for each factor should be above 0.7 [56]. Furthermore, there are two types of validity to account for—Convergent Validity and Discriminant Validity. The aim of convergent validity is to verify that the variables in each dimension are correlated with each other or share a high proportion of variance. For convergent validity, there are three values that need to be checked which are the standardised factor estimates, the average variance extracted and composite reliability. Discriminate validity aims to verify whether there is a distinct difference between latent constructs. To measure discriminant validity, the square root of the average variance extracted is compared to the estimated correlations between the constructs.

Briefly, correlation coefficients measure the amount of shared variance that occur between two variables. The value of this coefficient determines the relationship between the relative movement of the two variables. Clearly, correlation coefficients range in value between −1.0 to 1.0—whereby a positive value means that the two variables move in the same direction and a negative value means that they move in opposite directions. Altogether, this research will use the criteria defined by the *Multivariate Data Analysis* [57] for establishing reliability and validity of the factors.

5. Results

After a brief but concise explanation of the methodology in the previous section, we will now read what the outcomes of the survey were and determine if they justify the hypotheses that were formulated earlier in the chapter.

First and foremost, the results related to the descriptive statistics revealed that the majority of the respondents are in total agreement with the item of Customer Satisfaction. Therefore, it is concluded that customers are satisfied with the service provided by the Australian Consulting Engineers (ACE) consultant. In addition, their experiences with the ACE consultant are excellent and they are highly pleased with overall performance.

To clarify, let us explore the results of each of the variables that are the focus of this study. In terms of customer needs and expectations, the results confirm that mainstream participants revealed their total agreement. Hence, it is concluded that ACE consultants always offered as much information as needed. Moreover, employees spent an adequate amount of time sharing their needs with the ACE clients during the project. It is also found that ACE consultants surpassed the level of expectations and never disappointed their clients. The above is also related to timely delivery and entails that a large proportion of respondents agree that ACE consultants met deadlines and are highly punctual. In other words, ACE consultants made excellent choices and implemented time management. In short, ACE consultants are extremely dependable.

As for the quality of products and services, the data revealed that adequate agreement levels were observed. It is thus concluded that services provided by the ACE consultant are reliable and demonstrated overall quality. Additionally, the analysis of outcomes stressed that due to adequate training and development, ACE consultants are always expected to perform satisfactorily. This is because all in-house processes are instigated professionally so as to achieve successful project completions that meet client requirements.

Overall, the findings were clear to state that ACE consultants constantly kept their clients well informed about what was going on with relevant projects and were very transparent about all pertinent issues. In the same manner, engineering advice can be easily obtained from ACE consultants, and they are always available and present for the job. It is noteworthy at this point to mention that ACE services offer good value and so clients are satisfied with 'return on investment'.

Similarly, the data and results related to employee experience and knowledge confirmed that all respondents revealed the higher scale agreement with the statement that products and services that ACE consultants presented have demonstrated the expected quality for money. In addition, ACE consultants are well informed and knowledgeable affirming that they are proficient, competent and experts in their fields and execute their jobs both effectively and efficiently.

In particular, the findings regarding employee motivation confirm that overall greater motivational levels are found to be prevalent. The strong agreement level confirms that ACE consultants verified strong commitment and loyalty towards their jobs and fulfil their goals in the best possible fashion. Unquestionably, ACE consultants are found to be highly motivated.

Correspondingly, the survey findings regarding employee professionalism that was highlighted previously attested that strong agreement is evident for the demonstration of respect by ACE consultants. These consultants readily presume responsibility for decisions, choices, and errors. The results also confirmed that ACE consultants' clothing, grooming, and hygiene consistently portrayed a professional appearance. As such, the ACE consultant is considered be highly straightforward, sincere, truthful, and sincere, but the results did not reflect that professionalism influenced Customer Satisfaction.

The outcomes regarding the Employee Proximity confirms that higher agreement scale affirms that ACE Consultant connected to life and experiences and takes an account of other perspectives. The ACE Consultant knows how to construct a close relationship during a project. The results also confirm those ACE consultants are aware about the effective treatment of client.

Furthermore, the outcomes of the employee care factor especially confirmed that ACE consultants displayed extra effort to meet their clients' needs. For instance, they were always ready to comply by working late, extra hours or on weekends; they were always accessible. Obviously, this ascertains that ACE consultants were well equipped to provide a suitable resolution to any problem encountered during the project.

Finally, the data regarding customer retention and word-of-mouth revealed that clients are loyal to the consultants of ACE. In the forthcoming future, clients are more prone to utilise the ACE consultant for any engineering deliberation. Besides, most respondents confirm that ACE is the firm of choice when it comes to engineering consultancy services. Likewise, the result regarding word-of-mouth validates that respondents are more inclined to maintain a positive attitude where ACE consultants are involved. In sum, most respondents are in harmony and are adamant and keen to recommend the services of ACE consultants to others.

Accordingly, the eight hypotheses stated above are justified and fully supported except for the effect of Professionalism and Proximity on Customer Satisfaction.

Clients that have strong relationships with consultants are assured that they will continue to meet their expectations due to precedence. Switching to a new consultant means confronting uncertainty in the quality of the project. This explains why clients are more likely to do business with people they have a good relationship with. As such, employee relationship quality is an essential element in ensuring the repeat purchase decision by clients as compared to the price competition or contractual agreement. Thus, Employee Proximity is good, and the employee is aligned with the Client then this should lead them directly to repurchase decisions.

We therefore suggest a direct and positive influence of perceived employee proximity on client retention, and this hypothesis has been tested and successful.

6. Discussion

At this point, the authors would like to emphasise that this study was conducted for the requirements of a Doctorate in Business Administration dissertation. This chapter has aimed to present as fully and as concisely as possible the findings of the data analysis that contributed to the overall justification of context. It is worthy to note that all documentation-text clarifications, full interview transcripts, letters sent to participants to ensure ethical behaviour, descriptive analysis of data, quantitative and qualitative analysis in the form of matrixes, tables, graphs etc.-is available in the original dissertation.

This research conducted a Structural Equation Model (SEM) analysis on the influence of human resource determinants of engineering consultants on client satisfaction. In the initial proposed conceptual model, there were determinants such as experience and knowledge, motivation, proximity, professionalism, and the care factor. Namely, the results showed that experience and knowledge, motivation, and care led to a positive effect on client satisfaction, which in turn directly influenced client retention and word of mouth. More precisely, the results from the analysis revealed that employee professionalism had a negative effect on customer satisfaction which is a strange effect which needs to be explored further and that employee's proximity had no effect on customer satisfaction but rather on customer retention. It was further demonstrated and supported that client satisfaction leads to a higher likelihood of client retention and positive word of mouth.

Based on these basic findings, Engineering Consultancy Firms (ECFs) can now understand that a consultant's experience and knowledge are crucial behavioural dimensions. If ECF's are aiming to retain their clients and create a positive word of mouth, they must initially ensure client satisfaction and guarantee that their employees offer effective and efficient outcomes. Specifically, perceived employee experience and knowledge followed by perceived employee motivation and then followed by perceived employee care factor show a standardised direct effect of 0.47, 0.335, and 0.202 respectively. Moreover, the data analysis proved that the effect of client satisfaction on client retention was very high and positive with a standardised direct effect of 0.45. Following that, the effect of client satisfaction on word of mouth was also high with a standardised direct effect of 0.559. Furthermore, the effect of client retention on word of mouth was also positive but with a smaller standardised direct effect of 0.433. All the above-mentioned hypotheses were supported with significant p-values.

Whilst the engineering consultant's professionalism had no significant impact on client satisfaction, then this study failed to propose a mediating or moderating effect that is necessary to understand the underlying phenomena but must be addressed in the future. For clarification, the professionalism dimension asked clients about consultant attitude and whether they respected clients, assumed responsibility, were straight forward, truthful, and sincere. Furthermore, it is concluded that experience, knowledge, motivation, and care factor all had positive and significant influence on client satisfaction. Regardless of the results of the other behavioural dimensions, the results of the afore-mentioned factors are important and engineering consultants should consider and prioritise them.

What is more, the findings conclude that client satisfaction, client retention, and word of mouth are all strongly related to each other in the engineering consulting industry. That is, satisfied clients are more likely to continue to employ the services of an engineering consultant and recommend them to others, whilst the opposite is also true. This finding confirms the importance of client satisfaction for engineering consulting firms as it is directly related to the ability to better retain clients and enhance the firm's overall reputation in the industry.

From the results presented, it can be concluded that the engineering consultant's proximity did not have a direct impact on client satisfaction and was not supported in this study. This result suggests that Proximity is a relationship management construct that leads to intention to continue the relationship but is not necessarily related to client satisfaction towards the delivered services. This is a meaningful result that must be considered, as it tells that the loyalty is towards the employee him or herself and not towards the consulting company and its services.

Moreover, it can be concluded that through the alternative model, the engineering consultant's proximity has a positive and significant impact on client retention. The proximity dimension asked the clients about the consultant's connection to the client's life and experience, if the consultant was able to consider the perspective of the client, if the consultant was able to take into consideration the input from the client, and if the consultant knows how to treat the client.

In fact, 34.6% of the clients in the survey totally agree that ACE consultants connected to the life experience of the clients. Similarly, 44.7% of the clients totally agree that ACE consultants were able to incorporate the needs of the clients whilst 46.2% totally agree that ACE consultants were able to establish rapport with the clients. Overall, 55.8% of the clients were in total agreement that that ACE consultants understand and know how to treat clients. Equally important is that over 50% of the clients totally agreed that ACE consultants consistently showed respect for others, readily assumed responsibility for performance. In short, consultants are always well groomed and display professionalism at all levels.

Furthermore, evidence revealed 45.2% of the clients in the survey totally agree that ACE consultant always took initiative and went out of their way to meet their needs. In doing so, clearly 51.9% of the clients agree that ACE consultants provided satisfactory resolutions to any problems encountered throughout the job. In general, over 50% of the clients also agree that ACE consultants have very good client relationship management and regard them highly in the industry. This is not surprising knowing that ACE just celebrated 25 years of service to the construction industry and the majority were repeat clients and over 45% have been dealing with ACE for more than 10 years.

In summary, it is noteworthy that the real managerial implications and outcomes of the research are the importance of the HR determinants in reaching client satisfaction especially for experience and knowledge, motivation, and the care factor. For all the above reasons, almost 60% of the clients that were surveyed totally agree that they are likely to say positive things about ACE consultants, are likely to recommend the services of ACE consultants, and are likely to suggest ACE consultants if they are asked to recommend an engineering consultant.

6.1 Limitations and future research

Initially, the original scope of the study aimed to compare results based on different demographics such as gender, age, project type and size, engineering discipline, and prior experience with engineering consultants. However, an accurate analysis [50] should consist of a minimum of 100 samples per group but unfortunately, this was not possible. The study showed 88.1% of the respondents were males, 70.4% of the responses were from the structural discipline, and 74% of the responses were based on residential projects. Surely, the results from this study should be carefully applied universally. Thus, it is unknown if the data would differ for female clients, other disciplines, and different project categories. Moreover, when identifying the determinants of client satisfaction, this study focuses only on the human resource determinants of the consultant whereas client satisfaction has been found to also include non-behavioural determinants, such as time, cost, and quality.

Overall, the scope of this research focused on the context of Engineering Consulting Firms (ECFs) in Sydney, Australia. Indeed, client satisfaction is a broad concept and can be extended to various frameworks. However, to keep the scope of this study applicable and relevant to engineering firms in Australia as a whole, then a study should certainly focus on the industry of engineering consultancy in Australia. Nevertheless, caution must be taken when extending the findings of such a study to other fields or regions as it is possible that the results may vary depending on cultural differences and industry norms.

Hence, further research that can be conducted to heighten the findings for future implementation may include the following themes, to list a few:

1. A focus on employee professionalism and its underlying phenomena with specific research on its mediating and moderating effects.

2. The inclusion of non-behavioural controls such as price, time, and quality in the model.

3. Comparing results from different demographic groups, such as gender, age, project type, project size, engineering discipline, and prior experience with engineering consultants. As [50] recommends, a minimum of one hundred samples per demographic group would require a larger sample size of approximately 1000.

4. Expanding the scope of the study to observe client satisfaction in a variety of different contexts such as industry and region. In addition, further research may possibly attempt to perform this study for a universal context.

7. Conclusion

To summarise, this research achieved an analysis on 221 samples to investigate two outcomes:

1. To identify, measure, and compare the impact that the Human Resource determinants of an engineering consultant has on client satisfaction.

2. To measure the impact that client satisfaction has on client retention and word-of-mouth reputation of the Engineering Consulting Firms (ECFs).

For interpretation, the literature review identified the essential HR determinants of engineering consultants to be knowledge, experience, motivation, professionalism, and the care factor. Following this, a set of hypotheses were formulated to highlight that these factors had a positive and direct effect on client satisfaction. A questionnaire was designed and distributed to various clients of an engineering consulting firm in Sydney, Australia; the response rate was 35% with a sample size of 256 and 221 valid responses were collected. Namely, it was discovered that due to high squared multiple correlations of the endogenous variables, there was enough statistical power to support the significance found for each HR determinant's impact on client satisfaction.

Thereafter, an extensive Exploratory Factor Analysis (EFA) of the data found that the questionnaire items loaded as expected to the original HR determinants. After a review of the questionnaire items and the data results, a Confirmatory Factor Analysis (CFA) and a Path Analysis were performed on the model that exhibited sufficient-to-good reliability, validity, model fit, and squared multiple correlations. These findings correspond with [6] who stressed the importance of the "unwritten expectations" that clients desire but do not clearly state.

In conclusion, Engineering Consulting Firms that can guarantee client satisfaction are more likely to retain those clients as well as receive recommendations that help them gain new clients. Specifically, it should be highlighted that experienced, knowledgeable, and motivated employees will always have the greatest and most desired impact on client satisfaction. Without a doubt, these firms should not only be aware of but also acknowledge that when their employees are satisfied, this will translate into satisfied clients and ensure satisfactory end products.

Author details

Elham Said Hasham and Anthony Said Hasham*
A.C.E. (Australian Consulting Engineers), Sydney, Australia

*Address all correspondence to: anthonyh@aceeng.com.au

References

[1] Akroush MN, Abu El Samen AA. An empirical investigation of the mediating role of relationship marketing skills on the relationship between customer satisfaction and customer loyalty. International Journal of Internet Marketing and Advertising. Jan 2012;**7**(1):1-30. DOI: 10.1504/IJIMA.2012.044956

[2] Lam SY, Shankar V, Erramilli MK, Murthy B. Customer value, satisfaction, loyalty, and switching costs: An illustration from a business-to-business service context. Journal of the Accademy of Marketing Science. 2004;**32**(3):293-311

[3] Rashvund P, Majid MZA. Critical criteria on client and clients satisfaction for the issue of performance measurement. Journal of Management in Engineering. 2014;**30**(1):10-18

[4] Stefan MC, Dogaru M. Organizarea acțivitățiiantrenareaangajaților, coordo nareaşicontrolulactivităților. Dezvoltări Social-Economice. 2012;**2**:65

[5] Androniceanu A. Motivation of the human resources for a sustainable organizational development. Economia (Bucharest, Romania: Seria Management, Faculty of Management, Academy of Economic Studies). Dec 2011;**14**(2):425-438

[6] Chelliah J, Davis D. What clients really want from management consultants: Evidence from Australia. Journal of International Management Studies. 2011;**6**(1):22-30

[7] O'Donnell J. The Benefits of Building Long-Term Business Relationships. Frazer John Recruitment; 2017. Available from: https://frazerjohn.com/insights/the-benefits-of-building-longterm-business-relationships/

[8] Haverila M, Bateman E, Naumann E. The drivers of clients satisfaction in strategic consulting engagements. Management Decision. 2011;**49**(8): 1354-1370

[9] Tang SL, Lu M, Chan YL. Achieving client satisfaction for engineering consulting firms. Journal of Management in Engineering. 2003;**19**(4):166-172

[10] Hecker PA. Human resources strategies for successful consulting engineering firms. Journal of Management in Engineering. 1996;**12**(5):32-36

[11] Shih CY, Huang CY, Huang ML, Chen CM, Lin CC, Tang FI. The association of sociodemographic factors and needs of haemodialysis patients according to Maslow's hierarchy of needs. Journal of Clinical Nursing. 2019;**28**(1-2):270-278

[12] Ganesh A. Understanding the relationship between employee motivation and clients retention. Vilakshan: The XIMB Journal of Management. 2016;**13**(1):101-114

[13] Martinaityte I, Sacramento C, Aryee S. Delighting the clients: Creativity-oriented high-performance work systems, frontline employee creative performance, and clients satisfaction. Journal of Management. 2019;**45**(2):728-751

[14] Hill N, Alexander J. The Handbook of Customer Satisfaction and Loyalty Measurement. 3rd ed. London: Routledge; 31 Dec 2016

[15] Pantouvakis A, Bouranta N. The interrelationship between service features, job satisfaction and clients

satisfaction. The TQM Journal. 2013;**25**(2):186-201

[16] Creswell JW, Creswell JD. Research Design Qualitative, Quantitative, and Mixed Approaches. 4th ed. Newbury Park: Sage; 2017

[17] Oliver RL. A cognitive model of the antecedents and consequences of satisfaction decisions. Journal of Marketing Research. 1980;**17**:460-469

[18] Cardozo RN. An experimental study of consumer efforts, expectations, and satisfaction. Journal of Marketing Research. 1968;**2**:244-249

[19] Hill N, Brierley J. How to Measure Clients Satisfaction. London: Routledge; 2017

[20] Ennew CT, Binks MR, Chiplin B. Clients satisfaction and clients retention: An examination of small businesses and their banks in the UK. In: Proceedings of the 1994 Academy of Marketing Science (AMS) Annual Conference. Cham: Springer; 2015. pp. 188-192

[21] Mohsan F, Nawaz MM, Khan MS, Shaukat Z, Aslam N. Impact of clients satisfaction on clients loyalty and intentions to switch: Evidence from banking sector of Pakistan. International Journal of Business and Social Science. 2011;**2**(16):263-270

[22] Tucker M. How important is clients satisfaction to the services sector? Journal of Facilities Management. 2010;**8**(4):91-100. DOI: 10.1108/jfm.2010.30808daa.001

[23] Cheng J, Proverbs DG, Oduoza CF. The satisfaction levels of UK construction clients based on the performance of consultants: Results of a case study. Engineering Construction, and Architectural Management. 2016;**13**(6):567-583

[24] Hussnain A, Khan RA, Fatima M. Impact of service quality on customer satisfaction in banking industry of Pakistan: A case study of Karachi. Journal of Social and Administrative Sciences. Sep 2018;**5**(3):219-238

[25] Lang et al. Word of mouth: What we know and what we have to learn. Journal of Consumer Satisfaction, Dissatisfaction and Complaining Behaviour. 2013;**26**:1-18

[26] Shi W, Tang L, Zhang X, Gao Y, Zhu Y. How does word of mouth affect clients satisfaction? Journal of Business & Industrial Marketing. 2016;**31**(3):393-403

[27] Zainol N, Ahmad Rozali A, Fadhlina M, Akhir I, Nordin NA. The Influence of Customer Satisfaction Towards Positive Word-of-Mouth in Hospitality Industry. 11 Jan 2016. DOI: 10.2139/ssrn.2713613. Available from SSRN: https://ssrn.com/abstract=2713613

[28] Doney P, Cannon J. An examination of the nature of trust in buyer-seller relationships. Journal of Marketing. 1997;**61**:35-51

[29] Yoon MH, Suh J. Organizational citizenship behaviours and service quality as external effectiveness of contact employees. Journal of Business Research. 2003;**56**(8):597-611

[30] Organ DW. A reappraisal and reinterpretation of the satisfaction-causes-performance hypothesis. Academy of Management Review. 1977;**2**(1):46-53

[31] Brown KG, Sitzmann T. Training and Employee Development for Improved Performance. Washington, D.C.: American Psychological Association; 2011

[32] Wakefield N, Abbatiello A, Agarwal D, Pastakia K, Berkel A.

Leadership Awakened: Generations, Teams, Science. Deloitte Insights; 2016. Available from: https://www2.deloitte.com/insights/us/en/focus/human-capital-trends/2016/identifying-future-business-leaders-leadership.html

[33] Sharmila JJV. Employee engagement– An approach to organisational excellence. International Journal of Social Science & Interdisciplinary Research. 2013;**2**(5):111-117

[34] Mouawad M, Kleiner BH. New developments in clients service training. Managing Service Quality: An International Journal. 1996;**6**(2):49-56

[35] Zaleznik A. Managers and leaders: Are they different? Harvard Business Review. 1977;**55**(3):67-78

[36] Hasham ES. Organizational Behaviour: Arab World Edition. UK: Pearson: Prentice-Hall; 2012

[37] Yukl G. Managerial leadership: A review of theory and research. Journal of Management. 1989;**15**(2):251-289

[38] Wang Z, Singh SN, Li YJ, Mishra S, Ambrose M, Biernat M. Effects of employees' positive affective displays on clients loyalty intentions: An emotions-as-social-information perspective. Academy of Management Journal. 2017;**60**(1):109-129

[39] Lu L, Lu ACC, Gursoy D, Neale NR. Work engagement, job satisfaction, and turnover intentions: A comparison between supervisors and line-level employees. International Journal of Contemporary Hospitality Management. 2016;**28**(4):737-761

[40] Johnson M, Gustafsson A. The effects of clients satisfaction, relationship commitment dimensions, triggers on clients retention (study on telecommunications services). Journal of Marketing. 2005;**69**(4):210-218. DOI: 10.1509/jmkg.2005.69.4.210

[41] De Wulf K, Odekerken-Schroder G, Lacobucci D. Investments in consumer relationships: A cross-country and cross-industry exploration. Journal of Marketing. 2001;**65**(4):33-50

[42] Bratton J, Gold J. Human Resource Management: Theory and Practice. 6th ed. London: Palgrave Macmillan; 2017

[43] Gazzoli G, Hancer M, Park Y. The role and effect of job satisfaction and empowerment on clients' perception of service quality: A study in the restaurantindustry. Research of Hospitality and Tourism Research. 2010;**34**(1):56-77

[44] Sundaray B. Employee engagement: A driver of organizational effectiveness. European Journal of Business and Management. 2011;**3**(8):53-59

[45] Swarnalatha C, Prasanna T. Leveraging employee engagement for competitive advantage: Strategic role of HR. Review of HRM. 2013;**2**:139-148

[46] Hammarberg K, Kirkman M, de Lacey S. Qualitative research methods: When to use them and how to judge them. Human Reproduction. 2016;**31**(3):498-501

[47] Daniel E. The usefulness of qualitative and quantitative approaches and methods in researching problem-solving ability in science education curriculum. Journal of Education and Practice. 2016;7(15):91-100

[48] Hanaysha J. Examining the effects of employee empowerment, teamwork, and employee training on organizational commitment. Procedia-Social and Behavioral Sciences. 2016;**229**(2):298-306

[49] Hoogland JJ, Boomsma A. Robustness studies in covariance structure modeling: An overview and a meta-analysis. Sociological Methods & Research. 1998;**26**(3):329-367

[50] Kline RB. Methodology in the Social Sciences. Principles and Practice of Structural Equation Modeling. 2nd ed. New York, NY, US: Guilford Press; 2005

[51] Tabachnick BG, Fidell LS. Using Multivariate Statistics. 4th ed. Boston, MA, USA: Allyn and Bacon; 2001

[52] Thompson B. Exploratory and Confirmatory Factor Analysis: Understanding Concepts and Applications. 1st ed. Washington, DC: American Psychological Association; 2004

[53] Hox JJ. An introduction to structural equation modeling. Family Science Review. 1999;**11**:354-373

[54] Sauro J. The Difference between Observed and Latent Variables [Blog Post]. 2016. Available from: https://measuringu.com/latent-variables/

[55] Violato C, Hecker G. How to use structural equation modeling in medical education research: A brief guide. Teaching and Learning in Medicine. 2007;**19**(4):362-371. DOI: 10.1080/10401330701542685

[56] Bonett D, Wright T. Cronbach's alpha reliability: Interval estimation, hypothesis testing, and sample size planning. Journal of Organizational Behavior. 2015;**67**(3):3-15. DOI: 10.1002/job.1960

[57] Hair JF, Black WC, Babin BJ, Anderson RE, Tatham RL. Multivariate Data Analysis. 7th ed. Upper Saddle River, NJ, USA: Prentice-Hall, Inc.; 2014

Section 2

Applications of QA/QC in Industry

Chapter 6

Quality Planning, Assurance and Control in Manufacturing, Industrial Projects and Services Environments

Rajenlall Siriram and Chantelle du Plessis

Abstract

The role of quality planning, assurance and control is well established in manufacturing and production environments; however, the same cannot be said for industrial projects and the services environment. It is evident that the tangibility of the output is different and, as a result, evaluated differently. In a project environment, many projects exceed budget, schedule and performance. One may argue that quality planning, control and assurance have a significant role in ensuring projects meet their budget, schedule and performance objectives. In contrast, in a services environment, quality planning, assurance and control procedures are used to ensure that the set of quality criteria meets the customer's requirements so that the customer gets a better service experience. There are differences in how quality planning, assurance and control are implemented and perceived in these three environments. This chapter explores the differences and similarities in the implementation, perception and maturity of quality planning, assurance and control. This chapter provides interesting opportunities for quality planning, assurance and control in a manufacturing, projects and services environment.

Keywords: quality planning, control, assurance, manufacturing, projects services

1. Introduction

Let us consider different industry settings such as mining, construction, manufacturing and service. All these industry settings have operations. Operations consist of turning inputs into outputs. Inputs go through a transformation process where inputs are converted to outputs. This is referred to as the operations management input-transformation-output model. While each industry sector differs, the input-transformation-output model is the same. In each of these sectors, inputs are transformed into goods and services the customer pays for. In mining, construction and manufacturing, the products delivered are tangible; for example, in mining, it is the raw materials that are produced, this includes gold, platinum, copper, coal, etc. In construction, the end product is a residential building, a bridge, a dam, etc.

These are large physical infrastructure projects. In manufacturing, the end products include, for example, an automobile, a laptop, a mobile phone, etc. In services, on the other hand, the finished product is intangible; it is a service offering which involves the interaction between the service provider and the customer; let us consider the following example: you have had to enquire at a bank because you want a home loan or a mortgage for a vehicle. You will, first, either go on the bank's website and navigate your way around the website depending on the ease of use; this could be either an easy or difficult experience. The next step will be to complete some personal information online. After that, you will have to interact with another human being. This whole encounter is rather intangible; it depends on the technology you used to access the service platform, as well as the banking platform, and the service you experienced from the banking employee. It will also depend on the information you provide to the banking employee. It becomes each individual's perception. Your objective was to obtain a loan; while this is tangible, the service experience was intangible and either left you feeling served well or a feeling of mediocre or poor service; therefore, it is difficult to put some tangible quality measure to the service experience. As demonstrated by these examples, it is fairly evident that each of the industry sectors, while they appear different at a conceptual level, may be considered a complex system trying to utilize resources to maximize performance [1].

Complexity arises from the interactions within and between firms [2]. In manufacturing and production, different departments of the firm have to interact internally, and departments also have to interact externally with suppliers and other stakeholders to manufacture the final product. The same can be said for the manufacturing and services industries. These interactions bring about complexity. To ensure that firms can utilize resources to maximize performance, these systems must be subjected to planning, assurance and control activities. Performance in a manufacturing industry may refer to the Sandcone model [3, 4], which speaks to the manufacturing mix criteria of quality, dependability, speed, flexibility and cost. In terms of the projects industry, many authors have suggested an insufficient understanding of the relationship between project management success and project success [5, 6]. However, [6] points out that project success is determined by achieving the project's overall goals, and [7] argues that project management success is determined by the iron triangle of time, cost and quality. Hence, in this chapter, we will focus on project management and project success as we see both as important. In terms of the services industry, one may refer to the SERVQUAL instrument for measuring service quality [8], which gives 10 dimensions for measuring service quality. To better understand these activities (Quality planning, assurance and control), we look at the Body of Knowledge in quality management in this chapter.

In terms of quality management Total Quality Management (TQM) philosophy is about managing systems through the seven principles of TQM [9]. These seven principles will be discussed in more detail in *Section 2*. Quality management, among other things, focuses on quality planning, assurance and control. Quality management is well understood and embedded in the manufacturing and production environments. Manufacturers easily relate to TQM philosophies, which are well-rooted in the firm's culture. However, the same cannot be said for the projects and services environment. In project management, quality management involves, among others, three processes, namely quality planning, quality assurance and quality control [10]. These processes, while at a conceptual level, are the same; in manufacturing industries, they are well embedded, but in the project environment, they are not so well embedded. In manufacturing and production, there is a process of continuous measurement and control, and it is widely accepted every component is strictly quality controlled, which is governed by a quality assurance system and requires careful planning. However, in the project environment, quality planning, assurance and control are accepted as important but are, however, more difficult to implement and hence, this hampers project management success as well as project success. These processes aim to ensure that the project will satisfy the needs for which it was designed and built [9]. While these processes are well defined, they do not seem to have the desired effect as many projects are out of control and many report huge losses [11–13] moreover, others like [14–16] also reported that many projects are overdue and over budget. In terms of the services industry, there are fundamental differences compared to the manufacturing industry, which is predominantly responsible for the production of goods. In contrast, the services industry is predominately responsible for the delivery of services and is about the service experience [17]. Distinguishes four attributes in terms of goods and services as follows:

a. Intangibility: Services are much less tangible compared to physical goods. Services relate to experiences, whereas manufacturing is about physical objects, i.e., tangible things that one can see, feel and touch.

b. Inseparability of production and consumption. In manufacturing, goods are produced, delivered and consumed by the customer at some later stage. In services, there is simultaneous creation of the service and the consumption thereof, usually on an ongoing basis.

c. Heterogeneity: In this case, the quality of service varies from one service provider to another service provider, from customer to customer and from day to day.

d. Perishability: Services are about experiences rather than tangible physical objects, and therefore, it cannot be stored and retrieved for usage at a later stage; in other words, it cannot be inventoried; therefore, service providers find it difficult to synchronize supply and demand. Service is a deliverable in terms of the service experience consumed by the customer; however, continued ongoing service is only achievable if the customer has a good service experience.

Quality management in the services industry is strongly affected by these attributes [18]; this is mainly because, firstly, services are about the service experience, it is about how the customer experiences or the customers' perception of the service encounter, the service experience is about ongoing continuous satisfaction [19]. "Defines the service encounter as the "face-to-face" encounter between a buyer and seller in a service setting." It is about a win-win or benefit-benefit for the customer and service provider. Moreover, because services are about simultaneous production and consumption, services are not manufactured in a plant and delivered and consumed at some later stage; it is about the interaction between the service provider and the customer. Services have high heterogeneity; as a result, uniform quality will be difficult as it is dependent on the interaction between the customer, the platform of engagement and the service provider, more specifically, the service employee delivering the service [20], as well as the information provided by the customer. Perishability, that is, services cannot be stored and retrieved for usage at a later stage and hence, quality cannot be pre-determined; quality occurs at the point of delivery.

2. Systems thinking, TQM, quality planning, assurance and control

2.1 The firm as a system

Firstly, we recognize that a firm is connected, and there are many interaction points as we live in a networked society [2]. This interaction is necessary due to exchanging goods and services [2] within and between firms. Goods and service exchanges occur within firms as different sections (departments) of the firm interact with each other, and exchanges occur between firms as firms interact with other firms, like suppliers, customers, and other stakeholders. This interaction with other firms leads to connectivity [21]. Defines connectivity as "the property that enables effects to be transmitted through the system". In Ref. [2] positioned that firms do not exist in isolation; they are connected with other firms by buying and selling goods and services [2]. Therefore, taking a systemic view of the firm is well warranted. In addition, firms require inputs; these inputs are resources and aids. Resources include materials, customers, information, etc. and aids, which include human resources, equipment, facilities, technology, etc. These inputs are transformed through a process, and the raw materials are converted to finished products. This is referred to in operations management as the input-transformation-output model. **Figure 1** gives a graphical representation of the input-transformation-output model.

The operations management input-transformation-output model is a conceptual generic model and can be applied to different industry settings; as in the case of this chapter, it applies to the manufacturing, projects and services industries. In the manufacturing industry, let us consider an automobile manufacturing plant. Inputs include materials procured from suppliers in either unprocessed form or finished components from some suppliers; customers who have requirements and need an automobile to provide inputs regarding specifications, quantity, color, etc. Then, there is information on the general market, demand for other vehicles, and customers who may require a similar automobile. Aids will include the human resources required

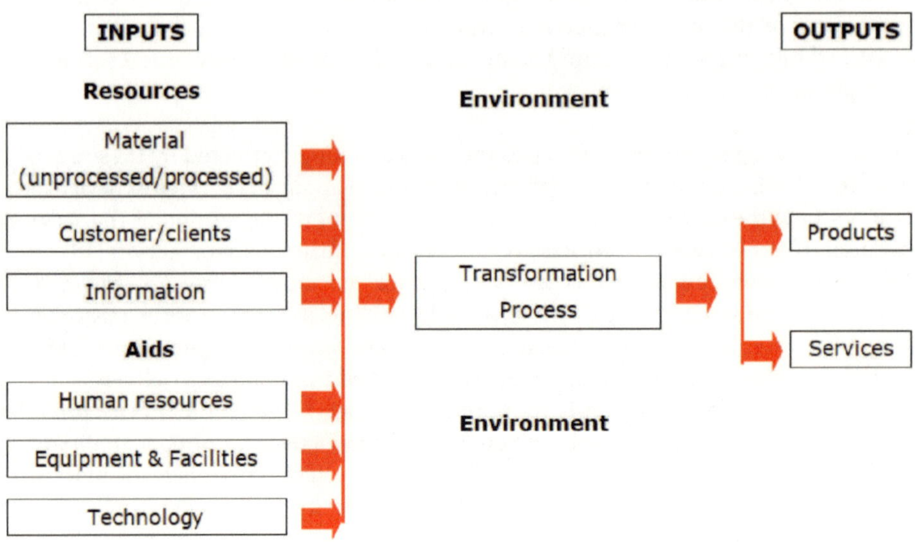

Figure 1.
Input-transformation-output model.

to manufacture, i.e. the engineers, the workforce, and other support staff like procurement, finance, maintenance, etc. Aids also includes the manufacturing facilities, equipment and technology required to produce the automobile. These inputs then go through a transformation process, which in this case is manufacturing, where all the inputs are transformed into the finished product, that is, the automobile, which is then delivered to the end customer.

In a project environment, there are also inputs: materials processed as in the manufacturing industry; there are customers who need and provide specifications, budgets and timelines, which the project firm has to adhere to. There is also other information about the external factors that influence the build of the project. In terms of the projects, industry aids also include human resources, as is in the case of the manufacturing industry. There are also more stakeholders and many different external factors like environmental concerns, weather, etc., which are also aids, but they have a high degree of uncertainty. Equipment and facilities, but these equipment and facilities may have to be transported to different or remote sites. This brings different complexities as opposed to manufacturing. Then, there is also the technology that is used to facilitate the project management process. In this case, the transformation process is project management. A project must be well managed to successfully deliver to the end customer. The delivery is the complete project's product to be built, be it a residential building, a bridge, a dam, etc.

Let us consider the banking example discussed in *Section 1* in the services sector. The inputs consist of information about the customer and the customer's needs; the materials include the documents provided by the customer, for example, proof of employment, proof of residence, etc. Information includes historical information about the customer's past performance regarding loan repayments, etc. In this case, aids also include human resources personnel from the bank and the customer, as without customer interaction, the service cannot be delivered. Aids also include the banks' platform for the interaction, the customer's technology that is being used to access the bank's platform and the banking technology that is used to process the loan application. In this case, the transformation process is accessing and either approving or rejecting the loan application. The final deliverable is when the customer receives a loan. In this case, the input-transformation-output model is also applicable.

In summary, we may conclude that in each case, manufacturing, projects and service the firms can be viewed from a systems perspective. There are differences in inputs, transformation process and output deliverables, but the conceptual generic operations input-transformation-output model applies to all three industry sectors. Now that we have established that we can look at these different industry sectors as a system, it will be prudent to take a TQM perspective on the transformation process for each sector.

2.2 Total quality management (TQM)

From a TQM perspective, it is important to first understand the concept of efficiency and effectiveness. These terms are well known in manufacturing they are often not given the right level of attention in the projects and services industries. Effectiveness is about doing the right thing; it is focusing on delivering value; it is about how well one is doing to achieve the target objectives. For example, the target revenue measure for a firm is $100 million per annum at the end of the year; the actual revenue achieved is $80 million, so then you are 80% effective. Considering efficiency, you can make many high-quality products; you do this well, but there is

Basis for comparison	Efficiency	Effectiveness
Definition or meaning	Doing things right	Doing the right thing
Focus is on	Measuring inputs and outputs	Measuring the outcomes resulting from the inputs
Time horizon	Short term	Long term
Approach	Introverted	Extroverted
Ascertainment	Strategy execution	Strategy formulation

Table 1.
Efficiency and effectiveness.

Principles	Description	Efficiency or effectiveness
Principle 1	Focusing on achieving customer value	Effectiveness
Principle 2	Continuously improve the system and its processes	Efficiency
Principle 3	Manage processes, not just people	Efficiency
Principle 4	Look for root causes to solve and prevent problems	Efficiency
Principle 5	Collect data and use science and analysis	Efficiency
Principle 6	People are the firm's primary resource	Efficiency
Principle 7	Work in teams to execute processes efficiently and effectively	Efficiency and effectiveness

Table 2.
Seven TQM principles (adapted from [10]).

not much market need for this product, so you have been very efficient. You have been efficient because you followed all the right processes and met all the quality requirements. However, you made a product that the market does not need, so while you were efficient, you were not very effective. Manufacturing industries have a good grasp of this concept; however, there is far too much focus on efficiency and insufficient focus on effectiveness in the projects and services industry. **Table 1** indicates some of the differences between efficiency and effectiveness.

Now that we have a better handle on efficiency and effectiveness, we refer to [10], who lists seven TQM principles for efficiently and effectively managing a system; these principles are depicted in **Table 2**. These TQM principles apply equally well to manufacturing, projects and service industries. Principle 1 refers to focusing on customer value. This is an effective principle to deliver what the customer wants. Principle 2 speaks of continuous improvement, that is, to continually improve the systems and processes is efficient. Principle 3, managing processes and not people, allows one to focus on team spirit and culture, steering away from being too hard on people. Principle 4 looks for root causes that are separate causes from systems, focusing on the factors causing the problem and not the symptoms of the problem. Principle 5 collects data and uses scientific principles, both quantitative and qualitative methods and techniques in problem analysis, staying away from the blame game, which can have a negative effect on the team and firm culture. Principle 6 recognizes that people are a core resource essential in delivering goods and services. Principle 7 focuses on teamwork to deliver efficiently, that is, doing things right and reducing scrap and waste, among other things and effectiveness by doing the right things by doing what the customer wants.

Now that we have established the groundwork in terms of viewing a firm as a system, adopting the input-transformation-output model, understanding the differences between efficiency and effectiveness and discussing the appropriateness of the TQM model for all three industry settings, our next focus is directed to quality planning, assurance and control in the manufacturing, projects and services industries.

2.3 Definition of the terms quality planning, assurance and control

These definitions are generic and apply to all three industry sectors. The differences and similarities stem from how these concepts are implemented. Quality planning is a proactive process where the role of planning is to establish the criteria so that the product or service that is being delivered will meet the customer's requirements. A quality plan consists of several documents that together form the specific quality standards, practices, resources, specifications and the sequence of activities to be performed in delivering a product or service [22]. In Ref. [22] emphasizes that a quality plan should include at least the following:

a. Objectives that are to be attained. This includes, for example, characteristics or specifications, uniformity, effectiveness, esthetics, cycle time, cost, natural resources, utilization, yield, dependability, etc.

b. Process steps include the processes that will be used in producing the final product or service.

c. Responsibilities and allocation thereof, as well as areas of responsibility and accountability.

d. Specific documented standards, practices, procedures, and instructions to be applied.

e. Testing and inspection plans, as well as auditing schedules at appropriate stages.

f. A change control procedure for design changes and other changes stemming from changes in customer requirements.

g. A procedure and method for measuring the achievement of quality objectives, including the cost of quality.

The highest-level quality goals and plans must be linked to the strategic objectives of the firm [22]. The essential elements of a quality plan must include quality assurance, quality control and quality management.

Quality assurance provides the overall guidelines and is a systematic process to ensure that the products or services meet the required specifications. Quality assurance is closely linked to international standards like ISO 9000, where many firms use ISO standards to ensure that their quality assurance system is in place and effective. There are many methods for quality assurance in the manufacturing and services industries. ISO Standards are quite widely accepted and used. In the project industries, firms are more likely to adhere to the guidelines in the Project Management Body of Knowledge (PMBOK) guide for effective project management.

Quality control is synonymous with inspection and testing, where inspection is more usually confined to indicators for fitness of use before actual use testing [23]. Testing and inspection are decision functions; they provide decisions concerning the fitness of use for a product or a manufacturing or production process [23].

2.4 Manufacturing: quality planning, assurance and control

As already pointed out, the concept of quality planning, assurance and control in manufacturing and production is well embedded. In manufacturing industries, many firms have adopted a TQM approach, which is a holistic view of quality where quality is essentially linked to the firm's overall strategy and is fundamental to the firm delivering value to the customer. Quality management is a core competence inclusive of quality planning, assurance and control. From a planning perspective, quality planning is well articulated, and it is a living plan subjected to continuous improvement. Quality assurance is closely coupled with standards like IS0 9000 and other such standards. Quality control is clearly visible with statistical process control (SPC) techniques. Initiatives like quality circles are practiced, and kaizen philosophies are a way of life.

Some techniques in no particular order of merit utilized for continuous process improvement (CPI) are depicted in **Table 3.** These techniques are widely used in the manufacturing industry and indicate the level of quality maturity in this industry.

2.5 Projects industry: quality planning assurance and control

Projects, by their very nature, are temporary operational structures which produce a unique product and are subjected to a liability of newness or upgrade and, as a result, have a propensity of failure [24, 25]. Moreover, [26] further argues that projects are temporary operational structures with a defined purpose and timeline. These temporary structures refer to the project organization in terms of how the project is structured or organized to build the project during the build phase. It does not refer to the end product that is being delivered, that is, the residential building,

#	Ingredients for CPI
1.	Top management drives CPI culture.
2.	Small, highly focused teams.
3.	Kaizen's attitude pervades the entire workforce.
4.	Qualitative analysis of problems.
5.	Focus on delighting the customer.
6.	In-depth analysis of problem causes.
7.	Consistent improvement of work processes.
8.	Statistical quality control.
9.	Systemic suggestion scheme with rewards.
10.	Quality circles.
11.	CPI is linked to other synergistic projects within the firm.

Table 3.
Ingredients for CPI.

bridge or dam. As already pointed out by [6], project success is related to achieving the project's overall goals, whereas project management success is related to the dimensions of time, cost and quality. Some authors, like [11, 26], have argued that project delivery quality affects project outcomes and the ability of the project to meet its success criteria. Therefore, projects must have quality incorporated from the outset when the project organization and structure are initially created [27]. Indicated that when setting up a project organization (structure), careful thought needs to be put in place on how the structure will facilitate coordination, communication and control [27]. Further articulates this is because the organizational structures create physical and operational boundaries for decision-making. These boundaries also reduce and increase complexity through choices made in terms of defining relationships and allocating resources, authority and tasks [27]. This is also especially important from a quality planning perspective, as there needs to be a clear demarcation of responsibilities and accountability. By incorporating quality planning in the project organization phase, we will ensure quality is inherent from the outset. The biggest project failures result from verticalizing the project's component parts. Then, not having an integral quality system to manage the risks on the interfaces between the component parts.

Moreover, [11] describes a project structure as "consisting of at least the following: a proper project plan (e.g. bar chart, network diagram, critical path analysis) and that is a realistic representation of the project; clear separated work packages with clear handover points; clear timelines of communication; an accurate representation of the stakeholders" [28]. Highlighted the importance of project structure, arguing that for a large project to be properly planned and executed, it must be decomposed into smaller segments, such as a work breakdown structure. In this way, it leads to smaller parts which are easier to manage. However, work breakdown structures bring a different level of complexity and risk: managing the interfaces between the different work breakdowns. It is essential to have a quality system incorporating a proper quality plan and a quality assurance and control system to efficiently and effectively manage these interfaces. Inspection and test plans (ITPs) are widely used and accepted in the project industry. ITPs may be viewed as follows: the creation of the ITP is part of the quality planning process, while ensuring that ITPs conform to certain standards and guidelines is the quality assurance component, and monitoring and control is part of the quality control component. These ITPs could, however, further be complemented by integrating them with some of the SPC techniques adopted in the manufacturing industry, as such risks can be better managed and mitigated. In this way, quality planning, assurance and control are built into every section or part of the project.

Projects being temporary organizations are subjected to high levels of uncertainty [29], and this, by its very nature, makes projects difficult to manage and, as such, requires careful organization, structuring, monitoring and control. The project, being a temporary structure by nature, is not static and evolves over the life of the project; therefore, projects must be subjected to proper organization, structuring, planning, monitoring and control. These aspects must be subjected to some kind of quality management. In the construction industry, most firms abide by PMBOK as a basis to guide project planning and execution. While PMBOK reflects and incorporates quality as a dimension, it does not sufficiently give attention to the philosophies of TQM. Projects are subject to risk, and as such, quality should be treated with a higher level of priority. Often, in projects, time and schedule take priority over quality. It would be beneficial to make quality a higher priority so as to reduce the risks associated with poor quality. Risk analysis must likewise be of paramount importance. It must also be emphasized that in a project environment, quality is a requirement in terms of the following:

a. Contract compliance reasons. Noncompliance can lead to expensive litigation.

b. Legal compliance. Noncompliance can lead to penalties, fines or even jail time.

c. Regulatory compliance. Noncompliance can lead to legal issues

d. Company rules. Noncompliance can lead to dismissal.

Therefore, quality is an essential issue. In a competitive world, the firm has to drive higher quality just to remain competitive in the market. So, to be better than your competitors, quality can be differentiated, and once this is recognized, it becomes non-negotiable.

Quality must be embedded in the project from the outset, which is at the project organization stage, which will then follow into the project structure. Quality will then cascade into the smaller sub-structures of the project. In this way, quality planning will be integral to the project organization and structure. Once quality planning is in place, quality assurance can follow, and thereafter, quality monitoring and control can be implemented. However, quality planning does not automatically ensure quality assurance, monitoring, and control. A concerted effort is required to ensure quality assurance is embedded; TQM philosophies can greatly assist in quality assurance as well as in creating a culture of quality. TQM will also help with quality monitoring and control.

2.6 Services industry: quality planning assurance and control

In the services industry, it is about the business experience that is the value derived from the collaborative co-creation in the production of services between the service provider and the customer. This business experience is dependent on value co-creation, relationships and service capability [30] as well as the perception of the customer. The quality dimension is catered for under service capability, where the quality of service is primarily seen from the perspective of the customer [30]. In services, it is about services-dominant logic, which is built on 10 fundamental premises. The 10 premises adapted for a services-dominant logic are quite well published, and the interested reader can refer to [30]. What is important is to understand the nuisances in services which are adapted from [30]; these are as follows:

a. Arrival variability: This implies that service me now is a need for immediate attention, and I do not want to wait in long queues. The difficulty here is in predicting the arrival rates of customers.

b. Requirement variability: This implies that service me the way I want to, that is, give me what I want, in other words, satisfy my needs. This is difficult because every customer is different.

c. Capability variability: Every customer is different; they have a unique set of skills and knowledge, which makes servicing such diversity of expectations challenging.

d. Effort capability: This relates to the level of effort in value co-creation. I, as the customer, do not want to put in too much effort; that is, service my needs and not expect me to be self-serving.

e. Subjective preference variability: This relates to factors that are associated with good service, for example, speed, accuracy, transparency, accessibility, friendliness, efficiency, and effectiveness. These can also be linked to the manufacturing mix criteria discussed earlier.

These nuisances make it difficult to deliver a good quality service; however, customers do not really appreciate these variabilities. Customers demand good service, or they have the option to switch to other service providers, especially in commoditized services, where there are not many barriers to switching from one service provider to the other, and there are low barriers to entry. Therefore, it is extremely difficult for service providers are able to provide good services to their customers. In ensuring that good services are provided, it is important that service providers measure the quality of service both from a quantitative manner and as well qualitative manner, including both lead and lag indicator levels. Each of these measures is next discussed:

a. Ensure that there are quantitative ways in which data can be collected so the demand for services can be predicated. For example, measure queue lengths, the average time customer spends in a queue waiting to be serviced, and when peak service occurs. This is dependent on data collection, and the more reliable the input data, the more reliable the prediction models. Prediction models also have an error factor, and these factors can be large, which makes prediction difficult. Therefore, it is important to combine quantitative and qualitative methods to improve the accuracy of prediction models. Prediction models help in building the capacity which is required to deliver a service.

b. Qualitative methods bring in the human dimension; while good service is measured from a customer perspective using surveys and other qualitative methods, it is important to understand that these qualitative measures are, to a large extent, based on customer perception, and perception is a lag indicator so customers will give you a view of their service experience but this view could also be influenced by a previous service encounter be it a good or bad service experience.

c. Lag indicators are based on customer perception; it is important to note that while these measures are important, it is firstly a perception, and secondly, it is a lag indicator. The customer is either satisfied or not satisfied, and the service encounter has occurred in the past. The next time a customer is serviced will be the time the service provider will have a chance to deliver better service, so whatever service improvements are made by the service provider, the benefits are in the future.

d. Lead indicators, on the other hand, could help predict future demand and can help build additional capacity which is consumed on service delivery; this allows for more resources to be deployed and increases service times through extended hours, all in an endeavor to better service the customer.

In order to ensure that the service provider delivers a good service, it is essential that service quality is inherent in the service systems. This requires planning for quality, assurance that the quality systems are in place and are sufficiently governed and ensuring that quality is controlled. There are tools such as the SERVQUAL instrument for measuring quality [8]. It is important that quality is built into the service processes. Just as is in the case of manufacturing, there are processes, systems, technology and people that also occur in services. While the services industry is different to the manufacturing industry, many of the TQM philosophies apply equally well in this industry. The TQM philosophies apply equally well here; the same can be said about the ingredients for CPI. These are all philosophies, techniques and tools that can be used to improve quality within the service industry.

In the service industry, it is about value co-creation where both the firm (service provider) and the recipient (customer) are involved in a mutual exchange to ensure that the desired service is delivered. The service exchange is about a win-win or benefit-benefit for both the service provider and the customer. This brings about variabilities and different types of measures that have to be considered, which, together with careful planning of quality, can be built and embedded through the entire service system, which, if correctly implemented, will benefit the service provider and the customer.

3. What are the implications for practitioners

3.1 Quality planning, assurance and control in context

Quality does not occur by default; it has to be carefully built into the firm, the project and the service being provided. While there are differences between the three industry sectors we have discussed, each with its own challenges, there is much cross-pollination that can be derived from the TQM philosophies, which, if correctly adopted, will benefit the end customer. The importance of quality planning early on cannot be underestimated. There needs to be a culture of quality that pervades throughout the firm, be it a manufacturing, project firm or services firm. First, one must establish a quality culture, which will then set the foundation for quality planning, assurance and control. Quality assurance and control will not be successful without addressing quality planning at the initial stages. Quality assurance and control can be greatly enhanced through the adoption of TQM.

In terms of TQM, the importance of change management and leadership cannot be over-emphasized. Continuous worker training in quality techniques and quality management will also aid in creating a quality culture throughout the firm. Creating a culture of continuous improvement and a culture of innovation, specifically in terms of how to do things better and, in so doing, improve efficiency and effectiveness. The same goes for leadership and team building; this is an integral part of embedding a quality culture; this becomes critical in environments that are prone to high risk and high uncertainty, like the project environment. Notwithstanding the high-risk and uncertain environments, leadership and team building is also essential in the manufacturing and services industry. In quality management, leadership is key as this sets the tone for quality throughout the firm.

3.2 Tools that can facilitate quality planning, assurance and control

In the manufacturing industry, there are many tools, systems and techniques that can be used to enhance quality planning, assurance and control. The manufacturing

DOI: http://dx.doi.org/10.5772/intechopen.113995

industry at large is fairly mature in this area. The manufacturing industry has embraced the ISO standards quite well in terms of ensuring that their quality systems conform to international standards, which are audited, monitored and controlled. In terms of quality control, tools like SPC are widely used and accepted.

In the projects industry, especially in construction projects, many firms have adopted the PMBOK as the Body of Knowledge to guide project management processes. Some project-driven firms have also adopted the ISO standards. We have, however, not seen much attention toward TQM as an approach to facilitate quality assurance and further to this, techniques for CPI can also be better adapted.

In the services industry, there does not appear to be any such standard that firms can use to guide the quality assurance function; however, it may be beneficial to look at capability maturity model integration (CMMI) for services. This will be a good toolset to use in terms of accessing a baseline maturity of services within services firms. CMMI creates the opportunity for continuous improvement and self-benchmarking. CMMI also has a wider toolset that can be used for manufacturing and project management. It will be worthwhile to at least explore the CMMI options. We have also not seen widespread adoption of ISO standards in the services industry.

4. Conclusions

In this chapter, we highlighted the maturity of quality planning, assurance and control in the manufacturing industry. We also highlighted that the projects industry is prone to failure, and many failures can be attributed to poor quality management processes and procedures. Based on this, it is extremely important that quality be given a high priority and that more attention is devoted to quality. There is also a high focus in the projects industry on implementing risk process management in all the project phases. The importance of TPs was also emphasized. There is also an opportunity to address risk as part of the quality, planning and assurance functions. Similarly, in the services industry, quality needs to be integrated into the services culture and be an integral part of the services process, so it is therefore essential that quality planning be integrated early when first developing the service systems and processes.

Quality cannot be an afterthought; the high-level quality objectives must be linked to the overall firm's strategy. Quality management requires both a top-down and bottom-up approach. Higher quality does not mean a higher cost; it means better control and monitoring of the process whereby faults in the manufacturing, construction or service can be picked up early and reduce the cost of rework.

The importance of quality cannot be overemphasized. Customers demand quality products and services. Customers pay for the quality; it is not an option to sell poor-quality products and/or services. To ensure that good quality products and services are delivered, it is essential that firms strive to integrate TQM philosophies within the wider firm. Finally, it must be recognized that people buy from people, and customers will come back if they are satisfied. There are different customer satisfiers in all three industry sectors; at the end of the day, it boils down to the customers getting what they paid for. Some questions to ponder:

a. Why do you shop at the same place every week?

b. Why do you go to the same hairdresser, man or female

c. Why do you buy petrol at the same petrol station

d. The main answer is you trust them, and you get service know your customer

Acknowledgements

The authors recognize the contribution made Jacobus Johannes Smit, Bohdan Pylypczak and Thomas Eichbaum for reading the initial drafts of this chapter and providing valuable comments on editing, as well as further insights, which helped improve the content and quality of this work. The authors are grateful for their valuable contributions.

Conflict of interest

"The authors declare no conflict of interest."

Author details

Rajenlall Siriram* and Chantelle du Plessis
North-West University, Faculty of Engineering, School of Industrial Engineering, Potchefstroom, North-West, South Africa

*Address all correspondence to: raj.siriram@nwu.ac.za; raj@alpha-concepts.com

References

[1] Jones A, Yih Y, Wallace E. Monitoring and controlling operations. In: Salvendy G, editor. Handbook of Industrial Engineering Technology and Operations Management. 3rd ed. New York: Wiley; 2001

[2] Siriarm R. Factors affecting the adoption of systems thinking. Systems Research and Behavioural Science. 2020;**37**:235-254. DOI: 10.1002/sres.2617

[3] Ferdows K, De Meyer A. Lasting improvements in manufacturing performance: In search of a new theory. Journal of Operations Management. 1990;**9**(2):168-184. DOI: 10.1016/0272-6963(90)90094-T

[4] Schroeder RG, Shah R, Ping X. The cumulative capability 'Sand cone' model revisited: A new perspective for manufacturing strategy. International Journal of Production. 2011;**49**(16):1-28. DOI: 10.1080/00207543.2010.509116

[5] Mir FA, Pinnington AH. Exploring the value of project management: Linking project management performance and project success. International Journal of Project Management. 2014;**32**(2):202-217. DOI: 10.1016/j.ijproman.2013.05.012

[6] Radujković M, Sjekavica M. Project management success factors. Procedia Engineering. 2017;**196**:607-615. DOI: 10.1016/j.proeng.2017.08.048

[7] Atkinson R. Project management: Cost, time and quality, two best guesses and a phenomenon, it's time to accept other success criteria. International Journal of Project Management. 1999;**17**:337-342. DOI: 10.1016/S0263-7863(98)00069-6

[8] Parasuraman A, Zeithaml VA, Berry LL. SERVQUAL: A multiple-item scale for measuring consumer perceptions of service quality. Journal of Retailing. 1988;**64**:12-40

[9] Shtub A. Project management cycle: Process used to manage project (steps to go through). In: Salvendy G, editor. Handbook of Industrial Engineering Technology and Operations Management. 3rd ed. New York: Wiley; 2001

[10] Bounds GM, Dobbins GH, Fowler OS. Management a Total Quality Perspective. Cincinnati, Ohio: International Thompson Publishing; 1995

[11] Siriram R. A project management investigative framework establishing links for better project outcomes. South African Journal of Industrial Engineering. 2019;**30**:100-114. DOI: 10.7166/30-2-2064

[12] Siriram R. Project management assessments (PMAs): An empirical study. South African Journal of Industrial Engineering. 2018;**29**:108-127. DOI: 10.7166/29-1-1675

[13] Miller R, Lessard D. The Strategic Management of Large Engineering Projects: Shaping Institutions, Risks and Governance. Cambridge, MA: Massachusetts Institute of Technology; 2009

[14] Evans M. Project Management: Overdue and Overbudget, Over and over Again [Internet]. The Economist; 2005. Available from: https://www.economist.com/business/2005/06/09/overdue-and-over-budget-over-and-over-again [Accessed: August 5, 2023]

[15] Morris P, Hough G. The Anatomy of Major Projects: A Study of the Reality of Project Management. Chichester: Wiley; 1987

[16] Shehu Z, Endut IR, Akintoye A, Holt GD. Cost overrun in Malaysian construction projects: A deeper insight. International Journal of Project Management. 2014;**32**:1471-1480. DOI: 10.1016/j.ijproman.2014.04.004

[17] Zeithaml VA, Parasuraman A, Berry LL. Problems and strategies in services marketing. Journal of Marketing. 1985;**49**:33-46. DOI: 10.1177/002224298504900203

[18] Shostack GL. Planning the service encounter. In: Czepiel JA, Solomon MR, Surprenant CF, editors. The Service Encounter: Managing Employee/Customer Interaction in the Service Businesses. Lexington: Lexington Books; 1985

[19] Solomon MR, Surprenant CF, Czepiel JA, Gutman EG. A role theory perspective on dyadic interactions: The service encounter. Journal of Marketing. 1985;**49**:99-111. DOI: 10.1177/002224298504900110

[20] Bitner MJ. Evaluating service encounter: The effects of physical surroundings and employee responses. Journal of Marketing. 1990;**54**:71-84. DOI: 10.1177/002224299005400206

[21] Checkland P. Systems Thinking, Systems Practice. West Sussex England: Wiley; 1999

[22] What is a Quality Plan [Internet]. Available from: https://asq.org/quality-resources/quality-plans [Accessed: August 6, 2023]

[23] Drury GG. Human factors and automation in test and inspection. In: Handbook of Industrial Engineering Technology and Operations Management. 3rd ed. New York: Wiley; 2001

[24] Bannerman PL. Smoothing innovation discontinuities. In: Proceedings of the IEEE International Conference on Communications (ICC '08). Engineering Management Mini-Conference (IEMC-China'08); 19-23 May 2008; Beijing. China: IEEE; 2008. pp. 5458-5462

[25] Bannerman PL. Structuring risk into projects. In: Paper Presented at PMI Research Conference: Defining the Future of Project Management. Washington, DC. Newton Square, PA: Project Management Institute. Available from: https://www.pmi.org/learning/library/structuring-risk-projects-propensity-fail-6400; [Accessed: July 21, 2023]

[26] Basu R. Managing quality in projects: An empirical study. International Journal of Project Management. 2014;**32**:178-187. DOI: 10.1016/j.ijproman.2013.02.003

[27] MacAskill K, Guthrie P. Organizational complexity in infrastructure reconstruction–A case study recovering land drainage functions. International Journal of Project Management. 2017;**35**:864-874. DOI: 10.1016/j.ijproman.2017.02.013

[28] Globerson S. The impact of various work-breakdown structures on project conceptualization. Interntaional Journal of Project Management. 1994;**12**:165-171. DOI: 10.1016/0263-7863(94)90032-9

[29] Atkinson R, Crawford L, Ward S. Fundamental uncertainties in projects and the scope of project management. International Journal of Project Management. 2006;**26**:687-698. DOI: 10.1016/j.ijproman.2006.09.011

[30] Fitzsimmons JA, Fitzsimmons MJ, Bordoloi SK. Service Management Opearations, Strategy, Information Technology. 8th ed. New York: McGraw-Hill International edition; 2014. p. 523

Chapter 7

Perspective Chapter: Quality Determination of Hydraulic Pumps with Adaptive Fuzzy Pattern Classifiers to Reduce the Risk for Quality Management

Amir Nemati, Elif Öztürk, Steffen F. Bocklisch and Welf-Guntram Drossel

Abstract

Automated production of complex assemblies such as hydraulic pumps also requires reliable detection of defects utilizing functional tests. In principle, this is a classification task in good/bad, which, however, is often not to be made sharply but should provide gradations for detailed error analysis. From this, conclusions can be drawn, for example, about the type or location of the defects, wear, or aging of components in the production chain. A high-dimensional vector of data from static or dynamic measurements including is generally available as the basis for the fault detection model. Modeling such complex nonlinear systems under various load conditions with dynamic test procedures leads to uncertainties that should also be reflected in the diagnostic model. For this, the design of the classification model (the classifier) should be largely automatic during the training phase for time and cost reasons. In addition, online updating under actual operating conditions is also often desired. These challenging goals can be met through the artificial intelligence (AI) methodology of fuzzy pattern classification. This chapter deals with the development of a fuzzy classifier for the application case of the final inspection of hydraulic axial piston pumps. The focus is on the automatic training of the classifier employing a new adaptation procedure and permanently (until termination) evaluates the resp. current classifier using performance measures. Using real experimental data, the procedure and the step-by-step adaptation results for different links between the current classification model and the new data are presented and compared.

Keywords: fuzzy pattern classification, adaptive fuzzy pattern classifiers, hydraulic pump, quality control, productivity improvement

1. Introduction

Today, in the various industrial sectors, the primary objective of each industrial unit is to improve the quality level of the products manufactured. More confidence in the equipment used is one of the requirements that the designer must consider as one of the basic objectives of any system, which is achieved through the timely detection of defects. Designers face the challenge of designing diagnostic systems that must be robust to faults within the system and resistant to uncertainties in the system. The uncertainties introduced by modelings, such as linearizing nonlinear systems and process parameter variations, cause the diagnostic system to fail and generate false alarms [1, 2].

Due to the numerous advantages of hydraulic drives over other drives, such as reliability, low power consumption, precise controllability, and high performance, such assembly systems are increasingly used. In the hydraulic industry, the hydraulic pump is the central part of hydraulic systems, responsible for generating flow and pressure in the process. In general, defects in hydraulic pumps can cause severe and irreparable damage, requiring costly repair and overhaul. Therefore, diagnosis, detection, and repair of defects in the pump components are crucial at the initial stage. The operation of hydraulic pumps is affected by various factors such as friction, adhesion, incompressible fluid, and internal and external leakage. Therefore, hydraulic pumps have nonlinear behavior and are complex nonlinear systems with multiple inputs and multiple outputs whose input and output behavior cannot be described by physical exact mathematical models. Therefore, theoretical modeling would only be an inaccurate model of the system. Since we need to simplify the input data, we are often forced to accept a certain degree of inaccuracy and uncertainty in mathematical models, which then cannot achieve acceptable results for the dynamic behavior of the systems [3, 4]. In such situations, a data-based controlled nonlinear black-box system provides a reasonable approximation to nonlinear systems. In these approaches, models are built based on the process's measured input and output data, requiring little or no physical or formal information [3]. The literature shows that black-box models, such as neural networks and fuzzy logic-based models, are widely used to build fault diagnostic models from measured input/output data [5–7]. These approaches perform better than statistical models [1] and mathematical equations [3].

For the prediction of hydraulic pump failure, the correct and accurate categorization of faults is essential because various factors affect the failure of the hydraulic pump. The fault categories can be understood as classification. Qualitative variables are considered, and the classification approach can be used for qualitative variables. Various approaches to classify data sets include regression, artificial neural networks, probability, and fuzzy approaches. However, these classification approaches are static, and the nature of static classification leads to several primary principle weaknesses on the user's side [8, 9].

1. The first issue is that these static classifications cannot adapt to the current user data stream that occurs in real systems. A comprehensive data set that covers all features of the system (big data), which is often required, is costly and time-consuming, complicated, and often impossible to achieve.

2. The second point is that the flow of new, unclassified, and unlabeled data requires a new classification that an expert or supervising supervised person has not provided. So the user's need may change, and the classifier should keep up with this need and automatically add a new class to its database.

Therefore, the goal is to develop a new method for automatic modeling in real-world situations and also with online data streams. The technical term for this category is Evolving Classification [10]. Evolving classifications can adaptively update their structures, components, and parameters according to the requirements of new process characteristics, system behavior, and operational and environmental conditions. These systems support modeling arbitrary scenarios of data flow, online measurements, and dynamic data whose nature and characteristics change over time. The main features of the evolving classifications are as follows [11]:

1. The learning process can be started from scratch, and the system can learn the classes required by the user without an initial learning phase.

2. Input samples can form new classes without losing previous knowledge, that is, or forgetting previous classes that the classifier has not currently seen.

3. Unlike traditional approaches, these evolving classification systems may have little or no learning phase.

The research work presented in this paper deals with axial piston pump. Specifically, a subset of these pumps of the variable displacement pump type was selected for the study. The axial piston hydraulic pump was designed and optimized for demanding heavy industry and shipping use. Thus, this type of pump, this axial piston pump in swash plate design in an open circuit with pressure ratings up to 420 *bar* and high speeds offers the user a high productivity and power density. The axial piston hydraulic pump is characterized by high conversion flexibility [12].

The main objective of this work is to provide a new fuzzy pattern classification approach [13] that works automatically, online, and without the presence of users. This classification method is applied to the final inspection and testing of hydraulic axial piston pumps.

2. Principles of pumps and axial piston hydraulic pumps

Nowadays, power transfer at low cost and high precision is desired in many processes. Hence, the use of pressure fluid in the transmission and control of power in all branches of the industry is expanding. Fluid power is separated into two categories: hydraulic and pneumatic. Pneumatic systems are used where relatively low forces (about one ton) and high-mobility speed are required (such as in systems that are used in drive robots). In pneumatic systems, compressible fluids such as air are used. In cases where high power and precisely controlled speeds (such as hydraulic jacks, brakes, and hydraulic steering) are needed, hydraulic systems are used. The advantages of hydraulic systems to mechanical and electronic systems are a simple design, power increase capability, simplicity and accuracy of control, flexibility, high efficiency, and reliability. Hydraulic systems are highly regarded in many industrial applications, such as industrial robot control, industrial machinery, automobile industry, hydraulic suspension systems, mechanical power transmission systems, and aerospace industries. Other advantages of hydraulic systems, compared to other mechanical systems, are the need for fewer actuator components and the achievement of high power and high-power movement at any point due to high-pressure flow in the tubes and hoses. In other mechanical systems, it is used to transmit power from components such as a cam, gear, lever, or clutch. As the heart of the hydraulic system, the pump converts the mechanical energy supplied by electric motors or internal combustion to hydraulic energy. The pump in a hydraulic cycle increases fluid energy to be used in the desired position. Hydraulic systems are based on Pascal's law. Like many mechanical systems, the disadvantages of hydraulic systems are their nonlinear and similar behavior. Several factors affect the operation of hydraulic systems, such as friction, adhesion, incompressible fluid, and internal and external leakage [14, 15].

The hydraulic axial piston pump is a type of hydraulic pump that has two fixed axial piston pumps and a variable axial piston pump. The multi-piston pump is used for mechanized hydraulic systems and can produce a more uniform flow of fluid. **Figure 1** shows the hydraulic axial piston pump. In the hydraulic axial piston pump type, the pistons are parallel to the central axis of the pump. The pistons are located around the central axis of the pump and in the circle's environment. The primary axial piston pump is based on the motion of the piston and the piston inside the cylinder that operates the suction and pressure of fluid in every move. In other words, the pump converts the rotary motion of the input shaft to the piston's back-and-forth linear motion. By embedding a mechanism for changing the angle of the back of the pistons, the displacement volume of a pump can be changed. This mechanism changes each piston movement's course, and the pump's discharge rate is adjusted by changing the angle between the two pump axes. There is no discharge at zero angles, and the maximum discharge is obtained at the maximum angle. Axial piston pumps are high-pressure pumps. Depending on the type of controller each pump has, the flow changes uniformly and continuously.

The pumps used in this work are axial piston pumps with a swash plate designed and optimized for demanding use in heavy-duty industrial and marine applications. With pressure ratings of up to 420 *bar* and high-speed ratings, this open circuit, swashplate-type axial piston pump provides its users high productivity and power density.

Figure 1.
Hydraulic axial piston pump [16].

As a standard, every swashplate-type axial piston pump comes with an integrated pre-compression volume, which ensures low ripple operation. A wide range of displacements and control options allows for various applications such as primary metal refining/processes, machine tools, marine, oil gas and power generation. The displacement for this pump ranges from 16 cm^3/rev to 360 cm^3/rev. These pumps are classified into five frame sizes according to the displacement. Each frame size can generate the discharge rate concerning the value of the swashplate angle. In **Table 1** and **Figure 2**, all types of frame size and their output flow are expressed.

The advantages of a swashplate-type axial piston pump include the following:

- Operating pressures of up to 350 *bar* (continuous) or 420 *bar* (intermittent) and high-power density.
- Accurate, high dynamic controls
- Outstanding response characteristics and productivity improvements.
- Excellent suction characteristics, high self-priming speeds, and increased productivity.

Frame size	Displacement $[cm^3/rev]$
1	16, 20, 23, 28
2	32, 40, 46
3	63, 80, 92
4	140, 180
5	270

Table 1.
Physical conditions and characteristics of the pumped fluid.

Figure 2.
Axial piston pumps with five different frame sizes [16].

There are several approaches to determining the quality of products or detecting defects and their classification. Product quality is diagnosed by processing multiple measurements using data analysis or logical reasoning links. It is also another way to compare real actual data and mathematical models of hydraulic pumps. However, creating a mathematical model is difficult because the pump system has considerable nonlinearities in parameters.

Another approach is to develop an intelligent system that learns the behavior of the hydraulic pump. Fuzzy pattern classification can be referred to as an approach to developing intelligent systems [13, 17].

3. Theoretical basis and methodology

Human intelligence has creativity, skill, awareness, intuition, and emotion. Intelligence is the ability to understand, think, and learn. The question is how to model this human ability and demonstrate it in a way that has computational efficiency. Traditional artificial intelligence and various mathematical and analytical tools for characterization, description, and analysis of systems attempted to simulate such intelligent behavior in systems requiring accurate and comprehensive representation of knowledge. With the increase in the complexity of dynamic systems, the traditional modeling techniques could not satisfy the needs any further and fall short in managing such processes, mainly because these methods were especially suitable for linear systems and could not solve problems in nonlinear systems. Hence, new tools for nonlinear systems should be developed that, on the one hand, can learn or deal with new and unknown situations and, on the other hand, can predict future events accurately enough for proper decision-making. This system, capable of understanding and predicting the environment or processes and its changes, can be called an intelligent system.

Computational intelligence means extracting intelligence, knowledge, and algorithms. Intelligent systems, in principle, provide free models for dynamic systems through the approximation of functions and/or mapping. Additionally, computational characteristics, such as accuracy, flexibility, and ease of implementation, are also of utmost importance in an intelligent system. Intelligent techniques have been applied, especially in uncertain, complex, nonlinear, time-varying, and randomly targeted processes. These models are not comparable to traditional mathematical modeling techniques mainly because [18]:

- The mathematical description of the processes is extraordinarily complex.
- Evaluation of these models is difficult and expensive.
- There is uncertainty in process operations.
- Nature often has a nonlinear process and random distribution.

The significant challenge in computational intelligence and one of the distinctive criteria for classification models is whether the system can continually learn from the incoming stream of data [19]. When human learning is considered, it is observed that the learning process is incremental. Human learns concepts from the environment and refines these concepts from facts, new observations, and new information in an increasing manner. For two primary reasons, humans need to learn continuously [20].

- The continuous flow of information.
- Limited memory and information processing.

Nowadays, with the massive increase in the speed of data collection, the goal of processing data is not only to achieve the least possible error, as it always was, but to do so in the shortest possible time. However, traditional approaches for training machine learning models fall short in this aspect since most of the learning algorithms in computational intelligence are batch learning or optimization, which are very costly and very slow in the data flow. The use of these algorithms is very challenging, especially in an online process, due to the sensitivity of the parameters in these algorithms, which makes it difficult in many cases to maintain accuracy.

Batch learning does not provide the possibility to learn continuously. Therefore, the system must be trained using all the available data, hence the requirement for a lot of computing time and resources. That is why this learning is done offline. In the presence of new data or facing an ever-changing environment, the models that are trained offline need to be trained again. Such a process, in which the system is stopped, trained again, and replaced as a new one, obviously cannot be done during system execution. Consequently, if the available data, a collection of the recently obtained and already existing data, are immense in size, the system should be trained from scratch and then replace the older version, which was trained by the previously available data. This, of course, would require a lot of resources, such as computational power from CPUs, memory space, disk space, etc., and would probably take hours to be done [19].

Because of the need for new techniques to continuously improve the performance of models with the flow of new data, incremental or online learning techniques are proposed. Incremental learning is, in fact, a method in advanced modeling [11, 21]. An incremental learning system is especially beneficial for computers with limited resources or systems that need to adapt quickly to the current data. In an online system, there is no need to keep the data after training; the data can be discarded and, as a result, save memory space. These systems are mostly suitable for large data, where storing them in the device's memory is neither a valid nor practical option.

Three important aspects of incremental learning are as follows [20].

1. How to adapt the previously learned knowledge to the currently received data and use the new raw data for learning.
2. How the accumulated experience and knowledge over time can support and further benefit future decision-making processes.
3. How classes can be produced, merged, or divided based on the dynamic nature of environmental changes.

3.1 Fuzzy computational

To counter complexity in modeling and solving new issues in physics, engineering, medicine, biology, and many other sciences, creating and developing new computational methods more closely aligned with human thinking is required. Artificial intelligence aims to match a computer system's behavior and response to the patterns in which humans behave and respond. In reality, many concepts which humans use are perceived as imprecise, unclear, and vague. Though words and concepts such as hot, cold, long, short, old, young, and so on do not refer to exact numbers, the human mind immediately understands everything with fantastic flexibility and uses them for decisions and conclusions. At the same time, the machine is precise and only understands numbers. The purpose of new methods in computer science is to learn these abilities from human beings and then to teach them to machines as closely as possible [22].

The scientific laws of Newtonian physics and mechanics are all based on old logic. Old logic can only express two attitudes: white and black, yes and no, bright and dark, one and zero, true and false. In general, variables in nature or calculations are of two types:

Quantitative values can be expressed in a given number, and qualitative values are expressed from a feature. These two values can be converted. Since the human mind operates with other logic and adapts decisions, formulating and developing new logic and the multivalued requirement, fuzzy logic is one of them. Following the first fuzzy set theory by Professor Lotfi Zadeh [23] in 1965, a new calculation appeared on the scene. This theory was developed to compensate for the inadequacy of the Boolean logic to describe many real-world issues. Classical math operations recognize only 0 and 1. Much real-world data is inaccurate and has uncertainties, so classical logic cannot cite the fact in this respect. Fuzzy logic presents a systematic concept for investigating uncertainty both quantitatively and qualitatively. Fuzzy theory is for applications with uncertain conditions and utilizes linguistic variables [18, 24].

One of the critical issues in fuzzy logic is distinguishing it from the theory of probabilities in mathematics. Often, the fuzzy theory is confused with the probability theory. However, these two concepts are entirely different from each other. Fuzzy logic deals with inaccurate facts and refers to the limits and levels of reality. However, probability theory is the theory of random events and discusses the chance of an event occurring. When the predicted event occurs, the theory is assumed to be accurate [25].

The main difference between fuzzy logic and neural networks is the data used. In other words, neural networks are based on data-driven modeling, but fuzzy models are knowledge-based. Data-driven models are any form of mathematical model extracted or learned from data. Knowledge-based models are extracted from the experience of experts, operators, and users working on the system [11, 18].

3.2 Fuzzy pattern classification (FPC)

Fuzzy pattern classification is suitable for complex issues, including technical and non-technical, and can provide support for control or decision support. Fuzzy patterns are modeled using multidimensional fuzzy membership functions, specified in a dimensional feature space representing the measured variables or characteristics. Compared to other approaches, the fuzzy pattern is a sign of parallel and non-sequential. One classifier can comprise some patterns, which are all semantically interpretable. Fuzzy pattern classifiers can be modeled using a knowledge-based approach or data-driven approach or as a combination of both [10, 13]. Fuzzy classification is a method for describing classification systems of observations defined as feature vectors in feature space of one or more dimensional (n-dimensional) Euclidean space. Compared to other methods, this model allows modeling interdependencies of the variables by rotating the multi-axis system within the entire coordinate system. Therefore, the correlation between input variables can be discussed. Fuzzy pattern classification can also be designed and modeled as a data-driven or knowledge-based approach or a combination of both [13]. The following is a discussion of the one-dimensional and multi-dimensional membership functions appropriate for the fuzzy pattern classification.

3.2.1 One-dimensional membership functions (MF)

Fuzzy pattern classification is based on an asymmetric membership function that differentiates between left and right function branches. This is a generalized potential function of the Aizerman's model and the Bocklisch model is similarly used to describe niches in ecology [16]. It has been proven that Aizermans modified function is very suitable for describing very high-dimensional membership functions (practical to 100 dimensions). Eq. (1) defines the one-dimensional asymmetric membership function type for the variable u.

$$\mu(u) = \begin{cases} \dfrac{a}{1+\left(\dfrac{1}{b_l - 1}\right)\left(\frac{r-u}{c_l}\right)^{d_l}} & u < r \\ \dfrac{a}{1+\left(\dfrac{1}{b_r} - 1\right)\left(\frac{u-r}{c_r}\right)^{d_r}} & u \geq r \end{cases} \tag{1}$$

Besides the mathematical model of this function, the following semantic meaning can be assigned to the parameters. This function is based on a set of eight parameters:

The parameter r denotes the representative position of the MF, which can be determined in various ways, for example, as the center of gravity of the objects constituting the class, as the arithmetic mean, or as a reference point. The maximum value of the membership parameter a of this unimodal MF is assigned to r. The membership parameter $a \geq 1$ can indicate the weight or authenticity of a class. The membership parameter can indicate the weight or authenticity of a class. The parameters c_l and c_r ($c_{l/r} > 0$) are the class information that contains the position of the farthest object from the center in projection to the axis. Therefore, parameter c represents the greatest distance of an observed object from r. As shown in **Figure 3**, c_l and c_r characterize a fuzzy pattern class's left- and right-sided expansions. Both parameters represent the range of a class. The parameters b_l and b_r ($0 < b_{l/r} \leq 1$) are factors that determine the value of the MF at the sharp boundaries $c_{l/r}$ of the fuzzy pattern class. The parameters d_l and d_r ($d_{l/r} \geq 2$) determine the form of the function and carry information about the object distribution in the corresponding class. In special cases the amount for $d_{l/r} \to \infty$ changes to a sharply described class (red color in **Figure 4**) [3, 13].

3.2.2 Multidimensional membership functions

The previous considerations only focused on the presentation of fuzzy pattern classification in one-dimensional (feature) space $\mathbb{R}^1$. The fuzzy potential function points to the theoretical possibility of multidimensional membership functions. Hence, one-dimensional membership functions define the membership function in N-dimensional space $\mathbb{R}^N$ Therefore, the exceptional advantage of the potential function is that a suitable conjunctive combination of several fuzzy sets/logic can be used to obtain a multivariate membership function in parametric form [3, 13]. The normalized function ($a = 1$) with N dimension is given by

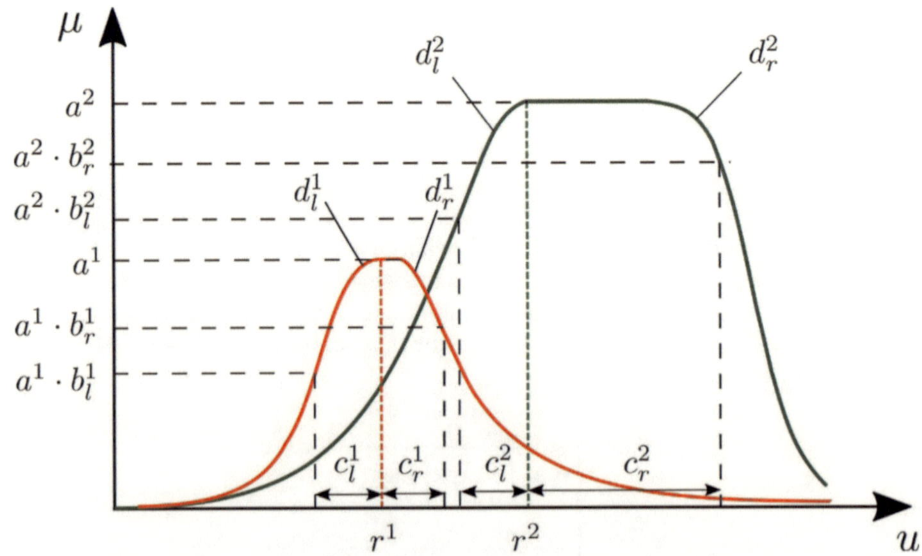

Figure 3.
Two one-dimensional asymmetric membership functions with different parameters [3].

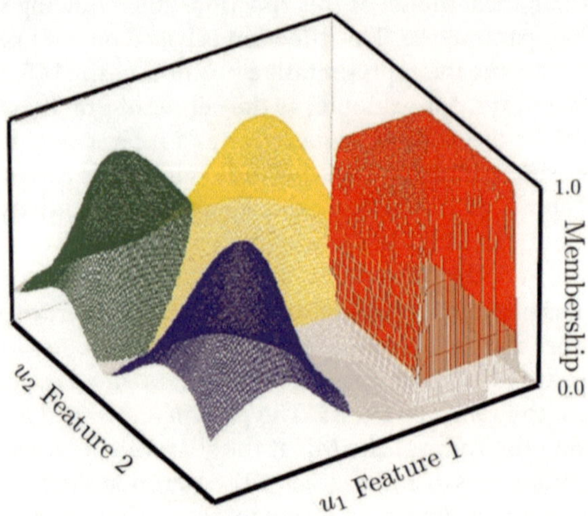

Figure 4.
Illustration of two-dimensional MFs for four classes [3].

$$\mu(\underline{u}) = \frac{1}{1 + \left(\frac{1}{N} \cdot \sum_{n=1}^{N} \sigma_n\right)} \tag{2}$$

where j belonging to number of features (n = 1,2,3,..., N; N = the number of dimension). Therefore, σ_j denoted by

$$\sigma_n = \begin{cases} \left(\frac{1}{b_{l,n}} - 1\right)\left(\frac{r_n - u_n}{c_{l,n}}\right)^{d_{l,n}} & u_n < r_n \\ \left(\frac{1}{b_{r,n}} - 1\right)\left(\frac{u_n - r_n}{c_{r,n}}\right)^{d_{l,n}}, & u_n \geq r_n \end{cases} \tag{3}$$

Figure 4 illustrates the basic join operation for two-dimensional fuzzy pattern classes.

3.3 Adaptive fuzzy pattern classification (AFPC)

As previously explained, some classification systems can be designed dynamically. These systems can modify classifications automatically during the operation phase and online. These systems can adjust their parameters continuously to improve system performance or change their structure for new classes. For these systems, two types of incremental learning algorithms can be presented: algorithms for incremental parameter learning and algorithms for incremental structure learning. The system is considered at the beginning of the learning process for incremental learning of the parameters, and the system parameters are learned according to the new data. Such algorithms are considered "adaptation" algorithms, and the classifications whose parameters change based on new data are called "adaptive classifiers." In incremental structure learning, parameters and classification structures change; they are called "evolving classifiers." Adaptive classifiers can modify their parameters based on the new process's demand characteristics, the system's behavior, and operational and environmental conditions. These systems can support modeling any scenario for data flow, online measurements, and dynamic data whose nature and characteristics change over time. Continuous adaption capability is essential for two reasons [10, 11]:

1. New data samples are related to the specific context in the real world, which can improve classification and differentiate it.

2. The storage of information and learning data is not required for classification consistency. Therefore, the classification can be set up with a small dataset and improve system performance.

*3.3.1 Adaption of parameter **a***

The parameter a, as already explained, represents a measure of the "weight" of a class. This "weight" is limited upwards by a value max which is determined individually ($a_{max} = 10$). The parameter a is a function of the number of N_k objects belonging to a class. For the total number of objects belonging to the classifier, a is defined as

$$a_k = a_{max} \cdot \left(1 - \left(\frac{a_{max} - 1}{a_{max}}\right)^{N_k}\right) \tag{4}$$

The dynamic of the evolvement can be changed using the weight of the transition patterns $N_k \geqslant 1$ depending on the membership function parameter a, which represents the number of data points that created the patterns in the past [10, 13].

*3.3.2 Adaption of parameter **r***

Adapting the r values leads to modification of the fuzzy pattern's position in the coordinate system. The new position of r, called $r_n^{(k,new)}$, with n representing the index of the dimension and with $k = 1, 2, 3, \dots, K$; whereby K is the number of classes, is defined for each additional data value x_n through the weighted arithmetic mean:

$$r_n^{(k,new)} = \frac{N_k}{N_k+1} \cdot r_n^{(old)} + \frac{1}{N_k+1} \cdot x_n \tag{5}$$

The dynamic of the evolvement can be changed using the weight of the transition patterns $N_k \geqslant 1$ with N_k depending on the membership function parameter a, which represents the number of data points that created the patterns in the past [10, 13].

3.3.3 Adaption of parameter $c_{l/r}$

Adapting the $c_{l/r}$ values leads to modification of the fuzzy pattern's boundary of the class. The new $c_{l/r}$ boundary $c_{((l/c),n)}^{(k,new)}$, with n representing the index of the dimension and k representing the number of classes, is defined for each additional data value x_n.

$$c_{(l/c),n}^{(k,new)} = \begin{cases} c_{(l/n)}^{(k,new)} = \frac{N_k}{N_k+1} \cdot c_{l/n}^{(old)} + \frac{1}{N_k+1} \cdot x_n \\ c_{(r/n)}^{(k,new)} = \frac{N_k}{N_k+1} \cdot c_{r/n}^{(old)} + \frac{1}{N_k+1} \cdot x_n \end{cases} \tag{6}$$

To adapt the parameter $c_{l/r}$, first, it must be determined that the new point in each dimension, in what position of the new parameter $r_n^{(k,new)}$, is in the same dimension.

After determining the position of the new point, the new parameter $c_{l/r}$ is calculated from Eq. (6). At this phase, we introduced two approaches for adaptation and evaluated their results.

I: To adapt the parameter $c_{l/r}$, first, in each dimension the new point must be determined, at what position of the old parameter c is in the same dimension. Then, if the corresponding object is out of range, that is, it is smaller than the left c parameter or greater than the c right parameter, then adaptation is done for the c parameter. II: The $c_{l/r}$ parameter is compatible with all objects (see scenarios II and III in Section 5.1 and results in Section 5.2).

4. Experiments on an axial piston pump and data collection

The level and complexity of product reliability and safety requirements have been increasing in many industrial fields. Complying with such demands requires various tests and evaluations at each stage of production, from research and development of materials to evaluating finished products. Dynamic testing is used to measure product behavior and response characteristics and structural members in response to environmental changes, for instance, through the internal analysis of excitation vibration, pressure, and the temperature source transfer path of the piston pump. In order to make the product work reliably and safely, it is essential to ensure that the system has excellent dynamic attributes. Therefore, dynamic analysis and design of mechanical products and equipment are necessary to satisfy the requirements of static and dynamic characteristics of the mechanical structure.

4.1 Hydraulic pumps testing machine

Any pump, regardless of its type and size, should be tested in different methods throughout the operation. If done otherwise, there is no way of knowing if all the requirements and needs of the user will be met. Therefore, the type of test and how to use it depends on the ultimate goal of the application. The performance test of the axial piston pump is done after assembly completion to prove that the pump has the required specifications based on requests. The performance test is performed after completing quality control inspections, such as mechanical and tolerance tests. These tests are a benchmark for the acceptability of the pump. Therefore, testing machine (see **Figure 5**) in laboratory has been developed with the ability to test rotational speeds up to 3000 *rpm*, a torque of 5000 *Nm*, a pressure of up to 450 *bar*, and a flow of 500 *liters/min*. The hydraulic pump is fixed with the help of fixators on the work table so that it does not move and vibrate during the hydrological pump test. After setting the values based on the pump controller and test program, the device is turned on through the touch screen. The main factors affecting the pump's operation are suction, discharge, and speed. Secondary factors affecting the performance of pumps include physical and climatic variables such as temperature, viscosity, density, and turbulence in the pumped liquid. At the pump outlet, a pressure gauge measures the output pressure, the average of the piston pressures in the pump. There are also sensors for measuring the outlet flow and rotational speed of the electromotor in the testing machine. The circuit of the test station is closed. This means the fluid is pumped from the reservoir and returned to the reservoir again after leaving the pump and passing through the equipment and filters.

In this work, the standard controllers from Parker Hannifin [16] are selected for testing, and in these controllers, there are two plans for the axial hydraulic pump, full stroke, and zero stroke. In the full stroke plan, four features are measured at three levels of pressure, and in the zero stroke plan, five features are measured at the three levels of pressure (see sections 4.1.1 and 4.1.2). The characteristics of the pumped fluid and the physical conditions are defined in **Table 2**.

4.1.1 Full stroke in axial piston pumps

In the axial piston pumps, the generated flow volume Q depends on the displacement swash plate. In other words, the displacement value is proportional to the

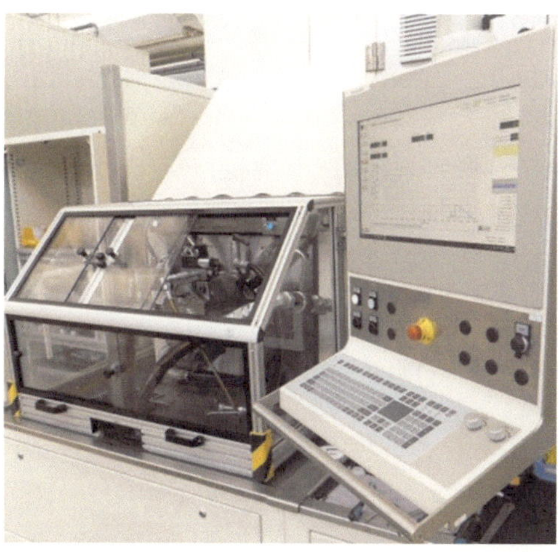

Figure 5.
Axial piston pumps test bench [16].

Hydraulic fluid	Hydraulic oil, Group HLP
Viscosity class	ISO VG 46
Viscosity class of the hydraulic fluid	$v = 46 \pm 5\ mm^2/s\ in\ 40°C$
Temperature of the hydraulic fluid	$t = 50 \pm 2°\ C$
Purity level of the hydraulic fluid	18/16/13 according to ISO 4406

Table 2.
Physical conditions and characteristics of the pumped fluid.

change in the angle between the swashplate and the axis of the pistons. In the full stroke, four features (see **Table 3**) are measured at three levels of pressure, 100, 200, and 300 *bar*. When the test machine is turned on, the hydraulic oil pressure is increased to 100 *bar* (set value) within four seconds, and it stays at this level for five seconds. Then, the actual feature values are determined by the sensors every second and stored. After the recording of these values, the hydraulic oil pressure increases to 200 *bar* (set value) in four seconds. At this step, the oil pressure stays at the value of 200 *bar*, and over five seconds, the values of the features are measured every second by the sensors. This process is repeated for a pressure of 300 *bar* (set value). When the measurement is finished, the pressure value is reduced from 300 *bar* to zero. **Figure 6** shows the pressure variations according to the time in the full stroke plan.

The measured features include:

Symbol	Unit	Feature
P	[*bar*]	outlet pressure
VM1	[*l/min*]	outlet flow
VM2	[*l/min*]	leak oil flow
NM1	[*rpm*]	electromotor rotational speed

Table 3.
The feature definition for the full stroke plan.

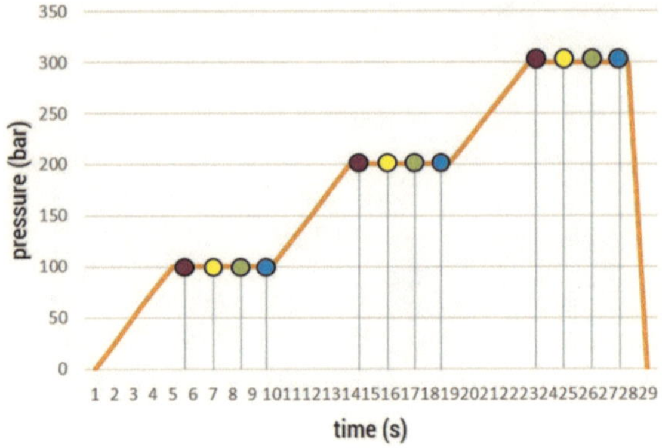

Figure 6.
Pressure vs. time curve for the full stroke plan.

4.1.2 Zero stroke in axial piston pumps

After completing the measuring in the full stroke plan, the test program enters a new phase called the zero stroke plan. The goal of the zero stroke plan is to measure five features (see **Table 4**) at the three levels of pressure 350, 200, and 100 *bar*. **Figure 7** shows the pressure variations over time in the zero stroke plan.

In this process, the hydraulic oil pressure is increased to 350 *bar* (set value) within four seconds and stays at this level for five seconds. Then, the actual feature values are measured by sensors every second and stored. After storing these values, the hydraulic pressure of the oil is decreased to 200 *bar* within four seconds. Afterward, the oil pressure stays at the value of 200 *bar* for five seconds, and the values of the features are measured by the sensors every second. This process is repeated for 100 *bar* oil pressure, too. In the end, the oil pressure is reduced to zero. At this point, the process of testing a hydraulic piston pump is finished.

The measured features include:

Symbol	Unit	Feature
P	[*bar*]	outlet pressure
ΔP	[*bar*]	Pressure difference between output and control
VM1	[*l/min*]	outlet flow
VM2	[*l/min*]	leak oil flow
NM1	[*rpm*]	electromotor rotational speed

Table 4.
The feature definition for the zero stroke plan.

4.2 Data collection and data labeling

In the database, there may be corrupt data such as I. Failure to store data. II. There needs to be more compatibility between the data. Therefore, the quality of these data should be checked. Preprocessing can eliminate inappropriate and corrupt data. In this way, the desired output can be obtained from the data. After data acquisition and

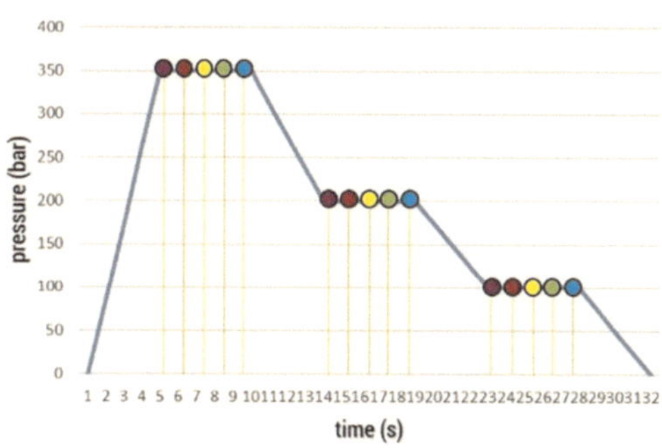

Figure 7.
Pressure vs. time curve for the zero stroke plan.

merging, data preparation was started. The next step was data cleaning. Collected data usually has three significant breaches: some data is noisy, there are some missing values, or in some cases, there is an inconsistency within the data, so data cleaning, to treat missing values and possibly inconsistent data, is necessary. After cleaning the data, selecting features was started. This process will discuss which plans are selected and how they can be prepared for the labeling phase. The full stroke and the zero stroke plans were selected among different plans. Then, the full stroke set values 100, 200, and 300 *bar* and the zero stroke set values 350, 200, and 100 *bar* were selected. In the end, the results were used in the labeling step. Due to the use of a supervised learning approach, it is essential to determine the appropriate label. At this step, labeling means evaluating data. The process of labeling receives data from the previous step includes the two plans: full stroke and zero strokes. It should be noted that all pumps tested in this work were in good condition. According to frame size, displacement, and the set values of this plan (see **Table 5**), the number of 18 classes was defined by the expert for full and zero stroke.

Depending on the available dataset, the classifier design can be considered as supervised learning for fuzzy pattern classification. This work has a set of labeled objects with known class membership. About 70% of this set is extracted and applied to obtain a classifier. These objects build the training set. The remaining objects, whose correct class assignments are also known, are referred to as the sets for adaptation and are used to validate the adaptive classifier's performance.

Class	Frame size	Displacement [cm^3/rev]	Set value full stroke	Set value zero stroke
1	3	60	100	350
2	3	80	100	350
3	3	92	100	350
4	4	140	100	350
5	4	180	100	350
6	5	270	100	350
7	3	60	200	200
8	3	80	200	200
9	3	92	200	200
10	4	140	200	200
11	4	180	200	200
12	5	270	200	200
13	3	60	300	100
14	3	80	300	100
15	3	92	300	100
16	4	140	300	100
17	4	180	300	100
18	5	270	300	100

Table 5.
The labeled dataset for the full stroke and zero stroke plan.

5. Results and discussion

5.1 Introduction

In this section, the fuzzy pattern classification and adaptive fuzzy pattern classification algorithms developed in this work are evaluated, and their performances are presented. For the fuzzy modeling, the axial piston pump test data was used. Two full stroke plans and zero stroke plans were used for tests on pumps, which aim to model the data obtained from these two test plans. The modeling is based on supervised learning. So, for each test plan, the 18 classes were defined (see **Table 5**).

Since the validity of any model computed in the research needs to be estimated, the division of the samples into the three training, fitting, and testing datasets was applied to measure and verify the accuracy of the models. For learning the fuzzy model, three scenarios are considered:

Scenario I: About 70% of the data was used for fuzzy modeling as a basic model. Therefore, 15% of the data (testing data respectively, the last 15% of saved data.) was used to evaluate the accuracy of the models.

Scenario II: Models obtained from scenario one were applied to an adaptive algorithm. Therefore, 15% of the data (adaption data) was used to adapt the model's parameters from the scenario I and to obtain the new pattern with the new parameters. In order to evaluate the accuracy of a new fuzzy pattern obtained from an adaptive algorithm, it has been used of the data (test data).

Scenario III: In this scenario, as in scenario II, 15% of the data is used in the basic model adaptation process, and 15% is used as test data to evaluate the adaption model. The difference in this scenario is that only the objects that are outside the scope of any classifier are included in the adaptation process.

The training phase was defined with initial values ($b_{l/r} = 0.5$, $d_{l/r} = 2$) and is used for the fuzzy pattern classification for all classes.

5.2 Evaluation of the fuzzy pattern classification and adaptive fuzzy classification

Figure 8 shows the membership values of the test data (15%) for full stroke, set value pressure 100 *bar*, and for different scenarios. As you can see, in class 1 and scenario I, the membership value, in other words, the accuracy of the basic model, has improved to 0.81 and after the adaptation process according to scenario.

II: to 0.84 and scenario III to 0.86. In class 6, these improvements and changes are more, and in class 3, these changes are much less. However, in these classes, it is more accurate after adapting the model. As it is clear from diagram 8, scenario II in classes 2, 4, and 5 has less compliance than the other two scenarios, but overall, scenario III shows a useful improvement considering the small number of data.

Figure 9 shows the membership values for different frame sizes for full stroke at a pressure of 200 bar. As can be seen in **Figure 9**, class 8 has the most changes to increase the accuracy of the base model after adaptation, and these results can be seen in classes 7 and 12. In classes 9 and 10, the changes after adaptation are almost negligible, while these classes display more than 95% accuracy after adaptation.

Figure 10 shows the membership values for full stroke and in 300 *bar*. As you can see, in class 14, in other words, in scenario I, the basic model has a low accuracy of about 10%. In contrast, after adaptation and with a limited number of objects, the model's accuracy has improved. Also, in other classes, except class 17, we see an

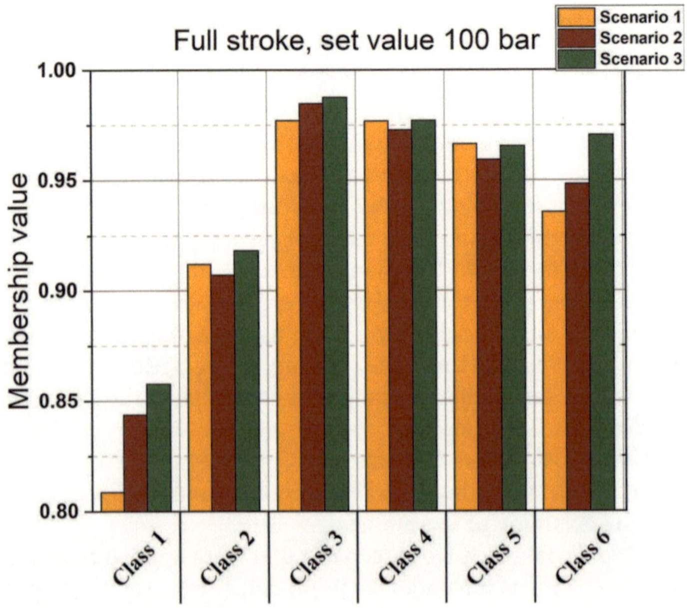

Figure 8.
Accuracy of different scenarios for full stroke in set value 100 bar *by testing data.*

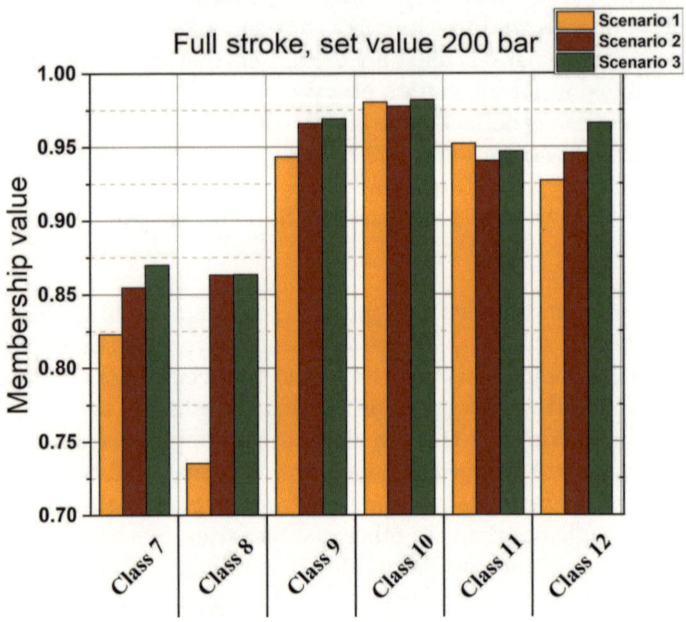

Figure 9.
Accuracy of different scenarios for full stroke in set value 200 bar *by testing data.*

increase in accuracy in the model after adaptation, and this accuracy is more than 80% in all classes except class 14.

Figures 11–13 show the membership values of test data for zero stroke at pressures of 350, 200, and 100 *bar*. As it is evident in these figures, in frame size 5 (see **Table 1**),

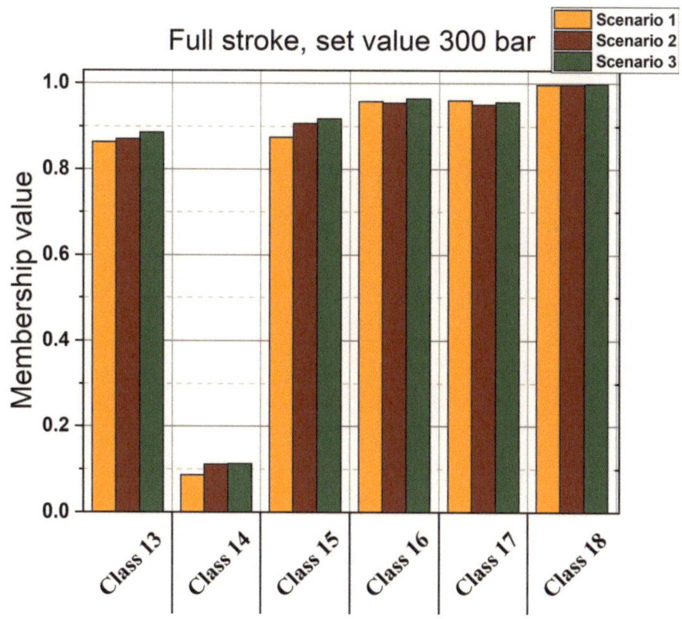

Figure 10.
Accuracy of different scenarios for full stroke in set value 300 bar *by testing data.*

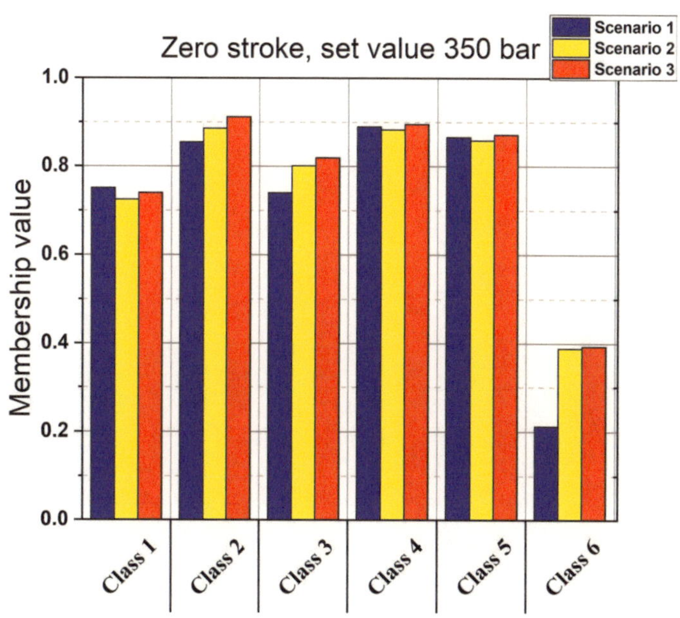

Figure 11.
Accuracy of different scenarios for zero stroke in set value 350 bar *by testing data.*

in other words, classes 6, 12, and 18, the accuracy of the basic model, in other words, scenario I, is in the range of 0.21 to 0.41, which after the adaptation of these models, there is a significant improvement that they show from 0.39 to 0.65. As discussed in the zero plane test and Section 4.1.2, the tested pumps are suddenly pressurized to

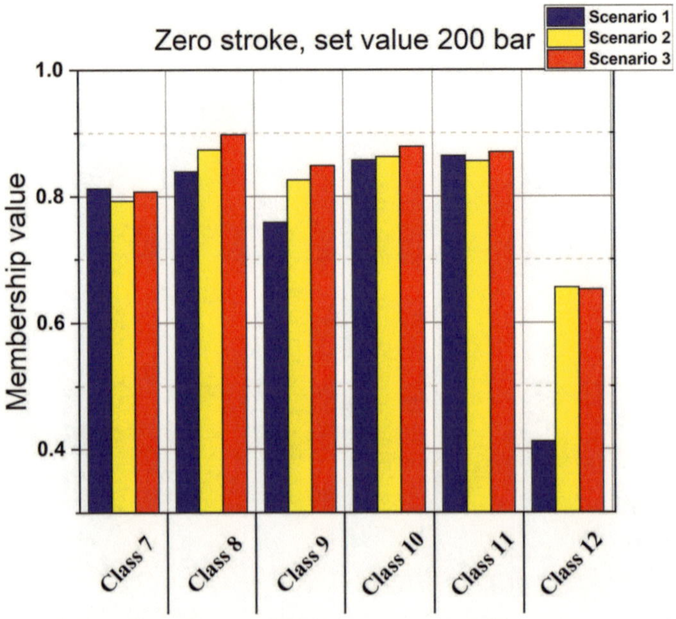

Figure 12.
Accuracy of different scenarios for zero stroke in set value 200 bar *by testing data.*

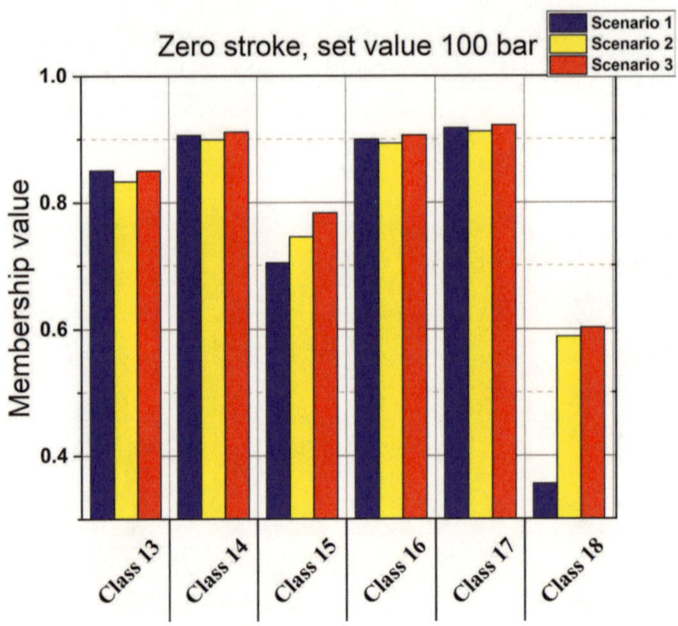

Figure 13.
Accuracy of different scenarios for zero stroke in set value 100 bar *by testing data.*

350 *bar* from a static state within a few seconds. In this case, the data dispersion is almost high, and the system faces uncertainty because the oil pressure needs time to reach equilibrium. The thing to think about in this process is that the fuzzy model has

displayed a stable state with appropriate accuracy concerning this uncertainty. Also, the accuracy of the model has improved after adaptation.

5.3 Evaluation of the adaptive fuzzy classification process

One of the questionable points in the adaptation process is how accurate the adapted model of each adaptation step is to the test data, whether the model's accuracy has continually improved during the adaptation process or exhibits other results.

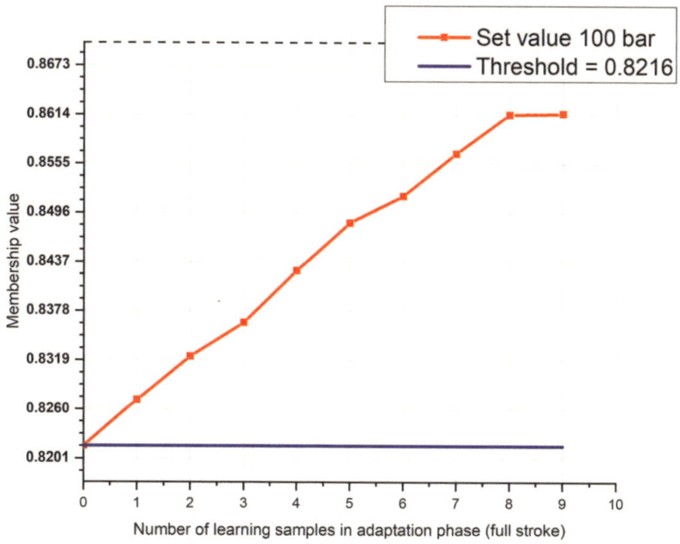

Figure 14.
Example of each adaptation step for full stroke of class 1 in set value 100 bar.

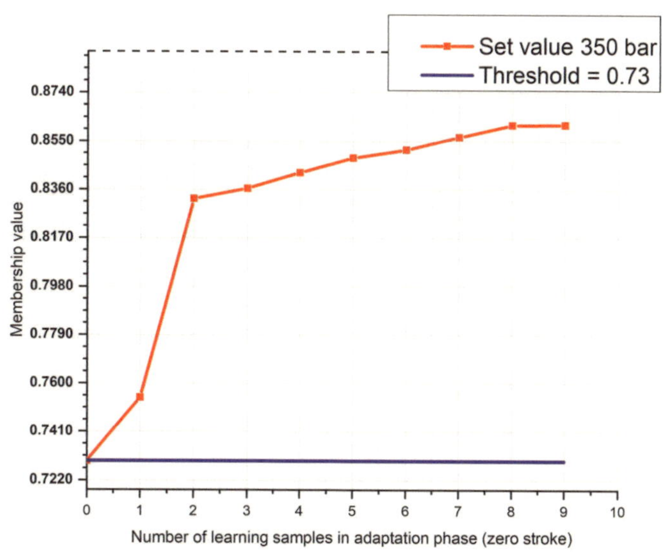

Figure 15.
Example of each adaptation step for zero stroke of class 2 in set value 350 bar.

To answer this question, the new model should be evaluated with test data for each object participating in the adaptation process and compared with the reference or threshold value. The result of this study is shown in **Figures 14** and **15**. In these graphs, the blue lines are the reference values and, in other words, the thresholds displayed for class 1 in plan five and class 2 in plan six based on strategy III. The red lines show each stage's adaptation process and the model's accuracy. It can be seen from these graphs that the adapted models have improved compared to the threshold. So that in plan five and **Figure 14**, the model's accuracy has improved from 0.82 to 0.86, and this value in plan six and **Figure 15** is from 0.73 to 0.86. It should be noted that the number of objects during the adaptation process was 9.

6. Conclusion and future work

The primary goal of this work is to provide a new method for the final inspection and testing of hydraulic axial piston pumps working automatically and online without the user's presence. Motivated by the increasing need for flexible classifiers that can be automatically adapted continuously to cope with dynamic work environments in the context of Industry 4.0 and 5.0, this work proposed an adaptive fuzzy pattern classification algorithm, which can model nonlinear and complex relationships between empirical input and output data with precise accuracy. This algorithm can adapt models to real situations and with online streams of data. The performance test of the axial piston pump is done to verify that the pump has the required specifications based on requests. These tests are a benchmark for the acceptability of the pump. In this work, the pump data from the good situation are tested for four months. In the standard controllers, there are two application plans: full stroke and zero stroke for the axial hydraulic pump. In the full stroke, four features are measured at 100, 200, and 300 *bar* pressure levels. In the zero stroke plan, five features are measured at 350, 200, and 100 *bar* pressure levels. The axial piston pumps range the displacement from 16 cm^3/rev to 360 cm^3/rev. According to frame size, displacement and the set values of each plan are defined as the number of 18 classes out of 448 pumps.

The parameters of a class are composed of position parameters and shape parameters. The strategy to update the initially static fuzzy pattern classifier to changes in the fuzzy pattern classifier parameter is to the classifier from the sequence of the object.

In the scenario I, to evaluate the accuracy of the models in both test plans, 70% of the data (learning data) and 15% of the data (testing data) were used. In scenarios II and III, 70% of the data was used as a basic model for fuzzy modeling. Therefore, 15% of the data was used to adapt and 15% to evaluate the accuracy of the models.

As can be seen from the results, the model's accuracy improves with the adaption process and increasing number of samples. These results show that the models can be improved significantly when using adaptation/ evolution based on new measurements.

Similar to the parts presented in this work, our standpoint can be divided into application and theoretical perspectives.

As future applications, we present here the essential points:

- In the tested pumps, two pumps were found with faults whose data can be defined and modeled as fault classes in the future.

- These pumps can be controlled online by defining classes in a microcontroller, connecting with axial piston pumps, and defining fault classes.

- It is possible to transmit this concept and approach to other technical applications, such as machine tools, forming and pressing machines, surface analyses.

As future theoretical works, we present below the direct signs that we consider to improve our approach:

- For a better classification of the complex features according to the analysis results, the model always requires an adjustment of the combination of all parameters. Therefore, adapting the remaining class parameters $(b_{l/r}, d_{l/r})$ is required.

- An evolving fuzzy pattern classification can update structural components on demand based on new system behavior and operating conditions. Evolving structure is a theoretical complex that should be researched in the future.

Author details

Amir Nemati[1], Elif Öztürk[1,2]*, Steffen F. Bocklisch[1] and Welf-Guntram Drossel[1,3]

1 Chemnitz University of Technology, Chemnitz, Saxony, Germany

2 Institute of Machine Tools and Manufacturing, ETH Zurich, Zurich, Switzerland

3 Fraunhofer Institute for Machine Tools and Forming Technology, IWU, Chemnitz, Saxony, Germany

*Address all correspondence to: oeztuerk@inspire.ethz.ch

References

[1] Helwig N, Pignanelli E, Schutze A. Condition monitoring of a complex hydraulic system using multivariate statistics. In: 2015 IEEE International Instrumentation and Measurement Technology Conference (I2MTC) Proceedings. Pisa, Italy: IEEE; 2015. pp. 210-215

[2] Helwig N, Pignanelli E, Schütze A. D8.1 - detecting and compensating sensor faults in a hydraulic condition monitoring system. In: Proceedings SENSOR 2015. Nürnberg, Germany: AMA Service GmbH, Von-Münchhausen-Str. 49, 31515 Wunstorf, Germany; 2015. pp. 641-646

[3] Nemati A, Jäckel M, Bocklisch SF, Drossel W-G. Fuzzy pattern modeling of self-pierce riveting for data from experiments and computer simulations. International Journal of Advanced Manufacturing Technology. 2022;**122**: 2203-2216. DOI: 10.1007/s00170-022-09914-6

[4] Baur M, Bocklisch SF. Similarity based local model approach for nonlinear modelling. In: 2001 European Control Conference (ECC). Porto: IEEE; 2001. pp. 3905-3910

[5] Chao Q, Gao H, Tao J, et al. Fault diagnosis of axial piston pumps with multi-sensor data and convolutional neural network. Frontiers of Mechanical Engineering. 2022;**17**:36. DOI: 10.1007/s11465-022-0692-4

[6] Xiao C, Tang H, Ren Y, Kumar A. Fuzzy entropy assisted singular spectrum decomposition to detect bearing faults in axial piston pump. Alexandria Engineering Journal. 2022;**61**: 5869-5885. DOI: 10.1016/j.aej.2021.11.019

[7] Tang S, Yuan S, Zhu Y, Li G. An integrated deep learning method towards fault diagnosis of hydraulic axial piston pump. Sensors. 2020;**20**:6576. DOI: 10.3390/s20226576

[8] Almaksour A. Incremental learning of evolving fuzzy inference systems: application to handwritten gesture recognition. Machine learning [cs.LG]. INSA de Rennes, (Doctoral dissertation). 2011

[9] Dougherty G. Pattern Recognition and Classification. New York, New York, NY: Springer; 2013

[10] Bocklisch F, Bocklisch SF, Beggiato M, Krems JF. Adaptive fuzzy pattern classification for the online detection of driver lane change intention. Neurocomputing. 2017;**262**:148-158

[11] Lughofer E. Evolving Fuzzy Systems – Methodologies, Advanced Concepts and Applications. Berlin, Heidelberg: Springer; 2011. DOI: 10.1007/978-3-642-18087-3

[12] Kayani OK, Sohaib M. Generic Simulation Model Development of Hydraulic Axial Piston Machines.

[13] Bocklisch SF. Prozessanalyse mit unscharfen Verfahren. Berlin: Verlag Technik; 1987

[14] Ivantysyn J, Ivantysynova M. Hydrostatische Pumpen und Motoren: Konstruktion und Berechnung. Vogel Buchverlag; 1993. Available from: https://books.google.ch/books?id=e1w7AAAACAAJ

[15] Eifler W, Küttner K-H. Küttner Kolbenmaschinen: [Kolbenpumpen, Kolbenverdichter, Brennkraftmaschinen] ; mit 40 Tabellen sowie zahlreichen

Übungen und Beispielen mit Lösungen, 7., neu bearb. Aufl. Wiesbaden: Vieweg + Teubner; 2009

[16] Parker Distributor. [Internet]. Available from: https://www.parker.com/pl/en/home.html

[17] Bocklisch F, Bocklisch SF, Krems JF. Sometimes, often, and always: Exploring the vague meanings of frequency expressions. Behavior in Research. 2012; **44**:144-157. DOI: 10.3758/s13428-011-0130-8

[18] Siddique NH, Adeli H. Computational Intelligence: Synergies of Fuzzy Logic, Neural Networks, and Evolutionary Computing. Chichester, West Sussex, United Kingdom: John Wiley & Sons; 2013

[19] Géron A. Hands-on Machine Learning with Scikit-Learn and TensorFlow: Concepts, Tools, and Techniques to Build Intelligent Systems. Sebastopol, CA: O'Reilly Media; 2017

[20] Ade RR, Deshmukh PR. Methods for incremental learning: A survey. International Journal of Data Mining & Knowledge Management Process. 2013;**3**:119-125. DOI: 10.5121/ijdkp.2013.3408

[21] Read J, Bifet A, Pfahringer B, Holmes G. Batch-incremental versus instance-incremental learning in dynamic and evolving data. In: Hollmén J, Klawonn F, Tucker A, editors. Advances in Intelligent Data Analysis XI. Berlin Heidelberg, Berlin, Heidelberg: Springer; 2012. pp. 313-323

[22] Buckley JJ, Eslami E. An Introduction to Fuzzy Logic and Fuzzy Sets. Heidelberg: Physica-Verlag HD; 2002

[23] Zadeh LA. Fuzzy sets. Information and Control. 1965;**8**:338-353. DOI: 10.1016/S0019-9958(65)90241-X

[24] Kacprzyk J, Pedrycz W. Springer Handbook of Computational Intelligence. Berlin Heidelberg, Berlin, Heidelberg: Springer; 2015

[25] Angelov PP. Handbook on Computational Intelligence. New Jersey: World Scientific; 2016

Chapter 8

Measuring the Service Quality of Mobile Smart Devices: A Framework for Best Practices

Abdulla Jaafar Desmal and Zainab Merza Madan

Abstract

This chapter presents a comprehensive framework for measuring and managing service quality of mobile smart devices across seven key dimensions: interactions, usability, efficiency, information quality, availability, security, and reliability. Grounded in established models like SERVQUAL and E-S-QUAL, the framework identifies specific metrics based on user perceptions for evaluating expertise, accuracy, responsiveness, and customization. The SERVQUAL model outlines five service quality dimensions: tangibles, reliability, responsiveness, assurance, and empathy. E-S-QUAL adapts these to the digital environment. The proposed framework draws on these seminal models while incorporating new factors relevant to interconnected, artificial intelligence (AI)-enabled mobile platforms and devices. The framework is designed to enable proactive service quality management throughout the customer journey with mobile smart devices—from product design and testing to postsales support. A continuous measurement, analysis, and improvement process is outlined involving quantitative and qualitative techniques. An implementation road map covers considerations for organizing, change management, and training to integrate the methodology into product development processes. The goal is to provide technology companies with actionable and evidence-based guidance for optimizing satisfaction and loyalty among mobile smart device consumers. As these devices continue proliferating, managing user expectations through integrated hardware, software, and service experiences will be key to competitive positioning in mobile ecosystems.

Keywords: service quality, mobile devices, user experience, satisfaction, loyalty, SERVQUAL, E-S-QUAL

1. Introduction

1.1 Background and importance of service quality for mobile smart devices

The widespread adoption of mobile smart devices like smartphones and tablets has fundamentally transformed how consumers interact with technologies and access information and services. As digitalization accelerates, mobile platforms provide ubiquitous connectivity and enable innovative applications across many aspects of life including entertainment, social connections, shopping, finances, health, education, and more. Consequently, mobile devices have become deeply integrated into daily routines for many people.

Service quality is critically important for providers of mobile platforms and applications to satisfy users, meet expectations, and encourage engagement. Perceived service quality directly influences customer satisfaction, loyalty, and long-term business success. However, evaluating and measuring quality can be complex for mobile contexts due to unique characteristics like intangibility, variability across users and contexts, and continuously evolving technologies and features. Service quality remains a high priority but also an ongoing challenge.

1.2 Challenges and opportunities of service quality for mobile smart devices

Delivering high-quality service through mobile platforms involves many potential complications. Device limitations, connectivity issues, security risks, and technology glitches can degrade experiences. Frequent feature updates change how people interact with apps and information. Diverse usage contexts like location, activity, or goals alter perceptions of quality. Personalization and customization enable more tailored experiences but also greater variability. Rapid evolution cycles continually reshape mobile technologies and ecosystems.

Despite these hurdles, mobile services also provide exciting possibilities to improve quality in innovative ways. Smart sensors and usage data analytics offer insights to understand users and contexts to optimize experiences. Machine learning (ML) and artificial intelligence (AI) can power intelligent assistance and interactions. 5G and enhanced mobile broadband will enable more immersive and seamless services. Leveraging these capabilities can enhance quality but requires adaptable strategies.

1.3 Purpose and outline of chapter

This chapter provides a comprehensive framework to evaluate and guide service quality management for mobile smart devices. Key objectives are to:

- Review established service quality theories and adapt principles to mobile environments.

- Identify core dimensions and metrics to assess mobile service quality.

- Develop implementation strategies for measurement, analysis, and improvement.

1.4 Defining service quality

Service quality refers to a customer's evaluation of how well a provided service meets expectations and satisfies needs [1]. It involves both subjective and objective factors across multiple dimensions. High-quality service directly leads to positive outcomes like customer satisfaction, loyalty, word-of-mouth promotion, and long-term business success [2]. The most widely used framework characterizes service quality along five key dimensions: reliability, assurance, tangibles, empathy, and responsiveness [3]. Reliability focuses on providing dependable and accurate service.

Assurance involves knowledge, competence, and courtesy of staff. Tangibles cover physical facilities, equipment, and overall presentation. Empathy relates to caring and individualized attention to customers. Responsiveness concerns promptness and ability to deliver timely service.

1.5 Mobile service quality

In the context of mobile platforms and applications, characteristics like intangibility, variability, and rapid evolution require adapting traditional service quality concepts [4]. Key dimensions include network quality covering connectivity and transmission; content quality focused on accuracy, completeness, and format; interaction quality dealing with operability, convenience, and customization; and contextual quality aligned to usage situations [5]. Both subjective perceptions and objective performance metrics are important for mobile service quality.

1.6 Strategic importance

Achieving high mobile service quality has become strategically vital for many businesses and organizations. Mobile channels are often the primary way companies engage with customers and provide services and support. With growing reliance on mobile platforms, any shortcomings in service quality can negatively impact customer acquisition, retention, loyalty, and profitability [6]. Organizations must continuously monitor, assess, and improve mobile service quality to enable positive experiences that fulfill user needs and expectations.

2. Service quality models

Service quality has been studied extensively across disciplines including marketing, management, and information systems science. Several foundational models that inform research on mobile service quality have been developed.

2.1 SERVQUAL model

One of the most widely adopted service quality frameworks is the SERVQUAL model developed by Parasuraman, Zeithaml, and Berry (1988). This model characterizes service quality along five key dimensions:

- Reliability – ability to perform the promised service dependably and accurately.
- Assurance – knowledge and courtesy of employees and their ability to convey trust and confidence.
- Tangibles – appearance of physical facilities, equipment, personnel, and communication materials.
- Empathy – caring, individualized attention provided to customers.
- Responsiveness – willingness to help customers and deliver prompt service.

The SERVQUAL model proposes that service quality depends not just on actual service delivery, but also the gaps between a customer's expectations and their perceptions of the service experience. By measuring both expectations and perceptions across these five dimensions through surveys or interviews, providers can identify areas where perceived service falls short of desired expectations. This framework equips organizations to understand customer perspectives and target improvements accordingly.

The SERVQUAL instrument has been applied in many industries and service contexts, making it one of the most widely utilized service quality models. However, SERVQUAL has faced some criticism. The expectation-perception gap approach can be difficult to implement reliably, with debate about whether customers can accurately recall preservice expectations [7]. The five dimensions may not fully capture all possible facets of service quality. Applicability across various service contexts has also been questioned [8].

Nonetheless, the SERVQUAL model pioneered the conceptualization of service quality as a multidimensional construct. It provides a useful foundation and starting framework that can be adapted or supplemented for specific service contexts. The model's focus on understanding customer perspectives through the gaps approach gives important insights for managing service quality. As such, SERVQUAL has strongly influenced subsequent service quality research and frameworks.

2.1.1 SERVPERF model

A widely employed adaptation of SERVQUAL is the SERVPERF model developed by Cronin and Taylor (1992). This streamlined model is based solely on perceived service performance rather than expectation-perception gaps. SERVPERF measures customer perceptions of quality across the same five SERVQUAL dimensions, without directly measuring prior expectations [7].

Cronin and Taylor (1992) empirically tested both SERVQUAL and SERVPERF approaches and found perceived performance alone accounted for more of the variation in customer satisfaction. They argued service providers should focus primarily on performance delivery rather than expectation management. The SERVPERF model simplifies data collection and analysis by relying only on performance measures. Subsequent studies have affirmed SERVPERF explains more variance in satisfaction and loyalty.

However, conceptual arguments for incorporating expectations remain strong. Customers often have preconceived service expectations that if unmet can undermine perceptions [9]. Measuring gaps can identify problem areas directly, rather than just low perception scores. Thus while SERVPERF offers advantages in execution, SERVQUAL provides richer diagnostics. Each approach has merits for developing mobile service quality frameworks.

2.1.2 IT alignment model

An important perspective for managing quality of technology-enabled services comes from the information systems (IS) success model developed by DeLone and McLean (2003). This model provides a comprehensive framework highlighting the interrelated dimensions of information quality, system quality, and service quality that drive intentions to use, user satisfaction, and net positive benefits. A key emphasis is the need to carefully adapt metrics and measures of IS success to the specific technology, application, and usage context [10].

As mobile platforms transform service ecosystems, enabling new forms of service delivery, consumption, and value creation, the DeLone and McLean framework provides critical guidance to reevaluate service quality assessment. Traditional models like SERVQUAL may not fully capture new dynamics, interactions, and outcomes made possible by smart mobile technologies and interfaces. More adaptable, context-aware approaches are required.

The DeLone and McLean model defines quality constructs as follows:

- Information quality – Considers content issues like completeness, accuracy, format, and timeliness of information. High information quality requires delivering current, reliable, and personalized content.

- System quality – Focuses on usability factors like accessibility, ease of use, system flexibility, and integration. High system quality provides simple, dependable, and intuitive technical interfaces.

- Service quality – Measures overall support delivered by service providers. It shares key dimensions with SERVQUAL including reliability, responsiveness, assurance, and empathy.

A key insight is distinguishing the separate but interdependent roles of information, system, and service quality for overall user satisfaction and system success. For mobile service ecosystems, advanced interfaces and content delivery may exceed traditional support services in dominating quality perceptions.

The model also emphasizes relationships between quality constructs, usage intentions, user satisfaction, and net benefits. Service quality alone is insufficient—positive outcomes depend on optimizing information and system quality. High information and system quality can also compensate for some service delivery shortcomings.

This model provides a more holistic, context-aware perspective on service quality. It underscores the need for flexible, multidimensional approaches adaptable to evolving mobile platforms and usage contexts. As technology-enabled services transform, so must measurement frameworks to capture new drivers of quality perceptions and system success.

2.1.3 Hierarchical model

Brady and Cronin (2001) proposed a multidimensional, multilevel conceptualization of service quality known as the Hierarchical Model. This model asserts that customers form perceptions of quality based on evaluations of performance at different levels [11].

The three primary levels in the hierarchy are:

- Interaction quality – The lowest level representing the interaction between a customer and service provider. It includes factors like attitude, behavior, expertise, and communication of employees.

- Physical environment quality – The second level covers ambient conditions, design, and social factors related to the servicescape. It ranges from intangible aspects like the atmosphere to tangible elements such as facilities and equipment.

- Outcome quality – The highest level focused on the results and what is accomplished through the service act. It encapsulates perceptions of what the customer obtains from the service experience.

A key premise is that customers make bottom-up quality judgments across these levels. Interaction quality lays the foundation. The environment and conditions surrounding service delivery form the next facet of evaluation. Finally, the ultimate outcome achieved determines perceptions of highest-level quality.

The model further proposes three subdimensions of quality for each level:

- Quality of service product – Core service functionality and offerings.

- Quality of service delivery – How the service is provided.

- Quality of service environment – Conditions under which service occurs.

By distinguishing between hierarchical levels and quality subdimensions, this model provides a more nuanced, multidimensional conceptualization of service quality. It moves beyond the SERVQUAL focus on just customer-employee interactions to also encapsulate broader environmental and outcome considerations.

For mobile service contexts, mapping quality perceptions to interaction, environment, and outcome levels can give greater diagnostic precision. Interaction may translate to human-device interactions and interfaces. The service environment consists of mobile hardware, operating platforms, and ecosystems. Outcomes depend on how technology-enabled services generate value across user contexts. A hierarchical approach allows linking specific quality issues to subdimensions within appropriate levels. This framework enhances understanding of multidimensional drivers to optimize mobile service quality.

2.1.4 Synthesis of models

The service quality models reviewed offer complementary conceptual perspectives for examining mobile service quality characteristics and requirements. While differing in their specific approaches, several key insights emerge across the major frameworks that can inform an integrative mobile service quality model:

- Multidimensional construct – Service quality is widely recognized across models as a multidimensional construct comprised of complex, interdependent subdimensions. Assessment requires a holistic perspective across key facets like interaction quality, environment quality, outcome quality, system performance, content quality, human-computer interaction, customer support, subjective perceptions, and value creation. No single dimension fully represents overall service quality. An integrative model must incorporate key dimensions specific to mobile service ecosystems.

- Adaptable measures – Models emphasize selecting appropriate metrics and adapting evaluation frameworks to fit different technologies, applications, and usage contexts. Mobile services have distinct characteristics necessitating customized, context-aware measures that capture new ecosystem dynamics, interactions, and outcomes enabled. Traditional models provide useful starting points but mobile-specific adaptations are required.

- Gaps modeling – Understanding gaps between customer expectations and perceptions provides powerful diagnostics to precisely identify problem areas. Assessing only perception of actual service performance is insufficient to fully determine drivers of quality shortfalls. Comparing user expectations to real-world service delivery gives greater precision in diagnosing issues. This vital gaps perspective must be incorporated to enable targeted mobile service improvements.

- Hierarchical levels – Layered quality models highlight that customers evaluate service quality at multiple levels, from discrete touchpoint interactions up to the overall environment and ultimate end results. Mobile services should be examined at interactive, environmental, and outcome dimensions to provide a complete perspective. This mapping of service quality levels gives greater insight into multidimensional drivers.

- Customer perspective – Incorporating subjective customer perceptions, rather than just objective performance metrics, provides crucial insights into quality judgments. For mobile services especially, quality hinges on how users experience and evaluate services within their particular contexts. Quantitative metrics must be complemented by qualitative insights from the user standpoint.

- Continuous improvement – Models recognize service quality as dynamic, necessitating ongoing monitoring, assessment, and enhancement. Mobile service providers must continually track shifting usage patterns, technology changes, perception evolutions, and outcome metrics to maintain quality as ecosystems rapidly transform. A continuous improvement mindset is essential.

These shared principles can guide synthesizing the strengths of each model into an integrative mobile service quality framework. Key tenets include:

- Assessing multiple dimensions spanning system performance, content delivery, human-computer interaction, customer support, and value creation.

- Adapting dimensional measures and evaluation models specifically for mobile service environments.

- Quantifying expectation-perception gaps across dimensions to pinpoint improvement opportunities.

- Mapping quality perceptions to interactive, environmental, and outcome levels.

- Bridging objective performance metrics with subjective user perspectives.

- Enabling continuous improvement closed-loop processes, given contextual dynamism.

No single model fully captures the complexity of assessing and managing mobile service quality. However, synthesizing elements of established models provides a robust foundation. The SERVQUAL model delivers core quality dimensions along with the vital gaps perspective. The DeLone and McLean IS Success Model emphasizes adaptable, context-specific measures spanning information, system, and service quality. Brady and Cronin's Hierarchical Model provides a multilayered view to map quality perceptions. The SERVPERF model focuses on performance metrics from the customer standpoint.

An integrative mobile service quality framework should incorporate synthesized elements:

- Customized quality dimensions spanning technology performance to experience to value.
- Gap analysis between expectations and perceptions across all dimensions.
- Levels ranging from component interactions to holistic service ecosystems.
- Objective metrics paired with subjective customer viewpoints.
- Continuous monitoring and improvement processes to track evolving contexts.

In essence, the integrative framework must be multidimensional, adaptable, hierarchical, gaps-driven, customer-centric, and dynamically managed to address complex and rapidly changing mobile service environments. This provides a robust platform to comprehensively evaluate and enhance mobile service quality from multiple crucial perspectives.

2.1.5 Challenges and tradeoffs

However, developing and implementing such an integrative framework also poses challenges and tradeoffs that must be considered:

- Model complexity – A multidimensional, layered model with adaptable metrics and continuous improvement processes will be more complex to conceptualize, operationalize, and manage. Simpler models may be easier to apply but provide less comprehensive insights.
- Measurement difficulty – Assessing expectations gaps, hierarchical levels, and subjective perceptions will require more extensive and specialized data collection through surveys, interviews, and analytics. Detailed contextual data are harder to gather consistently.
- Framework customization – Adapting measures and models for different mobile services and usage contexts demands significant upfront customization efforts and expertise. A one-size-fits-all approach will have less applicability.
- Data integration – Bridging diverse datasets like technical performance indicators, customer feedback, and usage metrics poses analytical and interpretive difficulties. Integrating insights is more challenging than isolated data streams.
- Resource requirements – Implementing continuous improvement processes and multidimensional monitoring requires more extensive technology, staffing, and management resources to coordinate. Budget limitations may constrain framework scope.

These challenges underscore why many providers apply only simplified service quality models. Yet, the integrative approach delivers significant benefits in providing complete, targeted, and actionable intelligence to drive mobile service excellence.

3. Toward an integrative model

Synthesizing established service quality models provides a strong conceptual foundation for an integrative mobile service quality framework. However, translating these concepts into an operational model requires further specification of dimensions, measures, and data integration strategies.

3.1 Potential dimensionality

Dimensions that could comprise an integrative model include:

- Interaction quality: Covers interactions between the user and the mobile device, app interfaces, virtual assistants, support channels, and any direct touchpoints. Key metrics include ease of use, convenience, customization, and communication effectiveness.

- Platform quality: Encompasses the underlying mobile operating system and capabilities including connectivity, speed, reliability, security, and device/app ecosystem integration. Focuses on performance measures.

- Content quality: Assesses the delivered information and services via the mobile platform including completeness, accuracy, relevance, timeliness, visual appeal, and personalization.

- Support quality: Measures availability, responsiveness, assurance, empathy, and communication of any customer support channels provided by the mobile service provider. Aligns with SERVQUAL dimensions.

- Contextual value: Subjective user-based assessment of how well the mobile services improve experiences or add value across usage contexts like work, leisure, shopping, travel, etc. Captures value-in-use.

These dimensions span technological and human elements of mobile service delivery, incorporating both performance metrics and user perceptions. The dimensions can be mapped to interaction, environment, and outcome quality tiers, with some spanning multiple levels.

3.2 Data collection approaches

Assessing these dimensions requires integrating diverse datasets:

- Technical performance metrics: Platform data on reliability, availability, throughput, latency, and other indicators from monitoring tools, network measurements, and operations records.

- User behavior analytics: Application usage patterns, journeys, and behaviors derived from mobile analytics platforms and app telemetry data. Provides insights into interaction quality.

- Subjective user feedback: Surveys, interviews, and self-reports to measure perceived quality across dimensions including contextual value added. Enables quantifying expectation-perception gaps.

- Support service metrics: Key performance indicators for customer support channels like call center response times, resolution rates, satisfaction scores, and usage volumes.

- Contextual data points: Situational details like user location, activity, goals, and other contextual variables that may affect quality perceptions and value. Can be collected via mobile sensors, surveys, or interviews.

Integrating these data streams provides a holistic, multidimensional dataset for model development and application. Advanced analytics techniques can help bridge quantitative and qualitative data. Contextual data add a crucial mobile-specific element.

3.3 Analysis and interpretation

Key analysis techniques include:

- Correlation analysis: Determine relationships and interdependencies between quality dimensions, technical metrics, perceptions, and value levels. Highlight the strongest drivers.

- Regression modeling: Estimate models predicting overall service quality and value based on dimension fulfillment levels. Identify the most significant dimensions for improvement prioritization.

- Gap scoring: Quantify gaps between user expectations and perceptions for each dimension. Diagnose shortfalls requiring attention, guided by gap magnitude.

- Context analytics: Profile variation in quality perceptions, gaps, and value across contextual variables. Adapt dimensions to usage situations.

- Journey mapping: Understand quality shortfalls across temporal user journeys encompassing multiple touchpoints. Identify problematic steps.

Interpreting integrated results requires a collaborative, cross-functional approach with stakeholders spanning technology, operations, marketing, and product teams. Statistical findings must be translated into tactical interventions personalized to each mobile service. Ongoing iteration and learning is critical for continuous enhancement.

3.4 Implementation architecture

A high-level architecture for an integrative mobile service quality platform includes:

a. Data ingestion infrastructure: Ingests technical metrics, user analytics, survey data, and other sources into a unified data lake, applying cleaning and transformation pipelines.

b. Analytics engine: Provides a toolkit to execute correlation analysis, regression modeling, gaps scoring, contextual analytics, journey mapping, and other techniques on the integrated dataset.

c. Interpretation tools: Enables collaborative filtering, visualization, and annotation of analysis results to extract insights. Provides capabilities to derive interventions.

d. Improvement processes: Closes the loop by linking identified issues to solution development workflows, postimplementation tracking to measure improvements, and capability building for ongoing enhancement.

e. Governance framework: Establishes policies, standards, controls, and processes to manage data collection, storage, access permissions, privacy, lifecycles, and overall platform security.

A modular architecture provides flexibility to start with foundational elements and expand capabilities over time. Open frameworks and interoperable components can integrate with existing mobile data and analytics investments, where possible. With thoughtful design, integrative mobile service quality platforms can deliver ongoing value.

This section has outlined potential strategies and components to progress from conceptual models to an operationalized integrative framework. Further elaboration and specificity would be required based on particular mobile service contexts. However, a multidimensional, data-driven approach following user-centric design principles demonstrates feasibility and value.

3.5 Critical evaluation and limitations

While an integrative model offers benefits, critical evaluation also reveals limitations that should be considered in implementation:

a. Generalizability vs. Contextualization Tradeoff

An integrative model aims to provide a generalizable framework by synthesizing findings across contexts. However, effective implementation requires extensive customization and contextual adaptation. Optimal dimensionality, data sources, analytics, and interventions will differ significantly across mobile service providers, market segments, use cases, and devices. A completely standardized approach lacks contextual precision. Yet, highly individual implementations sacrifice generalizability. Striking the right balance is challenging.

b. Subjective Perceptions vs. Objective Metrics Tension

Incorporating both subjective customer perceptions and objective performance metrics provides a more complete picture. But discrepancies between perceived quality and measured quality will inevitably arise, requiring interpretive caution. Relying too much on perceptions risks overlooking actual underlying issues. Focusing solely on technical metrics misses the customer angle. Determining appropriate weight between perspectives is difficult.

c. Static vs. Dynamic Equilibrium

An integrative model must be dynamically adaptable to changing expectations, technologies, and market conditions. But frequent model changes risk instability, inconsistent tracking, and initiative fatigue, if taken too far. The rate of change required to stay current, without introducing instability from continual changes, needs to be deliberately evaluated.

d. Theoretical Ideal vs. Operational Reality

As conceptualized, an integrative framework is thorough yet complex. But resource constraints, data limitations, and siloed teams may restrict implementation scope in reality. Adoption risks remaining superficial without proper organizational supports, resources, and buy-in. A simplified or phased approach may become necessary.

e. Upfront Investment vs. Realized return on investment (ROI)

Significant upfront investment is required to develop and launch an integrative model, with uncertain ROI realization. Short-term costs and resourcing needs could deter adoption, especially if leaders expect immediate returns. A proof-of-value pilot with projected returns backed by data may help secure buy-in for larger implementation.

These limitations warrant further examination to develop mitigation strategies. For instance, general frameworks could be created for common mobile contexts, then customized through rapid prototyping techniques. Change management and participation from both IT and business teams could bridge metrics and perceptions. Gradual rollout can allow stabilizing models before full launch. Despite limitations, an integrative approach carries major potential, if thoughtfully addressed.

4. Key dimensions of mobile device service quality

Multiple frameworks have been proposed for evaluating service quality of technology-based services and products [12–14]. Synthesizing key elements of these models, seven core dimensions emerge as most relevant for mobile smart devices: interactions, usability, efficiency, information quality, availability, security, and reliability.

4.1 Interactions

A primary way users evaluate service quality of mobile devices is through interactive experiences across various touchpoints [15]. This includes physical device interactions, customer service, online account management, and ecosystem integration. Metrics for assessing interaction quality focus on customization, convenience, responsiveness, and employee expertise [16].

Specific metrics include: personalized greeting, polite tone, agent knowledge, query resolution time, convenience of contact channels, proactive communications, and community forum responsiveness. Surveys, interviews, focus groups, and online reviews can collect user perceptions on these interaction metrics [17]. Data analytics on query volumes, channel usage, and response times also provide insights.

4.2 Usability

Usability evaluates how easy and satisfying a mobile device's user interface and hardware are to operate [18]. Key metrics focus on learnability, efficiency, memorability, error handling, and subjective satisfaction [16].

Usability testing and user experience research techniques like prototyping, think-aloud protocols, surveys, and beta testing provide qualitative insights [19]. Analytics on task completion rates and usage patterns supply quantitative usability data. Integrating usability studies throughout product development is crucial.

4.3 Efficiency

Efficiency measures a mobile device's performance in enabling users to accomplish goals and complete tasks [16]. Metrics evaluate processing speed, battery life, storage capacity, app performance, and connectivity quality [20, 21].

Standard benchmarks, usage tests, and technical diagnostics quantify efficiency. User surveys rate perceptions of speed and battery drain. Monitoring app crashes, storage usage, and network connectivity issues highlights problem areas. Efficiency must be balanced with other dimensions like usability and reliability.

4.4 Information quality

Useful, accurate, current, and personalized information is expected from mobile devices [10]. Metrics focus on relevance, completeness, clarity, accuracy, and customization of information presentation.

Surveys, focus groups, and online reviews provide user perspectives on information quality. Automated testing verifies accuracy. Analytics on search queries and content consumption patterns provide insights for improving personalization and recommendations.

4.5 Availability

Availability evaluates service access and uptime [16, 22]. Mobile devices are reliant on networks, servers, and integrations. Metrics assess percentage uptime, service reach, and continuity during roaming and network switching.

Technical monitoring tools track uptime and performance indicators. User feedback reveals availability issues. Location data analyzes network coverage and roaming quality. Availability is a top priority dimension needing continuous monitoring and improvement.

4.6 Security

Perceived security and privacy risks negatively impact user trust and satisfaction [23]. Mobile devices store sensitive personal data. Security metrics focus on vulnerability prevention, detection, and recovery [18].

Testing tools probe known threats and vulnerabilities. User surveys gauge security perceptions. Monitoring systems track fraudulent account access, suspicious network traffic, malware infections, and other threats. Promptly addressing identified issues maintains user trust.

4.7 Reliability

Reliability represents the consistent and error-free operation of mobile hardware, software, and services [16]. Key metrics include device failure rates, software crashes, complaint volumes, and user perceptions of consistency.

Reliability testing under diverse real-world conditions is essential during product development. Postlaunch monitoring of error logs, help desk tickets, returns/repairs, and online complaints guides continuous improvement. User surveys also rate reliability satisfaction.

These seven dimensions provide a comprehensive framework for managing mobile device service quality. The appropriate metrics and measurement methods will vary by product type, development phase, and brand objectives. Ongoing multichannel data collection, analysis, and improvement is critical for aligning with customer expectations.

5. Measurement of service quality

Robust measurement strategies are required to generate data and insights across the mobile service quality dimensions proposed. Integrating quantitative performance indicators with qualitative consumer perspectives provides a comprehensive assessment.

5.1 Quantitative metrics

Quantitative data enable objective assessment of certain service quality elements related to performance, accuracy, availability, and reliability. Useful metrics can be gathered through system monitoring, testing, and analytics.

Network speed and latency benchmarks provide indicators of service efficiency and availability [24]. Standardized tools like Ookla speed tests generate comparable connectivity data across locations. Real-world usage monitoring provides complementary insights into reliability and consistency.

Component reliability can be quantified through testing under diverse operating conditions and usage profiles. Accelerated life tests analyze failure rates and modes under environmental stress like temperature, vibration, and moisture [25]. These guide engineering improvements and also supply field reliability data.

Software quality metrics based on source code analysis techniques assess the maintainability, testability, reusability, and evolution of mobile apps [6, 26]. Automated static and dynamic analysis identifies vulnerabilities and establishes security benchmarks.

Analytics on app stability, battery and resource usage, crashes, and anomalies in large field data identifies optimization opportunities [27]. Online service uptime and response times are quantifiable through scripts and synthetic monitoring.

Usability metrics based on task times, clicks, conversions, learnability tests, and similar usage data offer objective efficiency indicators complementary to surveys [28]. Data logs provide visibility into usage patterns, while in-product telemetry tracks detailed flows.

5.2 Qualitative feedback

Despite useful insights from usage data, consumer perspectives remain important to fully assess user satisfaction across service dimensions [3]. Surveys, interviews, focus groups, and reviews reveal subjective perceptions difficult to capture through metrics alone.

Standardized rating scales allow statistical analysis and benchmarking. The System Usability Scale [29] and SERVPERF [9] provide validated instruments for measuring perceived service quality. Product-specific questionnaires customized to target contexts are also valuable.

Open-ended feedback through online reviews, interviews, and support channels provides details on pain points. Techniques like sentiment analysis assess emotions and extract common themes from unstructured feedback at scale [30, 31]. High-frequency concerns indicate systemic gaps needing priority action.

In-context user tests and observational studies reveal usability issues and interaction difficulties. Moderated sessions allow deeper probing through follow-up questions and task analysis. Remote synchronous tools have enabled more flexible and scalable qualitative testing [32].

5.3 Customer journey mapping

An emerging qualitative approach is documenting detailed customer journeys to map overall experience across channels and touchpoints [33]. This identifies emotional highs and lows, pain points and vulnerabilities throughout the user lifecycle.

Tools like experience maps, value chain diagrams, and blueprints systematically capture steps customers take before, during, and after transactions. Customer perspectives are integrated across sales, onboarding, engagement cycles, and support [6, 21]. This horizon view highlights priorities.

Journey mapping workshops, ethnographic observation, and longitudinal engagements/diaries provide immersive understanding [34]. Personas, scenarios, and storyboards make the narratives tangible. Comparison across customer segments reveals different needs.

Analytics enrich the qualitative story. Association rules analysis links emotions to touchpoints [30]. Predictive modeling identifies likely pain points and vulnerable moments [35]. Clustering classifies journeys for targeted improvements.

This cross-channel perspective across the lifecycle complements episodic surveys and transactional data with a more holistic view. Customer journey mapping integrates quantitative metrics with rich qualitative insights for driving mobile service enhancements.

6. Managing service quality

Realizing improvements requires processes linking measurement insights with strategic decisions and development prioritization. A culture valuing customer data guides mobile brands to proactively address experience gaps.

6.1 Product design and testing

Service quality focus must begin well before launch through research, prototyping, and design iteration. User needs analysis and usability testing ensure product-market fit and ease of use [19]. Incorporating metrics and feedback into requirement reviews and feature prioritization promotes satisfaction.

Developer communities and public beta testing prerelease enable crowdsourced improvements [21, 32]. Regular usability testing postlaunch identifies adoption barriers. Monitoring app store ratings and social media sentiment guides incremental enhancements.

Experience analytics and in-product telemetry provide granular visibility into painful journeys and optimization opportunities. Continuously built integration and controlled rollouts facilitate data-driven improvements [36].

6.2 Postsales support

Despite best efforts, some defects and quality gaps will remain. Analyzing incoming issues for trends highlights systemic problems versus one-offs. Monitoring channels like app reviews, call centers, online communities, and social media provides voice of the customer insights [37].

Case routing and resolution tracking based on root causes rather than symptoms drive effective diagnosis and prevention. Knowledge bases codify workarounds while development focuses on permanent fixes for common problems.

Over-the-air updates should provide fixes with minimal user effort. Push notifications and in-app messaging inform customers of solutions. Proactive alerts when usage data indicate emerging pain points also boost satisfaction [30].

Continuous improvement of support experience—through tools, training, and community engagement—is as vital as resolving technical issues. Poor service recovery compounds product frustrations.

6.3 Continuous improvement process

Sustaining mobile service quality requires institutionalizing measurement and improvement as an ongoing capability versus isolated initiatives. Regular monitoring of metrics, journey mapping, and cross-functional reviews maintain visibility [33].

Quantitative analytics inform trends and benchmarks. Qualitative insights reveal human impacts. Technical and customer teams should collaborate closely on issues spanning software, hardware, design, and communications.

Improvement goals and projects related to reliability, usability, efficiency, and other dimensions drive progress. Results are validated through sustained metric improvements and user feedback. Enablers like knowledge management, communication rhythms, and continuous education sustain gains.

With rigorous measurement, systematic processes, and cross-functional coordination, mobile brands can deliver service quality on par with innovations in smart devices and applications. The following section concludes with key takeaways.

7. Conclusion

Mobile smart devices deliver capabilities that are unprecedented yet intricately woven into everyday life. As user dependence and spend increases, managing service quality is vital alongside introducing advances. This requires a comprehensive framework spanning technical and human elements.

Key dimensions like interactions, usability, availability, and reliability were outlined. Quantitative metrics and qualitative inputs enable multifaceted measurement. Customer journey mapping provides cross-channel insights. Closing gaps requires continuous processes integrating analytics, consumer data, and multidisciplinary improvement projects.

Further research can refine techniques for specific mobile services and use cases. Comparative benchmarking across demographics and device types would offer additional nuance. As technologies evolve, new dimensions may emerge around interfaces like augmented reality and brain-computer integration.

Nonetheless, the frameworks and best practices presented offer a robust starting point for mobile brands seeking to match service quality with product innovation. In the growing data economy, competitive advantage will be defined by experience delivery as much as smart features. By instilling user-centric service quality across the mobile customer journey, companies can establish durable bonds amid fickle consumers and fleeting technologies.

Author details

Abdulla Jaafar Desmal[1*] and Zainab Merza Madan[2]

1 University of Technology Bahrain, Bahrain

2 Bursa Uludağ University, Bursa, Türkiye

*Address all correspondence to: a.desmal@outlook.com

References

[1] Parasuraman A, Zeithaml VA, Berry L. A conceptual model of service quality and its implications for future research. Journal of Marketing. 1985;**64**(1):41-50. DOI: 10.2307/1251430

[2] Zeithaml VA, Berry LL, Parasuraman A. The Behavioral consequences of service quality. Journal of Marketing. 1996;**60**(2):31. DOI: 10.2307/1251929

[3] Parasuraman A, Zeithaml VA, Berry LL. SERVQUAL: A multiple-item scale for measuring consumer perceptions of service quality. Journal of Retailing. 1988;**64**(1):12-40. DOI: 10.1016/S0148-2963(99)00084-3

[4] Desmal AJ, Hamid S, Othman MK, Zolait A. A user satisfaction model for mobile government services: A literature review. PeerJ Computer Science. 2022;**8**:e1074. DOI: 10.7717/PEERJ-CS.1074/SUPP-3

[5] Desmal AJ, Othman MK, Hamid S, Zolait A. Exploring the information quality of mobile government services: A literature review. PeerJ Computer Science. 2022;**8**:e1028. DOI: 10.7717/PEERJ-CS.1028/TABLE-2

[6] Desmal AJ, Othman MK, Hamid S. The uniqueness of Mobile government service quality: A review on quality drivers. In: 2021 International Conference on Computer Science and Engineering (IC2SE), Padang, Indonesia. 2021. pp. 1-6. DOI: 10.1109/IC2SE52832.2021.9791955

[7] Babakus E, Boller GW. An empirical assessment of the SERVQUAL scale. Journal of Business Research. 1992;**24**(3):253-268. DOI: 10.1016/0148-2963(92)90022-4

[8] Buttle F. SERVQUAL: Review, critique, research agenda. European Journal of Marketing. 1996;**30**(1):8-32. DOI: 10.1108/03090569610105762

[9] Parasuraman A, Zeithaml VA, Berry LL. Reassessment of expectations as a comparison standard in measuring service quality: Implications for further research. Journal of Marketing. 1994;**58**(1):111-124. DOI: 10.1177/002224299405800109

[10] Delone WH, McLean ER. The DeLone and McLean model of information systems success: A ten-year update. Journal of Management Information Systems. 2023;**19**(4):9-30. DOI: 10.1080/07421222.2003.11045748

[11] Brady MK, Cronin JJ. Some new thoughts on conceptualizing perceived service quality: A hierarchical approach. Journal of Marketing. 2001;**65**(3):34-49. DOI: 10.1509/jmkg.65.3.34.18334

[12] Brynjolfsson E, Hu YJ, Rahman MS. Competing in the age of omnichannel retailing. MIT Sloan Management Review. 2013;**54**(4):1-7

[13] Rafiq M, Fulford H. Measuring internet retail service quality using E-S-QUAL. Journal of Marketing Management. 2012;**28**:1159-1173. DOI: 10.1080/0267257X.2011.621441

[14] Zeithaml VA, Parasuraman A, Malhotra A. Service quality delivery through web sites: A critical review of extant knowledge. Journal of the Academy of Marketing Science. 2002;**30**(4):362-375. DOI: 10.1177/009207002236911

[15] Choi KS, Cho WH, Lee S, Lee H, Kim C. The relationships among quality,

value, satisfaction and behavioral intention in health care provider choice: A south Korean study. Journal of Business Research. 2004;**57**(8):913-921. DOI: 10.1016/S0148-2963(02)00293-X

[16] Parasuraman A et al. A conceptual framework for understanding E-service quality: Implications for future research and managerial practice. Journal of Service Research. 2005. pp. 1-46 [Working paper]

[17] Massey AP, Khatri V, Montoya-Weiss MM. Usability of online services: The role of technology readiness and context. Decision Sciences. 2007;**38**(2):277-308. DOI: 10.1111/j.1540-5915.2007.00159.x

[18] Desmal AJ, Hamid S, Othman MK, Zolait A. Exploration of the usability quality attributes of mobile government services: A literature review. PeerJ Computer Science. 2022;**8**:e1026. DOI: 10.7717/PEERJ-CS.1026

[19] Kuniavsky M. Smart Things: Ubiquitous Computing User Experience Design. San Francisco, CA: Morgan Kaufmann; 2010. DOI: 10.1016/C2009-0-20057-2

[20] Lu Y, Zhang L, Wang B. A multidimensional and hierarchical model of mobile service quality. Electronic Commerce Research and Applications. 2009;**8**(5):228-240. DOI: 10.1016/J.ELERAP.2009.04.002

[21] Desmal AJ, Othman MK, Hamid S. Evaluation of Government Service Quality from the Prospective of Reliability. In: International Conference Technology and Entrepreneurship in Digital Society. Russia; Jan 2020. pp. 5-14. DOI: 10.17747/TEDS-2019-5-14

[22] Desmal AJ, Othman MK, Hamid S, Zolait AH. EGovemment service quality measurement scales: A literature review. In: 2019 International Conference on Innovation and Intelligence for Informatics, Computing, and Technologies, 3ICT 2019. Sakhier, Bahrain: Institute of Electrical and Electronics Engineers Inc.; 2019. DOI: 10.1109/3ICT.2019.8910293

[23] Chen SCGSD. Interpreting dimensions of consumer trust in E-commerce. Information Technology and Management. 2009;**4**(3):303-318

[24] Desmal AJ, Hamid S, Othman MK, Zolait AH. Proposed a model to evaluate mGovernment service quality from the perspective of Perceived Ease of Use (PEU). International Conference on Innovation and Intelligence for Informatics, Computing, and Technologies, 3ICT 2022. 2022. DOI: 10.1109/3ICT56508.2022.9990856

[25] Teng X, Pham H. Considerations for combining the uses of accelerates life testing and field service data in reliability and warranty analyses. Reliability Engineering and System Safety. 2006;**91**(9):1014-1021

[26] Jaafar A. Service Quality Framework for Mobile Government Services in Kingdom of Bahrain/Abdulla Jaafar Mohamed Hasan Ali. Kuala Lumpur, Malaysia: Universiti Malaya; 2022

[27] Ciurumelea A, Schaufelbühl A, Panichella S, Gall HC. Analyzing reviews and code of mobile apps for better release planning. In: 2017 IEEE 24th International Conference on Software Analysis, Evolution and Reengineering (SANER), Klagenfurt, Austria; 2017. pp. 91-102. DOI: 10.1109/SANER.2017.7884612

[28] Sauro J, Lewis JR. Quantifying the User Experience: Practical Statistics for User Research. 2nd ed. Morgan Kaufmann, UK: Morgan Kaufmann; 2016

[29] Brooke J. SUS: A 'quick and dirty' usability scale. Usability Evaluation In Industry. 2020;**189**(194):207-212. DOI: 10.1201/9781498710411-35

[30] Liu Y, Teichert T, Rossi M, Li H, Hu F. Big data for big insights: Investigating language-specific drivers of hotel satisfaction with 412,784 user-generated reviews. Tourism Management. 2017;**59**:554-563. DOI: 10.1016/j.tourman.2016.08.012

[31] Zhang Y, Fang Y, Wei KK, Ramsey E, McCole P, Chen H. Repurchase intention in B2C e-commerce - a relationship quality perspective. Information Management. 2011;**48**(6):192-200. DOI: 10.1016/j.im.2011.05.003

[32] Desmal AJ, Alsaeed M, Hamid S, Zolait AH. Feedback for Excellence: Improving mGovernment Services Through User Insights. In: 2023 International Conference on Innovation and Intelligence for Informatics, Computing, and Technologies, 3ICT 2023. 2023. DOI: 10.1109/3ICT60104.2023.10391312

[33] Rosenbaum MS, Otalora ML, Ramírez GC. How to create a realistic customer journey map. Business Horizons. 2017;**60**(1):143-150. DOI: 10.1016/j.bushor.2016.09.010

[34] McColl-Kennedy JR, Zaki M, Lemon KN, Urmetzer F, Neely A. Gaining customer experience insights that matter. Journal of Service Research. 2019;**22**(1):8-26. DOI: 10.1177/1094670518812182

[35] Desmal AJ, Bin Othman MK, Hamid SB, Zolait AH, Kassim NBA. Proposing a service quality framework for Mobile commerce. In: Lecture Notes of the Institute for Computer Sciences, Social-Informatics and Telecommunications Engineering. Bahrain: LNICST; 2019. pp. 203-212. DOI: 10.1007/978-3-030-23943-5_15

[36] Rahman AA, Helo P, Hao Y, Weldeghiorgis A. T. Cloud Manufacturing Provided Selection. Ideal value co-creation a DEMATEL-ANP based approach. 2015;90:338-349

[37] Maalej W, Kurtanović Z, Nabil H, Stanik C. On the automatic classification of app reviews. Lect. Notes Informatics (LNI), Proceedings of - Series Gesellschaft fur Informatik. 2017;**P-267**(3):61-62

Chapter 9

Optimising the Quality Management System in Dairy Processing

Niamh Burke and Mark Southern

Abstract

Milk has a solid reputation as a staple food since time immemorial. It is a complete food in its raw form, high in fat, protein, vitamins and minerals, including calcium. While the most beneficial first food for mammals is mammalian milk until weaning, cow's milk and dairy derivatives are considered significant nutritional components in the human diet. While milk consumption has in fact sharply declined in recent decades, the consumption of liquid milk derivatives and dairy products has steadily increased. Quality in terms of product, process and the environment in a milk production plant can be measured through performance, reliability and durability. The quality management system, in whatever form that may take within a plant, is the pinnacle in ensuring how one organisation can differentiate from its competitors. Quality systems and analytical testing protocols, especially in the dairy industry, are seldom quantified or fine-tuned to guarantee their efficiency. Furthermore, the impacts of quality systems on process, product, and environmental optimisation are frequently overlooked. This chapter reviews the activities that allow for the optimisation of quality systems in a dairy processing environment. The outcomes of which highlight the importance of process based quality systems.

Keywords: dairy processing quality systems, quality optimisation, continuous improvement, process optimisation, process control

1. Introduction

Definitions of quality within the literature are manufacture and user based. W. Edwards Deming believed the definition of quality has a user based perspective in that the product fulfils the needs and expectations of the customer [1]. Philip Crosby believed the system of quality, was prevention, and conformance to requirements and was manufacturing based ([2], p. 116). A mixture of both concepts derived from Deming and Crosby can be highlighted through David Garvin's perspective on the quality concept. He identified five approaches: *The transcendent,* a perspective that states quality cannot be defined exactly. *The product based view*, a perspective that states quality can be measured exactly through the desirable characteristics a product should possess. *The user based view,* stating that the customer judges quality. *The manufacturing based approach,* conveys production requirements and finally *the value-based approach*, whereby quality is defined in relation to cost and price [3]. All of these principles are easily applicable to the dairy industry.

Increasing consumer demands for high quality dairy products puts a burden on dairy producers to manufacture top grade product that is free from microbiological, chemical and physical contaminants. To comply with regulatory guidelines routine inspections are carried out from milk intake to finished product [4]. While milk composition is nutritionally important to consumers, its composition and safety is equally important to milk producers and processors [5], as the quality demands of the product will influence processing parameters and final yield [6]. Inspections are based on randomised sampling and process control samples, where analysis can occur hourly on the production line and on bags of finished product, all of which are outlined in a company's SOP and analytical test matrices. Test matrices are simple tools for visualising how often and for what analysis, a sample of product should be tested.

Total quality management (TQM), is an application of quality management principles that are applied to all aspects of a business. The food safety management system (FSMS) is an integral part of the TQM in any food manufacturing company [7], including dairy manufacturing plants [8], examples of which include, HACCP, FSSC (Food Safety System Certification Standard), IFS (International Food Safety Standard), SQF (Safe Quality Food) and ISO (International Standards Organisation). HACCP [9], is a global quality assurance system, ensuring products are produced in the safest way possible and are compositionally stable for product formulations [10]. The HACCP system is widely used across the Irish dairy industry. TACCP (Threat Assessment Critical Control Point) and VACCP (Vulnerability Assessment Critical Control Point) are relatively new concepts. Where HACCP prevents food safety issues [11], TACCP identifies deliberate infidelities to a food product [12], finally VACCP assesses how vulnerable various points are along the process chain [13]. The general scope of VACCP is that raw material supply chains are protected, and TACCP protects all aspects contributing to raw materials from intake to consumer; both go hand in hand [13].

Significant gaps however, can be seen across the dairy FSMS globally. For example, melamine is a crystalline compound made by heating cyanimide [14], in 2008, the Sanlu Dairy Company in China were responsible for adding melamine to watered-down milk. Melamine is rich in nitrogen and the addition of the poisonous substance was to increase the apparent protein content in the product. More than 290,000 people, mostly infants ingested the contaminated product, six infants died, before the Chinese government issued a product recall [15]. More recently, Lactalis, a multinational French dairy corporation, was involved in a contaminated milk scandal at the end of 2017, when infants ingested formula which was contaminated with *Salmonella agona* at their production plant in Craon [16]. Both of these scandals involved flawed quality management systems whereby production operators, laboratory technicians and management were aware of contamination issues, without taking steps to control production.

To ensure high yield and exceptional quality [17], quality systems in any food industry must be established from raw materials right through to finished product [18], with sampling inspection plans [19]. This is achieved through analytical testing and sampling strategies [20] which are governed by legislative and regulatory guidelines and standards. Generally, within the Irish dairy industry inspection is based on EU Directives and Regulations. The EU Regulation (EC) No. 178/2002, states that it is necessary to consider all aspects of the food chain as a continuum to ensure food safety, furthermore, it states scientific risk assessment alone cannot guarantee food safety along the production chain and societal factors, environmental factors and *'feasibility of controls'* should also be considered [21]. Guidelines on sampling can be product or process based [22] and are routinely carried out from milk intake to finished product.

2. Dairy processing and quality

Quality control for a safe product must be the objective of any food processing industry. In reality, the universal objectives to any process are minimising cost and maximising throughput, and/or efficiency ([23], p. 205). A network of interrelated activities that are repeated over time can be defined as a process ([24], p. 456), adjusting the process so as to improve specific parameters is termed 'process optimisation'. Additionally, many factors can relate to the performance of a process, for example the condition of a machine/instrument, processing parameters, environmental variables and material quality ([23], p. 211; [25]). Food processing is defined as any procedure undergone by food commodities, after they have left the primary producer and before they reach the customer [26]. In terms of dairy production, processing can take many forms and vary in the degree of complexity, i.e. from fluids to solids. In its simplest form, processing may involve no more than controlled storage such as refrigeration. At more complex levels, commodities may be processed to yield ingredients that are later combined to yield foodstuffs, for example, the conversion of liquid skim milk to skim milk powder. The revolutionary changes in manufacturing and the advancements of the dairy industry over the past 30 years have increased the need for optimised quality assurance and quality control methodologies. 'Total quality control (TQC) is a thought revolution in management' [27] and is deemed as a continual practice [1]. It is a flexible system whereby both processes and methods can be easily transformed. Jagdev et al. [23] explored a range of techniques to enhance process optimisation and concluded that a knowledge based system, analysing historical and real-time events on an on-going basis, enables the identification of any variations likely to present within a process, which would contribute to overall process optimisation. Process optimisation and enhanced quality, affects the success of a manufacturing plant in terms of higher productivity and reduced costs due to out of specification (OOS) product and downgrades [28].

3. Techniques to optimise quality

Research findings suggest the basis for quality improvement needs to be understood by everyone at all levels of production ([24], p. 460; [17, 28]), for instance in a milk production plant, from milk intake all the way through to finished product. Dr. Kaoru Ishikawa defined *'the seven improvement tools'* in response to the realisation that everyone in a company had to become involved in improvement works, thus they needed to be simple and effective tools [27]. The set of tools within the toolbox methodology can vary across the literature; however, the end objective is always the same whereby they are systematically designed for problem solving and optimisation. As the 'Define' phase of the Six Sigma methodology states, the collection of data is a fundamental step in the road to quality improvements [29], Ishikawa's Seven Improvement Tools also demonstrated the importance of this step [27]. Histograms and Pareto charts, can be used as visualisation tools, as an important part of data analytics is to represent data efficiently and that can help in the decision of which problem should be solved first [30]. Once data has been analysed the root cause of problems within a process has to be defined. This can be achieved through the use of a cause and effect diagram [27, 29]. A tool that allows the description of causes that can possibly lead to quality problems and allows for the investigation of each cause, one problem at a time. One common problem however, with the use of cause and effect diagrams, is that a quality team can start with the major problems while the root causes of the minor ones become harder to solve [24, 31]. Stratification is another problem solving tool within the seven improvements toolbox, based on classifying data from different sources into subgroups and obtaining important information for improvement work [27]. Scatter plots can be used when stratification cannot, to identify process or product variations due to explanatory variables [32]. Finally, the use of control charts in quality optimisation has two main objectives, the first being to identify causes of variations and secondly to quickly detect when variations occur in a stable process [31].

4. Laboratory activities

A vital part of any quality system are laboratory quality standards [33, 34], and an increasing number of laboratories within the dairy industry are operating to accreditation standards. Calibration and testing activities can either be partially or fully accredited [35]. In Ireland, accreditation can be granted by the Irish National Accreditation Board (INAB), a voluntary scheme open to laboratories within the country, which offers accreditation to the international standard outlaying requirements on testing, personnel, confidentiality, laboratory facilities and environmental conditions [36]. In recent times, clinical laboratories have considered laboratory turnaround time (TAT) an important key performance indicator (KPI) of laboratory performance [37], defined as *the time it takes for a test to be ordered to the time it takes for a result to be delivered*. This is a topic however, that has not been widely researched in relation to food, and more so dairy manufacturing. Activities are carried out within dairy manufacturing plants with little consideration to downtime or the effect of sampling and analysis on personnel labour hours [34]. A study carried out in 2014, on methods of identifying unnecessary laboratory testing in a tertiary care hospital deduced that a user approach as well as a systems approach would ultimately reduce over-testing [38]. Within Dairy manufacturing plants however, over testing is more times than not, due to out of specification test results (OOS), defined as a test result that falls outside of the established testing criteria [39]. While an initial OOS result does not mean a failed batch, the concept of 'testing into specification' rather than applying an investigative approach to determine the root cause of such a result, often leads to unnecessary over analysis, ultimately negatively affecting cost and laboratory TAT. Investigating OOS test results should be thorough and timely and include the use of scientific decisions, justifications and risk based analyses [40]. Furthermore, applying a strategic in-process sampling system would minimise the level of OOS in final product. In-process sampling should always start at the beginning of the process to allow for early detection of processing issues. Finally, managing test results can be a complex process if it is not performed in an accurate manner. Improving how test results are recorded and communicated can achieve savings in staff time and improve the overall reliability and communication of the results. OOS results can be due to a number of reasons including failed equipment or reagents, in process issues and human error.

5. Advancing quality in dairy processing

The global dairy industry is fast evolving, as are trends in consumption patterns. The issues surrounding environmental sustainability and increased production demands are placing pressures on the dairy manufacturing industry to produce product in adherence to regulatory bodies and consumer requirements, as well as keeping costs to a minimum. Manufacturing product that is chemically and microbiologically safe for human consumption however, is the core of any production plant and ensuring that efficient quality systems are put in place to capture this is vital. The literature suggests however, that flaws can, and do exist within quality systems and this is evident within the dairy manufacturing industry. There are areas for optimisation in quality systems within the dairy industry. Trend analytics of historic data and making scientific and industry contrasts concerning quality systems will enable dairy manufacturers to become more in control of processes, which will effectively drive costs down.

To ensure high quality products are delivered to the consumer, the dairy manufacturer must evaluate milk quality throughout the value chain and be strategic in their approach to minimising product contamination and upholding quality. De Silva et al. [41] strongly recommended that following good management practices (GMP) will uphold product quality. GMP is described by the Codex Alimentarius; hygiene in the primary production, hygienic design of equipment and facilities, control of operations, maintenance and sanitation practices, personal hygiene, transportation, product information and consumer awareness and training [42]. Based on the PDCA (Plan, do, check, action) cycle, the implementation of GMP is a continuous process [43]. The roadmap to implementing GMP is led by strategic management where there are three basic elements; (i) The formation of a strategy, (ii) the implementation of a strategy and (iii) the control and evaluation of a strategy [44]. Considering the many variables involved in the milk processing and the effects they can have on quality and composition, having a strategic quality management system in place is critical. Moving from a product based testing regime to a process focused approach, has been widely adopted by the food and pharma industry in the last decade. An exclusive product based inspection system has proven to be an inefficient method of regulation, however a poorly monitored process based level of inspection can be just as incompetent.

An effective quality management system (QMS) will document processes, procedures and responsibilities, that will enable a company to meet customer and regulatory requirements [17]. Additionally, the effective management of data will facilitate a quality team's decision making surrounding process and product variability. According to Strong et al. [45], the barriers to data accessibility are poor systems, the timeliness of inputting data and accessibility to data systems.

The successful management of quality and traceability is dependent on a company's ability to demonstrate product quality and safety to regulators, customers and consumers. Influences on process optimisation and quality within the dairy industry can be multi-factorial including: seasonality; the quality of incoming milk and herd health; the level of skilled laboratory technicians; the level of production and the digital capabilities of a dairy processing company can all have an impact on optimising the quality management system.

6. Conclusion

This chapter critically assessed the factors that affect optimal dairy quality systems. The basic role of continuous improvement is – there is always a way to get improved quality. This review highlighted the necessity for heightened awareness needed among dairy processors to allow for a synergistic affect that all aspects of the quality system have on product quality and process performance. It also highlighted the value in-process based sampling has on final product output within a dairy plant as well as monitoring the efficiency of the process in terms of waste and energy usage. Finally, this review encourages any dairy manufacturer to consider all inputs and outputs of their entire product line to achieve high quality product and uphold brand reputation.

Funding

This research was funded by the Irish State through funding from the Dairy Processing Technology Centre's program - Grant Number TC/2020/028.

Author details

Niamh Burke* and Mark Southern
Dairy Processing Technology Centre (DPTC), University of Limerick, Limerick, Ireland

*Address all correspondence to: niamh.burke@dptc.ie

References

[1] Shewhart WA, Deming WE. Statistical Method from the Viewpoint of Quality Control. Courier Corporation. New York: Dover Publications; 1986

[2] Crosby PB. Completeness: Quality for the 21st Century. Penguin. United States: Dutton Publications; 1992

[3] Garvin DA, Quality WD. Really mean. Sloan Management Review. 1984;**25**:25-29

[4] Burke N, Zacharski K, Adley CC, Southern M. A comparison of analytical test methods in dairy processing. Food Control. 2020;**107637**

[5] NRC. Designing Foods; Animal Product Options in the Marketplace. Washington DC: National Academies Press (US); 1988

[6] Miciński J, Kowalski IM, Zwierzchowski G, Szarek J, Pierożyński B, Zabłocka E. Characteristics of cow's milk proteins, including allergenic properties and methods for its reduction. Polish Annals of Medicine. 2013;**20**(1):69-76. DOI: 10.1016/j.poamed.2013.07.006

[7] Chen E, Flint S, Perry P, Perry M, Lau R. Implementation of non-regulatory food safety management schemes in New Zealand: A survey of the food and beverage industry. Food Control. 2015;**47**:569-576. DOI: 10.1016/j.foodcont.2014.08.009

[8] Devitt C, McKenzie K, More S, Heanue K, McCoy F. Opportunities and constraints to improving milk quality in Ireland: Enabling change through collective action. Journal of Dairy Science. 2013;**96**(4):2661-2670

[9] FSIS U. The Hazard Analysis and Critical Control Point (HACCP) System and the Food Safety Inspection Service: Concept Paper. Washington DC: US Department of Agriculture; 1989

[10] Vega C, Roos Y. Invited review: Spray-dried dairy and dairy-like emulsions—Compositional considerations. Journal of Dairy Science. 2006;**89**(2):383-401

[11] Al-Kandari D, Jukes DJ. Incorporating HACCP into national food control systems-analyzing progress in the United Arab Emirates. Food Control. 2011;**22**(6):851-861

[12] Bogadi NP, Banović M, Babić I. Food defence system in food industry: Perspective of the EU countries. Journal für Verbraucherschutz und Lebensmittelsicherheit. 2016;**11**(3):217-226

[13] Manning L, Soon JM. Food safety, food fraud, and food Defense: A fast evolving literature. Journal of Food Science. 2016;**81**(4):R823-R834

[14] Venkatasami G, Sowa JR Jr. A rapid, acetonitrile-free, HPLC method for determination of melamine in infant formula. Analytica Chimica Acta. 2010;**665**(2):227-230

[15] Xiu C, Klein K. Melamine in milk products in China: Examining the factors that led to deliberate use of the contaminant. Food Policy. 2010;**35**(5):463-470

[16] Reuters. The Baby Milk Scandal Stalking France's Lactalis. 2018. Available from: https://www.reuters.com/article/uk-france-babymilk-lactalis-explainer/the-baby-milk-scandal-stalking-frances-lactalis-idUKKBN [Accessed: July 17, 2018]

[17] Trienekens J, Zuurbier P. Quality and safety standards in the food industry, developments and challenges. International Journal of Production Economics. 2008;**113**(1):107-122

[18] Aung MM, Chang YS. Traceability in a food supply chain: Safety and Quality perspectives. Food Control. 2014;**39**:172-184

[19] Santos-Fernández E, Govindaraju K, Jones G. A new variables acceptance sampling plan for food safety. Food Control. 2014;**44**:249-257. DOI: 10.1016/j.foodcont.2014.03.051

[20] Truchaud A, Le Neel T, Brochard H, Malvaux S, Moyon M, Cazaubiel M. New tools for laboratory design and management. Clinical Chemistry. 1997;**43**(9):1709-1715

[21] EU (2002) 'No 178/2002 of 28 January 2002 Laying down the General Principles and Requirements of Food Law, Establishing the European Food Safety Authority and Laying Down Procedures in Matters of Food Safety.' European Union Regulations, European Parliament and the Council of the European Union

[22] Torkko M, Katajavuori N, Linna A, and Juppo A. Quality KPIs in pharmaceutical and food industry. European Union Regulations. European Parliament and the Council of the European Union; 2013;**8**:205-211. DOI: 10.1007/s12247-013-9159-9

[23] Jagdev HS, Browne J, Keogh J. Manufacturing process optimisation — A survey of techniques. In: Davies BJ, editor. Proceedings of the Twenty-Eighth International: Matador Conference. London: Macmillan Education UK; 1990. pp. 205-215. DOI: 10.1007/978-1-349-10890-9_29

[24] Bergman B, Klefsjö B. Quality from Customer Needs to Customer Satisfaction [Online]. 3rd ed. Lund: Studentlitteratur AB. Available from: http://urn.kb.se/resolve?urn=urn:nbn:se:ltu:diva-16592; 2010 Accessed: 29 September 2016

[25] Davídek J. In-process quality control. Food Quality and Standards-Volume II. 2009;**10**:48

[26] Chawla R, Arora JS, Dubey RK, Mukhopadhyay CS. Omics approaches and applications in dairy and food processing technology. In: Omics Technologies and Bio-Engineering. Amsterdam: Elsevier; 2018. pp. 271-295

[27] Ishikawa K, Loftus JH. Introduction to Quality Control. 3A Corporation Tokyo. United Kingdom: Chapman & Hall; 1990

[28] Juran JM. Juran on Leadership for Quality. Simon and Schuster. New York: Free Press; 2003

[29] Pyzdek T, Keller PA. The Six Sigma Handbook. New York, NY: McGraw-Hill Education; 2014

[30] Juran JM. Juran on Planning for Quality. New York: Collier Macmillan; 1988

[31] Hung H-C, Sung M-H. Applying six sigma to manufacturing processes in the food industry to reduce quality cost. Scientific Research and Essays. 2011;**6**(3):580-591

[32] Smith B. Six-sigma design (Quality control). IEEE Spectrum. 1993;**30**(9):43-47

[33] Barbé B, Yansouni CP, Affolabi D, Jacobs J. Implementation of quality management for clinical bacteriology in low-resource settings. Clinical Microbiology and Infection. 2017;**23**(7):426-433

[34] Burke N, Southern M, Ryan A, Adley CC. Investigating the current skim milk powder inspection strategies for improvements in process optimisation. Food Control. 2018;**94**:17-21

[35] Karthiyayini N, Rajendran C. Critical factors and performance indicators: Accreditation of testing-and calibration-laboratories. Benchmarking: An International Journal. 2017;**24**(7):1814-1833

[36] ISO. ISO/IEC 17025:2017 General Requirements for the Competence of Testing and Calibration Laboratories. Geneva, Switzerland: International Organisation for Standardisation; 2017

[37] Goswami B, Singh B, Chawla R, Gupta VK, Mallika V. Turn around time (TAT) as a benchmark of laboratory performance. Indian Journal of Clinical Biochemistry. 2010;**25**(4):376-379. DOI: 10.1007/s12291-010-0056-4

[38] Khalifa M, Khalid P. Reducing unnecessary laboratory testing using health informatics applications: A case study on a tertiary care hospital. Procedia Computer Science. 2014;**37**:253-260

[39] Kuselman I, Pennecchi F, Fajgelj A, Karpov Y. Human errors and reliability of test results in analytical chemistry. Accreditation and Quality Assurance. 2013;**18**(1):3-9

[40] Devi NG, Chandramouli R. Real time release testing-a review. Journal of Pharmaceutical Research. 2018;**16**(4):314-318

[41] De Silva S, Kanugala K, Weerakkody N. Microbiological quality of raw milk and effect on quality by implementing good management practices. Procedia Food Science. 2016;**6**:92-96

[42] Codex (2003) 'Recommended International Code of Practice General Principles of Food Hygiene. Italy: Codex Alimentarius Commission; 2003. CAC/RCP 1-1969, Rev. 4-2003'

[43] Costa Dias MA, Sant'Ana AS, Cruz AG, Faria J d AF, Fernandes de Oliveira CA, Bona E. On the implementation of good manufacturing practices in a small processing unit of mozzarella cheese in Brazil. Food Control. 2012;**24**(1):199-205. DOI: 10.1016/j.foodcont.2011.09.028

[44] Houben G, Lenie K, Vanhoof K. A knowledge-based SWOT-analysis system as an instrument for strategic planning in small and medium sized enterprises. Decision Support Systems. 1999;**26**(2):125-135. DOI: 10.1016/S0167-9236(99)00024-X

[45] Strong DM, Lee YW, Wang RY. Data quality in context. Communications of the ACM. 1997;**40**(5):103-110

Section 3

Applications of QA/QC in Healthcare

Chapter 10

The Importance of Quality Assurance in Rural Public Healthcare Facilities

Takalani Rhoda Luhalima

Abstract

This chapter aims to discuss the importance of quality assurance in public rural healthcare facilities. In developing countries, there is a massive problem regarding poor health services. Inadequate resources, coupled with the waning of the Ubuntu philosophy in public rural healthcare facilities, led to a lot of complaints and litigations. However, improved healthcare quality can lead to fewer errors, shorter waiting times, higher efficiency, increased market share, and cheaper costs. Besides, Ubuntu as a reflection of a person's willingness to share, to give generously, to care, and to be full of compassion enhances quality assurance. A positive work environment and staff's positive attitude when providing quality care are effective approaches that augment patient satisfaction. The importance of providing quality patient care is nonnegotiable in all healthcare facilities. Quality assurance and quality control using the fundamental standards in place are the responsibility of leadership management. The art of leadership and the organizational culture determine and strengthen quality in healthcare facilities. Therefore, the chapter covers the significance of compliance in service standards applications to uphold quality assurance. Furthermore, the chapter addresses the assessment of staff capabilities, describes healthcare needs, resources, and maintenance approaches, explores strategies for Ubuntu restoration, identifies challenges, barriers, and poor services, and recommends improvement interventions. A literature review or desktop approach was used.

Keywords: healthcare facility, public health, rural healthcare, quality, quality of care, quality assurance

1. Introduction

This chapter aims to explain the importance of quality assurance in public rural healthcare facilities. By closely observing each step of the delivery or production process, quality assurance (QA) maintains the required standard of quality for a service or product [1]. Hence, in this study, quality assurance means providing quality patient care in a comprehensive manner, using available resources in place, and complying with the nursing norms and standards set. It involves administrative and procedural activities to achieve requirements and goals, ensuring systematic measurement, comparison with standards, and monitoring processes for error prevention. However, healthcare facilities are faced with several quality challenges such as underuse, overuse, preventable mistakes, differences in services, a lack of resources, inadequate diagnosis and treatment, inefficient resource usage, inadequate information, referral systems, disrespect for human dignity, a lack of drugs, poorly maintained records, and poor delivery systems.

All these mentioned challenges lead to various lawsuits that healthcare facilities are faced with. Hence, the author is interested in the importance of QA in public rural healthcare facilities. The objectives of the chapter are outlined below.

2. Objectives of the chapter

This chapter:

- defines key concepts.
- assess the capabilities of staff members who can manage the QA office.
- describes healthcare needs, resources, and assessment approaches for quality maintenance.
- explores strategies for restoring Ubuntu in healthcare facilities.
- Identifies challenges faced by relatives, patients, and quality assurance managers handling complaints.
- explains barriers for maintaining quality care in healthcare facilities.
- describes the effects of poor healthcare services.
- explains the improvement interventions.

3. Background

In many instances, this question is raised. "Why is QA important in healthcare facilities"? A robust healthcare QA program enables immediate identification of patient care and safety issues. Besides, high-quality health services provide appropriate care, timely delivery, and responsiveness to users' needs, minimizing harm, and resource waste. Two components of quality management are quality assurance (QA) and quality control (QC), with QA activities spanning the full quality system and QC being a subset of QA activities [2]. Seven measurable qualities of these services exist efficacy, safety, people-centeredness, timeliness, equity, integration, and efficiency. Additionally, "Ubuntu" should be accepted as a fundamental attribute in healthcare, representing humanity, hospitality, generosity, and empathy; all of which are necessary for providing patients and others with high-quality care [3]. An African traditional concept known as "Ubuntu" emphasizes the importance of taking patients' cultural backgrounds into account when practicing ethical medicine [4]. It encourages collective moral principles and unity. According to Oleribe et al. [5], the healthcare system comprises six core components: service delivery, healthcare workforce, information system, medicines and technology, financing, and leadership. All these need to be provided in line with the service standards that promote quality improvement in all healthcare facilities globally. However, due to constraints of poorly resourced public rural healthcare facilities mostly in developing countries, quality care is compromised [6].

Addressing the strain on African hospitals for critically ill patients requires capacity building and early interventions to reduce organ dysfunction progression and reduce strain on poorly resourced hospitals. The poor resources compromise the quality of patient care. Studies have shown that challenges faced by nurses working with constrained resources, erratic staffing schedules, and unpredictably high patient volume and acuity levels are common in rural settings [7, 8]. Similar problems affecting several states have been reported on the quality of care and patient safety at hospitals in under-resourced situations; according to a study conducted in Uganda by Kakyo [9], including issues patient-staff ratio is incredibly low, a lack of necessary medications and equipment, low salaries, and a high level of workload. Besides, the study conducted by Manyisa and van Aswegen [10] revealed that a lack of administrative equipment and well-trained professionals negatively impacts healthcare quality. The definitions of the concepts will be presented below.

4. Definition of concepts

4.1 Health facilities

Health facilities encompass various healthcare establishments such as hospitals, clinics, outpatient care centers, and specialized hospitals [11]. In this chapter, health facilities refer to all healthcare facilities that provide healthcare services.

4.2 Public health

All organized efforts, whether governmental or private, to prevent illness, enhance health, and increase lifespan are referred to as public health [12]. In this chapter, public health refers to measures used to promote quality of care.

4.3 Rural health

Rural health refers to the health of individuals living in rural areas, which are typically further from medical facilities and amenities compared to urban areas [13]. In this chapter, rural health refers to the health of individuals residing in rural areas, often located further from healthcare facilities.

4.4 Quality

According to Ref. [14], quality is described as measurable, effective, efficient, accessible, acceptable, equitable, and safe services. In this chapter, quality refers to achieving goals with available resources.

4.5 Quality of care

Quality of care refers to the effectiveness of health services in achieving desired health outcomes and aligning with current professional knowledge [15]. In this chapter, quality of care refers to achieving clients' needs with available resources. This includes equity, accessibility, comprehensiveness, continuity, privacy, and confidentiality when providing quality patient care.

4.6 Quality assurance

Quality assurance is any systematic process of determining whether a product or service meets specified requirements [16]. In this chapter, QA in healthcare encompasses the activities and series of measures that are designed to ensure continuous improvement in the quality of patient care.

4.7 Quality control

Quality control is the process that ensures that the stipulated standards are adhered to always [17]. In this chapter, QC means monitoring healthcare provided, healthcare providers, and resources to improve or maintain quality standards.

The author used desk research methodology technique for compiling and analyzing data from secondary sources that are easily accessible such as books, articles, reports, academic publications, and other materials that were obtained online [18]. The author employed a desk study as an initial appraisal process that gathers and reviews existing site information to guide the subsequent site search to discuss the importance of quality assurance in rural public healthcare facilities.

5. Content of the chapter

5.1 The assessment of the capabilities of staff members who can manage the quality assurance office

The assessment of the staff members who can manage the QA office depends on the culture and the goals or objectives of the organization. Being appointed as AQ manager is very important because this is attached to greater responsibility and accountability regarding the reputation of the organization.

Therefore, the person should have experience in clinical practice and quality assurance and be passionate about working with all stakeholders both internal and external. Furthermore, the manager should embrace Ubuntu's philosophy to work with all stakeholders with Ubuntu as a global concept derived from a South African Zulu word, promoting love, caring, sharing, respect, kindness, compassion, and cooperation. It encourages equal interaction without feeling less human.

Managers are crucial in maintaining control and strategic decisions in the quality assurance department. They ensure that the department operates at high standards, complies with corporate rules, and meets quality requirements. Their primary tasks include client satisfaction, collaboration with subsidiary companies, and maintaining project quality standards [19].

Quality assurance manager ensures successful implementation in:

- Coordinating meetings and interviews with employees.
- Analyzing project standards and ensuring technical solutions of the departments are aligned with QA standards.

- Communicating effectively with employees and technical implementers.
- Resolving issues.
- Addressing project and design issues.
- Allocating test budgets and preventing losses.
- Establishing quality standards.
- Monitoring operations.
- Assessing customer expectations.
- Conducting inspections and collecting data for proactive solutions.

5.2 Healthcare needs, resources, and assessment approaches for quality maintenance

Despite the importance of quality, managers must be active in patient care delivery. Managers must prioritize numerous quality aspects, take into account the demands of providers and patients, and manage resources given to them by the government, commercial entities, and the community. It is critical to address resource distribution, fee schedules, personnel patterns, and management procedures.

Quality, as a multidimensional notion, is especially useful for managers who prioritize access, effectiveness, technical competence, and efficiency. The QA, on the other hand, involves monitoring healthcare professionals' provision of patient care using cost-effective resources and motivating nurses to deliver excellence. It involves nurses taking accountability for their care and ensuring services are regulated by professional members.

Quality assurance aims to create an environment that produces consistent, reliable, and high-level products, detects challenges, and corrects problems that may affect quality. It is a proactive process that prevents quality failures by setting core standards and ensuring compliance. Procedure and policy manuals, healthcare standards, and the South African Nursing Council's scope of practice guide healthcare staff in providing quality patient care. However, the strategies were put in place with limited positive results. Patient satisfaction is not met. There are a lot of complaints in most rural public healthcare facilities that lead to various litigations. The study by Madaka [20] reveals that despite numerous quality improvement initiatives, South Africa's healthcare facilities have not achieved high-quality service delivery, posing a challenge for the government to ensure long-lasting improvement. Below are the challenges faced by relatives, patients, and quality assurance managers handling complaints.

5.3 Challenges faced by relatives, patients, and quality assurance managers handling complaints

The two components of quality management are quality assurance and quality control, with QA activities encompassing the full quality system and QC being a subset of QA activities [21]. The evaluation of healthcare quality is becoming more and more important in developing nations, with outcomes serving as a major indicator. Some African nations are creating national quality programs and policies, such as South Africa's National Policy on quality in Healthcare and Zambia's National Quality Assurance program [22, 23]. Ghana is putting into practice a national health quality program, and it is anticipated that quality assurance systems will become essential to the regular provision of health services. Through the implementation of quality assurance systems, Nigeria's health sector reform seeks to increase access to high-quality healthcare. All reform plans and programs must include quality improvement measures according to the Federal Ministry of Health. To establish a quality culture in healthcare systems, health professionals must participate in the quality program and be aware of the required behaviors. An example was the study conducted by Romdhane et al. [24], which showed three key problems in Africa's healthcare system:

- Insufficient human resources.
- Insufficient financial allocations.
- Poor management and leadership.

More than two-thirds of the perceived problems are attributed to these obstacles, which focus on leadership and governance, the healthcare workforce, the provision of health services, and financing [5]. Furthermore Broadhurst et al. [22], revealed that inadequate training facilities, brain drain, and adverse governmental policies are overt problems. Another significant issue is inadequate funding with low budgetary allocation and poverty being the two main financial obstacles that African healthcare systems must overcome. The approaches for restoring Ubuntu in healthcare facilities are described below.

5.4 Restoring Ubuntu in healthcare facilities

Ubuntu relates to humanity. It is a worldwide concept derived from a South African Zulu word. The concept is based on a saying in Zulu, which states that "umuntu ngomuntu Ngabantu" which means I am because you are Um and Lau [25]. In this chapter, "Ubuntu" refers to embracing love, caring, sharing, respect, kindness, compassion, and cooperation in promoting quality assurance. It is very difficult to ensure that QA is maintained in rural health facilities without embracing the Ubuntu philosophy. Therefore, incorporating Ubuntu in daily activities by health workers is very important.

Caring in nursing and Ubuntu have many parallels because both emphasize caring and other caring ideals, such as respect and dignity. The notion that "a person is a person through other people," which is a direct translation of practically all African languages, is at the center of Ubuntu moral theory, and is known in Tshivenda as "Muthu ndi Muthu Nga Munwe." Ubuntu ethics encourage people to share their belongings for the benefit of all and to care for all people.

The study conducted by Maphumulo and Bhengu [18] mentions Ubuntu as what is ethically decent for individuals and communities, and it is assessed by dignity, respect, and contentment to promote the well-being of others, all of which are frequently shared fairness ideas. Hence, the following Ubuntu principles need to be of value. Ubuntu principles can help clinicians gain a deeper understanding of patients and others in their daily lives. These include humaneness, empathy, compromise, learning, change, renewal, restorative justice, love, spirituality, and hope. By incorporating these concepts, clinicians can develop relationships with patients, and harnessing their energy to be the healers, they are meant to be. By embracing these principles, clinicians can create a more compassionate healthcare system.

On the other hand, QA integrated with Ubuntu is a crucial aspect of quality management, ensuring the fulfillment of quality requirements for both management and external stakeholders. It involves systematic activities within the quality system to demonstrate confidence in a healthcare service. For instance, [19] found that Ubuntu can affect how nurses provide care. It involves nurses taking accountability for their care and ensuring services are regulated by professional members. Furthermore, QA involves monitoring healthcare professionals' provision of patient care using cost-effective resources and motivating nurses to deliver excellence showing Ubuntu throughout the caring practices.

Quality assurance aims to create an environment that produces consistent, reliable, and high levels of quality patient care, detects challenges, and corrects problems that may affect quality. It is a proactive process that prevents quality failures by setting standards and ensuring compliance providing quality patient care using autonomy in whatever decision is taken.

5.5 Barriers to maintaining quality care in healthcare facilities

Financial obstacles to healthcare services with high rates of out-of-pocket spending are further issues with the continent's healthcare system that are common [20]. These issues are caused by an ineffective national health insurance system and insufficient service integration. According to some research, nurses working in rural locations frequently deal with few resources, erratic staffing schedules, and unpredictably high patient census and acuity levels [5, 8].

Inadequate resources: A particular level of output is unachievable with inadequate infrastructure, tools, funding for ongoing costs, and employees.

Personnel issue: The standard of treatment is impacted by a lack of properly trained, skilled, and motivated staff members.

Patients and staff who are unreasonable: Illness, anxiety, delayed response to therapy, unreasonable, and uncooperative behavior, which, in turn, impacts the quality of care.

Poor maintenance: To be used effectively, buildings and equipment need to be properly maintained.

Lack of an informed public: In order to increase the quality of nursing care, it is important for individuals to become informed and assert their legal and ethical obligations to receive high-quality care.

Lack of accreditation regulations: Hospitals are not adequately inspected to make sure that the minimum requirements are met: Investigate significant instances of neglect and punish the negligent health providers.

Lack of incident review procedures: Several occurrences that may occur, while a patient is hospitalized and have an impact on their care and eventual recovery, include wrong medicine, burns caused by substandard practices, or equipment.

Delayed attendance by doctor or nurse: A patient dies in a hallway without a doctor or nurse present.

Not enough effective healthcare information systems: An effective management information system is crucial for evaluating the quality of care.

Patient satisfaction surveys not conducted: surveys that will be conducted by social workers, hospital management students, and consultant groups using questionnaires, interviews, etc.

Lack of nursing care records: Nurses should consistently record the care they provide using the nursing process or a problem-oriented record system.

Lack of oversight: Absence of supervision, ignorance of the philosophy of nursing care, absence of a policy and administrative manual, absence of a process manual, inadequate quality and quantity of professionals, and subpar education and training.

Poor management: Lack of evaluation methods, ineffective departmental and cross-departmental collaboration, absence of formal job descriptions, and job specifications.

Inadequate programs for nurse education: Lack of continuous education and in-service training opportunities.

5.6 The effects of poor healthcare services

Poor healthcare services are increasing illness burdens and health costs globally, hindering progress in all income levels [12]. Besides, the study conducted by Mosadeghrad [26] showed that poor healthcare services failure endangers patients' health and safety. Therefore, QA is crucial to measure quality patient care and put quality control in place to mitigate deviations. For example, inadequate resource distribution, particularly inadequate medication supply, increases the risk of malaria, tuberculosis, and HIV/AIDS, which leads to more complications in patients.

The main causes of poor healthcare, which may be prolonged waiting times for treatment and surgery, negative outcomes, poor cleanliness, inadequate infection control procedures, limited resources, and poor record-keeping if well addressed can ensure quality care. In addition, inadequate training of staff members leads to inadequate knowledge and a lack of competency. As a result, it is critical to explain the interventions for the purpose of raising the standard of quality of patient care, see below.

5.7 The improvement interventions

Quality is every health worker's responsibility and accountability. This means that in every healthcare facility, the developed policies and procedures in place should be defined to reinforce compliance to the standard of service. Therefore, remedial actions should be suggested in case of deviations between the performance and the standards for QC. QA strategy integrates organizational approach, management, staff, and quantitative techniques to meet client needs and expectations. It is an orderly managerial improvement strategy aimed at addressing the demands and prospects of all organizations as they attempt to deal with rising change, complexity, and pressure within their organizations. Hence, the following interventions are implemented to improve the quality. Healthcare managers should contact non-returning patients for information and implement generic measures to improve service quality and patient well-being. Furthermore, suggestion boxes for survey questionnaires are used to determine patient satisfaction with services rendered.

Healthcare managers need to understand the importance of service quality dimensions in patient satisfaction and how dissatisfied patients react to service failures in order to intervene further to improve the quality of care. Ref. [27] affirmed that healthcare is dependent on patient-provider collaboration and supportive surroundings. Furthermore, [28] demonstrated that change can result from visionary leadership, sound planning, education, training, resource availability, proficient management, and provider collaboration. For example, peer reviews and audits must be completed within the agreed-upon time frame, and sometimes unannounced audits are required to improve the quality of care. Furthermore, the buildings must be evaluated in order to increase the security and safety of both clients and health personnel. To avoid unintended consequences, risk officers and occupational health safety personnel should conduct risk assessments in collaboration with quality and infection control practitioners.

The study conducted by Polizzi et al. [28] showed that quality improvement must be a long-term, ongoing activity aimed at reducing errors and creating a trusting atmosphere for both health workers and patients. The healthcare nursing service manager is not solely responsible for quality assurance. All staff members are responsible for assessing and controlling all accidents or hazards. In addition, stated that QA and improvement program is an ongoing assessment of internal audit and consulting work, aiming to enhance the quality and commitment of auditors. As a result, the purpose of the chapter was to explain the significance of quality assurance in public rural healthcare facilities.

6. Conclusion

Quality assurance (QA) is crucial in public rural healthcare facilities to ensure patient care quality and compliance with nursing norms and standards. QA involves administrative activities, measurement, comparison, and error prevention. Healthcare facilities face challenges such as under-use, overuse, errors, service variation, resource scarcity, inadequate diagnosis, treatment, and poor information. Poorly resourced facilities, especially in developing countries, compromise quality care. Capacity building and early interventions are needed to address these issues and improve quality of patient care.

Acknowledgements

The author of this paper sincerely acknowledges the support of my colleagues to encourage me to write the chapter. No funding.

Conflict of interest

The author declares no conflict of interest.

Notes/thanks/other declarations

Nothing to declare.

Author details

Takalani Rhoda Luhalima
University of Venda, Thohoyandou, South Africa

*Address all correspondence to: takalani.luhalima@univen.ac.za

References

[1] Busari JO. Comparative analysis of quality assurance in health care delivery and higher medical education. Advances in Medical Education and Practice. 2012:121-127. DOI: 10.214/AMEP.S38166

[2] Saville N, McElwee S. Quality management in test production and administration. In: The Routledge handbook of language testing. Routledge; 2021. pp. 597-621. DOI: 10.4324/9781003220756

[3] Ewuoso C, Hall S. Core aspects of Ubuntu: A systematic review. South African Journal of Bioethics and Law. 2019;**12**(2):93-103. DOI: 10.7196/SAJBL.2019.v12i2.679

[4] Sambala EZ, Cooper S, Manderson L. Ubuntu as a framework for ethical decision making in Africa: Responding to epidemics. Ethics & Behavior. 2020;**30**(1):1-3. DOI: 10.1080/10508422.2019.1583565

[5] Oleribe OO, Momoh J, Uzochukwu BS, Mbofana F, Adebiyi A, Barbera T, et al. Identifying key challenges facing healthcare systems in Africa and potential solutions. International Journal of General Medicine. 2019:395-403. DOI: 10.2147/IJGM.S223882

[6] Petersen I, Marais D, Abdulmalik J, Ahuja S, Alem A, Chisholm D, et al. Strengthening mental health system governance in six low-and middle-income countries in Africa and South Asia: Challenges, needs and potential strategies. Health Policy and Planning. 2017;**32**(5):699-709. DOI: 10.1093/heapol/czx014

[7] Twigg DE, Cramer JH, Pugh JD. Nurse staffing and workload drivers in small rural hospitals: An imperative for evidence. Online Journal of Rural Nursing and Health Care. 2016. DOI: 10.14574/ojrnhc.v16i1.370

[8] Smith S, Sim J, Halcomb E. Nurses' experiences of working in rural hospitals: An integrative review. Journal of Nursing Management. 2019;**27**(3):482-490. DOI: 10.1111/jonm.12716

[9] Kakyo TA, Xiao LD. Challenges faced in rural hospitals: The experiences of nurse managers in Uganda. International Nursing Review. 2019;**66**(1):70-77. DOI: 10.1111/inr.12459

[10] Manyisa ZM, van Aswegen EJ. Factors affecting working conditions in public hospitals: A literature review. International Journal of Africa Nursing Sciences. 2017;**6**:28-38. DOI: 10.1016/j.ijans.2017.02.002

[11] Sabbah N, Carles G, Demar M, Nacher M. Diabetes in French Guiana, adapting national standards of therapeutic education and care to the Amazonian challenge. World Journal of Diabetes. 2021;**12**(2):98. DOI: 10.4239/wjd.v12.i2.98

[12] Weine S, Eisenman DP, Kinsler J, Glik DC, Polutnik C. Addressing violent extremism as public health policy and practice. Behavioral Sciences of Terrorism and Political Aggression. 2017;**9**(3):208-221. DOI: 10.1080/19434472.2016.1198413

[13] Cyr ME, Etchin AG, Guthrie BJ, Benneyan JC. Access to specialty healthcare in urban versus rural US populations: A systematic literature review. BMC Health Services Research. 2019;**19**(1):1-7. DOI: 10.1186/s12913-019-4815-5

[14] World Health Organization. Delivering Quality Health Services: A Global Imperative. OECD Publishing; 2018

[15] Patil DK, Patil DR, Pati SA. A review on introduction to quality assurance. Research Journal of Pharmacology and Pharmacodynamics. 2023;**15**(2):73-76. DOI: 10.52711/2321-5836.2023.00015

[16] Evans AM, O'Donovan C, Playdon M, Beecher C, Beger RD, Bowden JA, et al. Dissemination, and analysis of the quality assurance (QA) and quality control (QC) practices of LC–MS based untargeted metabolomics practitioners. Metabolomics. 2020;**16**(10):113. DOI: 10.4239/wjd.v12.i2.98

[17] Carey RB, Bhattacharyya S, Kehl SC, Matukas LM, Pentella MA, Salfinger M, et al. Practical guidance for clinical microbiology laboratories: Implementing a quality management system in the medical microbiology laboratory. Clinical Microbiology Reviews. 2018;**31**(3):10-128. DOI: 10.1128/cmr.00062-17

[18] Maphumulo WT, Bhengu BR. Challenges of quality improvement in the healthcare of South Africa post-apartheid: A critical review. Curationis. 2019;**42**(1):1-9. DOI: 10.4102/curationis.v42i1.1901

[19] Murove MF, Murove MF. African ethics in a nutshell. African Politics and Ethics: Exploring New Dimensions. 2020:7-30. DOI: 10.1007/978-3-030-54185-9

[20] Madaka NC. Public Health Policy in Resource Allocation: The Role of Ubuntu Ethics in Redressing Resource Disparity between Public and Private Healthcare in South Africa (Doctoral Dissertation. Stellenbosch: Stellenbosch University;

[21] Havenga Y, Temane A, Ramukumba T, Nolte A. Ethics in nursing–A South African perspective. African Perspectives on Ethics for Healthcare Professionals. 2018:175-188. DOI: 10.1007/978-3-319-93230-9

[22] Broadhurst D, Goodacre R, Reinke SN, Kuligowski J, Wilson ID, Lewis MR, et al. Guidelines and considerations for the use of system suitability and quality control samples in mass spectrometry assays applied in untargeted clinical metabolomic studies. Metabolomics. 2018;**14**:1-7. DOI: 10.1007/s11306-018-1367-3

[23] Awoonor-Williams JK, Bawah AA, Nyonator FK, Asuru R, Oduro A, Ofosu A, et al. The Ghana essential health interventions program: A plausibility trial of the impact of health systems strengthening on maternal & child survival. BMC Health Services Research. 2013;**13**:1-2. DOI: 10.1186/1472-6963-13-S2-S3

[24] Romdhane HB, Tlili F, Skhiri A, Zaman S, Phillimore P. Health system challenges of NCDs in Tunisia. International Journal of Public Health. 2015;**60**:39-46. DOI: 10.1007/s00038-014-0616-0

[25] Um KH, Lau AK. Healthcare service failure: How dissatisfied patients respond to poor service quality. International Journal of Operations & Production Management. 2018;**38**(5):1245-1270. DOI: 10.1108/IJOPM-11-2016-0669

[26] Mosadeghrad AM. Factors influencing healthcare service quality. International Journal of Health Policy and Management. 2014;**3**(2):77. DOI: 10.15171/ijhpm.2014.65

[27] Alamantariotou K. Improving quality of care through risk management knowledge transfer and quality Assurance in Nursing Care. In: Quality Assurance in Healthcare Service

Delivery, Nursing and Personalized Medicine: Technologies and Processes. IGI Global; 2012. pp. 233-241. DOI: 10.4018/978-1-61350-120-7.ch014

[28] Polizzi S, Lupo F, Testella S. Quality assurance and improvement program: Some considerations for central banks. The TQM Journal. 2022. DOI: 10.1108/TQM-05-2021-0128

Chapter 11

Perspective Chapter: Quality Assurance in Diagnostic Ultrasound

Raffaele Novario

Abstract

Ultrasound techniques have become a gold standard, both as a complement to conventional diagnostic methodologies and as a technique of choice. This has led to the need for quality control procedures, as happened in other sectors of diagnostic imaging. The aim of this chapter is to propose a series of physical parameters related to quality, both in B-mode and in Doppler velocimetry that could be measured with the use of commercial or "homemade" phantoms, following a protocol that considers both innovative proposals and indications published in the literature. In order to do this the different parameters are described and their physical meaning is discussed. Tests on different equipment are performed to evaluate the robustness of the protocol and the chosen parameters. The main results obtained with high clinical significance are presented. This allows both the acceptability of the equipment from a clinical ultrasound point of view and the consistency of diagnostic performance over time.

Keywords: ultrasound, quality control, patient protection, optimization, radiological imaging

1. Introduction

Ultrasound is a very popular diagnostic imaging technique. The technology has been known since the eighteenth century, and its use in medicine was first suggested by Austrian physician Karl Theodore Dussik in 1941. Usually, however, obstetrician Ian Donald and engineer Tom Brown are cited as pioneers of ultrasound, as in the 1950s they developed a prototype of an ultrasound device, called ultrasound or sonography scanner, to be used for medical purposes, mainly to identify any fetal malformations.

Ultrasound has spread throughout the world since the 1970s. Unlike methods such as X-rays, it does not use ionising radiation but rather creates an image of the inside of the body using high-frequency mechanical waves, which cannot be picked up by our ears: ultrasound. The sounds that we can normally hear have a frequency ranging from 20 to 20,000 Hz, whilst ultrasound for medical diagnostics usually has frequencies from 1 to 20 MHz.

To carry out an ultrasound scan, the sonographer uses a small probe: this is held in hand, contains many transducers (order of magnitude 10^1, 10^3) and is placed on the skin in correspondence with the volume to be examined. This device emits high-frequency sound impulses into the body, and acts both as a transmitter and a receiver, picking up the impulses reflected by the organ or tissue to be examined. More specifically, every time the sound waves encounter a separation surface between different types of tissue (i.e. acoustic impedance), part of them is reflected backwards and is picked up by the transducer; the unreflected part of the wave continues its travel.

To "seal" the space between the probe and the skin, and thus ensure better propagation of the mechanical waves, a dense water-based gel is spread on the skin immediately before the exam.

Each probe contains numerous piezoelectric crystals, i.e. ceramic or other materials that vibrate in response to the passage of electricity, producing ultrasonic waves. Vice versa, when the probe captures the waves reflected from the examined volume (echoes), the opposite happens: the transducers convert the pressure of the waves into current, and this current forms a signal which is transformed into an image by the sonographic scanner.

The instrumentation is actually able to calculate the position (depth) of the different tissues starting from the time it takes for the waves to return to the probe (being the speed of sound through the body's tissues constant, 1540 m/s). Thanks to this information the sonographic scanner is able to reconstruct a digital tomographic image of the volume being analysed obtained by merging the several lines representing the amplitude of the single line waves received (**Figure 1**, real-time brightness mode, B-mode). Today, with this technology, we have the opportunity to view images in great detail [1–5].

The frequency of the sound waves used varies depending on the type of analysis. For example, for superficial structures such as the thyroid, muscles, tendons, ligaments and breast glands, higher frequency ultrasounds with high resolution but low penetration are used (7–18 MHz), whilst to examine deeper organs such as the liver, kidneys and the heart, or to visualise the inside of the abdomen, a probe that produces lower frequency waves with low resolution, high penetration is more suitable (1–6 MHz).

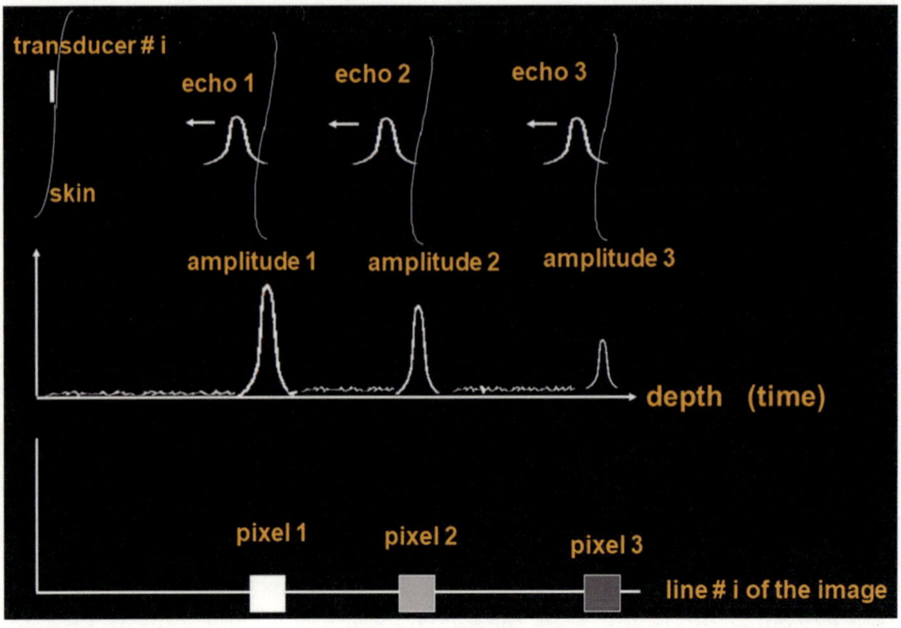

Figure 1.
From ultrasound reflected echoes to pixels of a line of the image.

Ultrasound is also used to monitor the flow in the blood vessels, through an analysis method called Doppler ultrasound or echo-Doppler. The underlying physical principle is the Doppler effect: when the receiver and the source are moving relative to each other, the sound waves perceived by the former have a different frequency than that of the sounds emitted by the latter.

In particular, if the wave received by the probe has a higher frequency than the one emitted, the receiver and source are getting closer, whilst if they are moving away, the received frequency will be lower. Doppler ultrasound is widely used in the analysis of blood flow and can provide information regarding the speed and resistance to blood flow, allowing the highlighting of vessel narrowing (stenosis) or enlargement (aneurism), increasing (narrowing, stenosis) or slowdowns (enlargement, aneurism) of the regular flow velocity.

This information can be displayed either through a velocity spectrum referring to a sampled volume (**Figure 2**, spectral Doppler) or by colouring the approaching and receding parts of the flow differently in the panoramic image (typically the former red and the latter blue), (**Figure 3**, colour Doppler).

The use of ultrasound techniques has become increasingly widespread in recent years, both as a complement to conventional diagnostic methodologies and as a technique of choice. This has led to an ever-increasing diffusion of ultrasound equipment, with the consequent need to apply quality control procedures on the performance of these instruments by the radiologist and medical physicist responsible of the scanner, as has happened in other sectors of diagnostic imaging.

However, it is necessary to point out an important and peculiar characteristic of ultrasound techniques: the diagnostic performance is never linked only to the characteristics of the equipment but is strongly influenced by the operator. The large number of variables involved, the numerous possibilities for different machine setups, and the direct interaction of the sonographer-probe-patient determine the impossibility

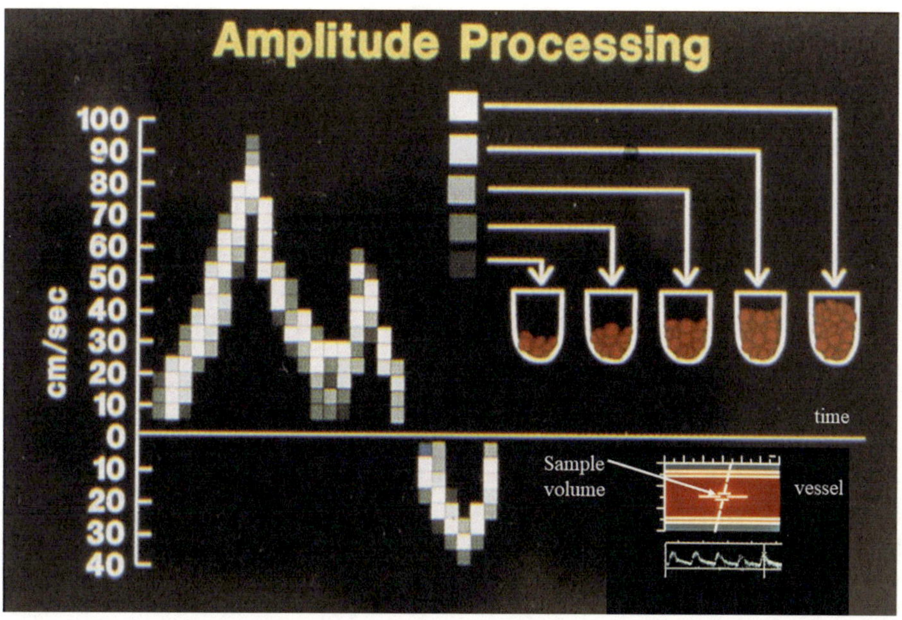

Figure 2.
Doppler velocity spectrum as a function of time referring to a sampled volume in a vessel. Pixel brightness = blood mass. Positive speed = moving towards the probe. Negative speed = moving away the probe.

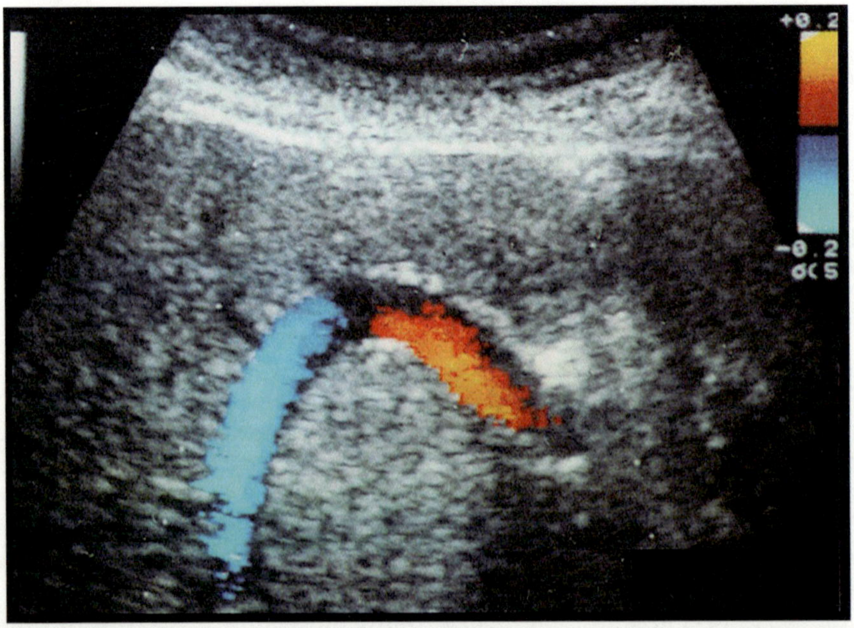

Figure 3.
Colour Doppler image. Red pixel = moving towards the probe. Blue pixel = moving away the probe. Brightness = higher speed.

of separating the quality component linked to the machine from the one linked to the operator. Therefore, the results of quality control must be interpreted in this light.

A complete quality control programme allows to:

- Make comparisons between different equipment and/or probes when choosing to purchase;
- Check the correspondence of the parameters certified by the manufacturer upon acceptance of the supply;
- Carry out status tests after each significant intervention (extraordinary maintenance, implementation of new modules, etc.);
- Regularly evaluate the constancy of the functional performance of the entire system in order to guarantee a minimum standard of diagnostic quality.

It is therefore necessary to draft a complete quality control protocol which includes, in addition to the definition of the parameters being evaluated, the identification of the limits of acceptability of these parameters. The extreme variability of the setup configurations, the different technological quality of the scanners (and cost) and the conditions of use have not allowed, at present, to agree in setting these values. We will refer to the limits proposed by the AAPM (American Association of Physicists in Medicine) working group protocol, keeping in mind that the approach commonly accepted in centres with more experience in the field of quality controls on ultrasound equipment is to set their own acceptability limits relating to the individual equipment in every foreseen configuration.

Quality checks on ultrasound equipment can be classified into clinical checks and physical checks.

Clinical checks are based on a subjective evaluation carried out by the sonographer/radiologist on a series of clinical images, which can therefore also be performed during the examination.

Physical checks are based on assessments carried out on physical parameters that can be obtained using test objects and/or phantoms (**Figures 4–8**).

This last type of control can also be divided into two further categories: a first level, which usually produces qualitative indications, and a second, which provides quantitative information. In the first case, we proceed only with the visual investigation of the image obtained, whilst in the second level, it is necessary to have access to the numerical content of the image, downloading it to obtain numerical values of parameters. This second type of approach is advisable as it minimises errors and is also much simpler and quicker to perform. There is also software on the market that allows you to completely automate the management of second-level controls. The parameters subject to physical quality control will be listed below: each of these parameters can therefore be evaluated with a first or second-level procedure.

Clinical quality control consists of a subjective evaluation by the operator of a clinical image. The resulting judgement is obtained through the qualitative evaluation of some parameters, described below, applicable to both real-time two-dimensional ultrasound (B-mode) and Doppler velocimetry.

In this scenario and considering that sonography exams performed in 2022 as stated by the World Health Organisation were 2,8 billion worldwide (vs 3,6 billion exams using X-rays), a quality assurance programme becomes mandatory also with this technology. A set of clinical and physical parameters to be used to understand their meaning and weight has to be proposed, a test of them performed in order to assess the feasibility of their use, and some comments about acceptance limits, so different in their setting from diagnostic X-ray, discussed.

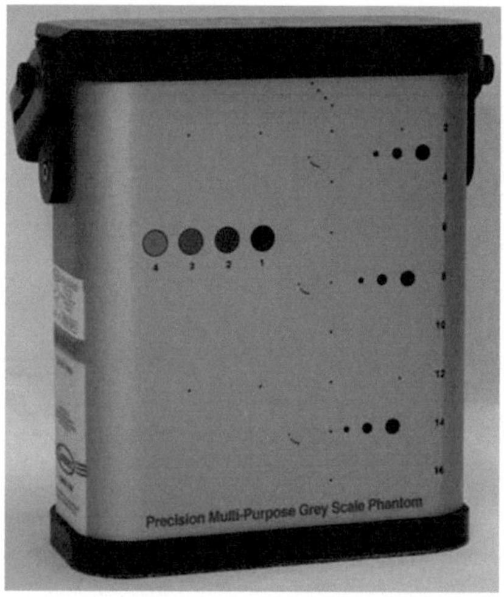

Figure 4.
B-mode quality control phantom.

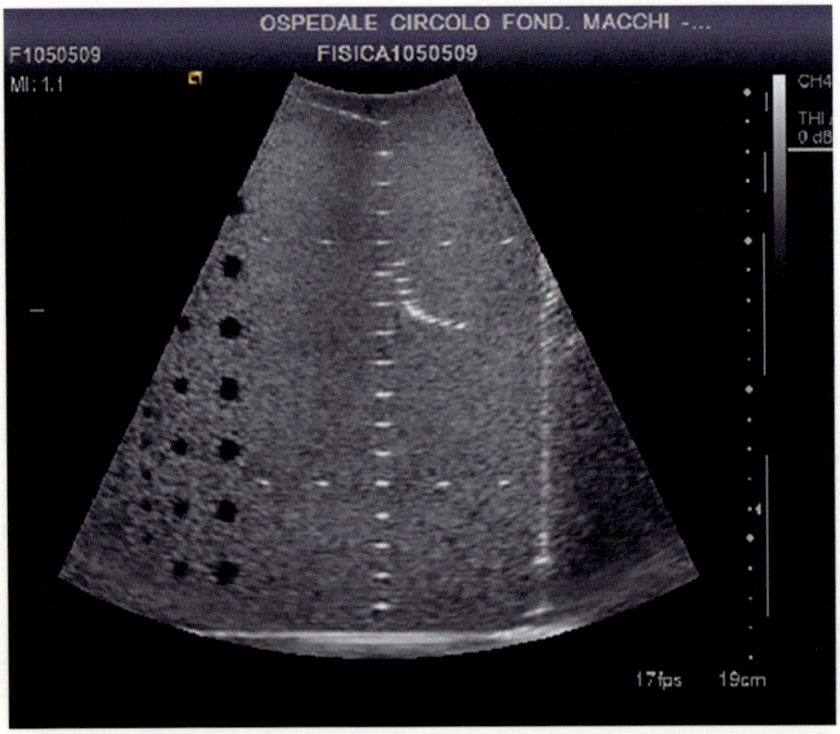

Figure 5.
B-mode quality control phantom image.

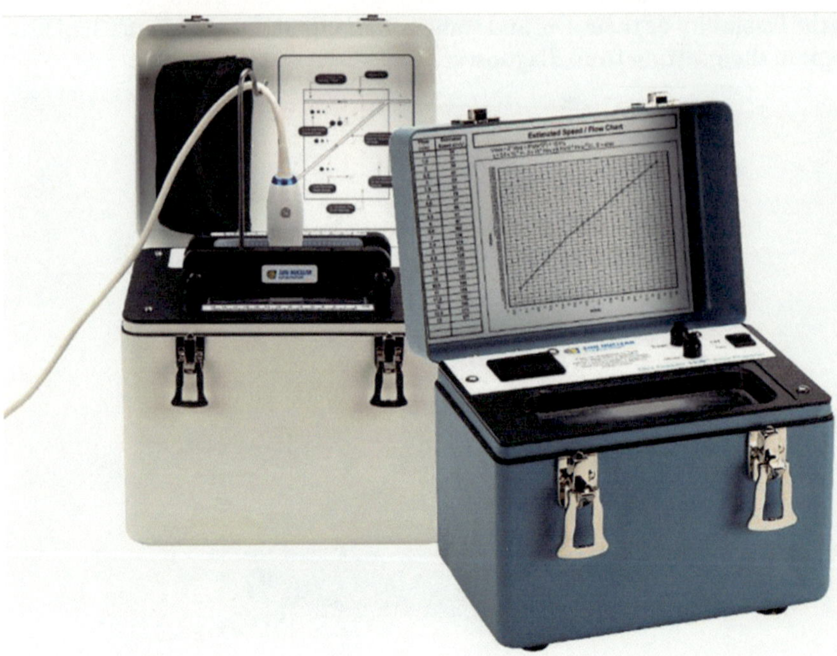

Figure 6.
Doppler quality control phantom.

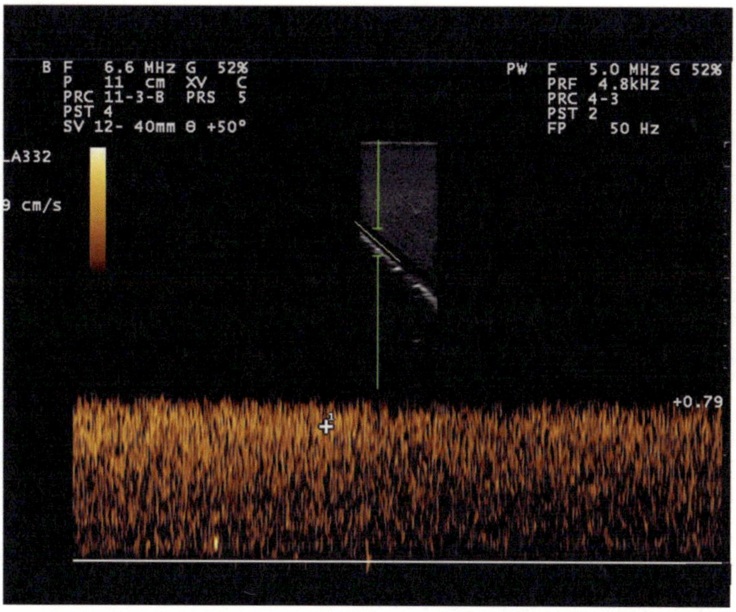

Figure 7.
Doppler velocity spectrum of a Doppler phantom with laminar flow directed towards the probe.

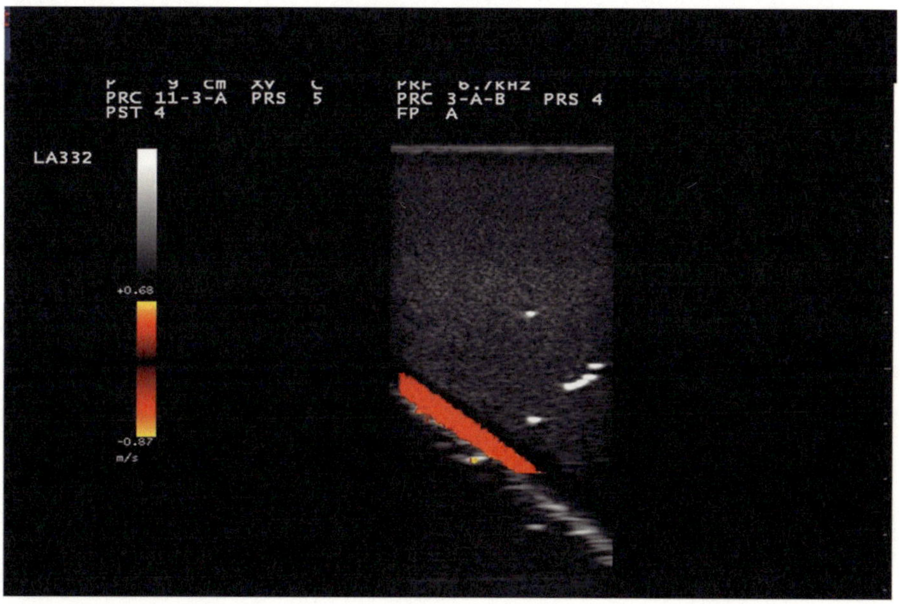

Figure 8.
Colour Doppler image of a Doppler phantom with laminar flow directed towards the probe.

2. Diagnostic image quality (DIQ)

It is a parameter linked to the ability to represent the normal and the pathological. For this purpose, it is necessary to identify variable "targets" depending on the different clinical use (e.g. representation of the thyroid and its pathology in the evaluation of the equipment configured for the study of superficial organs; representation of superficial vessels and their pathology for the vascular study; representation of the gallbladder in the study of the upper abdomen, etc.). The DIQ is naturally linked to the next parameters.

2.1 Overall image quality (OIQ)

It is a subjective parameter that evaluates:

- The definition of the details represented (solid-liquid and solid-solid interfaces), linked to the presumed resolution capabilities of the system;
- The presence of artefacts, linked to the spatial localization capacity of the system;
- The sensation of breadth of the grey scale, linked to the dynamic range;
- Anechoic behaviour in the representation of liquid structures, linked to the signal/noise ratio of the system;
- The sensation of "real time", linked to the frame rate allowed by the system.

2.2 Exploration efficiency (EE)

It is a parameter that depends on the dimensions of the field represented and above all on the morphological and structural characteristics of the probe (support surface, size, weight) and therefore on the ergonomics of the system. This parameter also affects the speed of execution of the exam and often the operator's feeling of safety.

2.3 Flexibility of use (FU)

It is a related parameter:

- The amount of adjustments that are necessary when moving from one patient to another or from one probe to another;
- The simplicity of use and therefore the potential of the keyboard and measurement systems, as well as image reproduction systems. This parameter is also linked to the ergonomics of the system.

An arbitrary numerical scale can be attributed to each of these subjective parameters (for example with a score from 1 to 5: poor, sufficient, fair, good, excellent). An evaluation table of the equipment can thus be constructed for the different clinical "targets" and an overall one obtained from the sum of the partial results, with the aim of both comparing different equipment with each other and checking the stability of the characteristics over time.

3. Real-time two-dimensional ultrasound physical QC

To carry out periodic physical quality control on ultrasound equipment, it is necessary to define a minimum set of significant and measurable parameters and compare them with parameters measured upon acceptance of the machine itself, or with appropriate levels of acceptability.

To do this it is necessary to proceed with the measurement starting from a known machine state that can be reproduced in subsequent measurements. The standard configuration set for the test is called the initial setup and consists of a series of settings and adjustments to be recorded on the test protocol [6–10].

The parameters tested are the following:

- Accuracy and spatial linearity of measurement
- Spatial resolution
- Contrast resolution
- Uniformity
- Dead zone
- Focus
- Sensitivity (or depth of maximum penetration)
- Dynamic range
- Size, shape and filling of pseudo-cysts
- Size, shape and shadowing of solid pseudo-masses

3.1 Accuracy and spatial linearity of measurement

Most of the equipment allows the using electronic callipers, to carry out measurements of depth, dimensions of objects, areas and volumes, necessary to ensure a correct diagnosis.

Accuracy is a parameter that provides an indication of the error made in the measurement of distances and is evaluated by comparing the distance measured between two targets in the phantom with the known distance declared by the manufacturer of the phantom. By determining the relationship between the true dimensions and the measured ones, spatial linearity is obtained: this can be evaluated both horizontally and vertically. Several test objects and phantoms available on the market are equipped with a horizontal series (at a given depth) of equidistant points and a vertical series. By representing the measured distances on a Cartesian plane, on the x-axis and the true distances on the y-axis, it is possible to evaluate both the linearity of the calliper's response (statistical p-value of regression) and its accuracy (angular coefficient of the straight line close to 1 i.e. 45°). No particular problems were found in performing this test and the results obtained were quite often excellent (p very low, angular coefficient very close to 1, typical 0.98–0.99).

3.2 Spatial resolution

Spatial resolution represents the system's ability to distinguish point nearby objects under high contrast conditions.

The spatial resolution can be evaluated with first-level quality controls using phantoms or test objects in which point targets are placed at gradually decreasing and known distances, horizontal or vertical respectively, (e.g. 4, 3, 2, 1, 0,5 mm) and evaluated at what level the points no longer appear distinguishable in the image (first level visual inspection). These checks can be carried out at different depths and at different lateral positions with respect to the axis of symmetry of the image.

The second level of controls involves the processing of the digital image of a point-like object in an area of the field of view identified by selecting the depth and lateral position with respect to the axis of symmetry of the image. The full width at half maximum (FWHM) of the object's two-dimensional point spread function in the axial and transverse directions is the absolute numerical estimate of the spatial resolution.

In ultrasound imaging, the spatial resolution can be axial (parallel to the direction of propagation of the beam, generally better especially for linear probes) or lateral (perpendicular to the direction of propagation, generally worse especially for convex probes). There is also a third resolution, called "elevation resolution", which represents the possibility of separating two adjacent layers in a direction perpendicular to the displayed plane.

Axial spatial resolution is defined as the ability to distinguish two adjacent objects along the direction of propagation of the ultrasound beam. This parameter, from a physical point of view, is determined primarily by the emission frequency of the transducer. In fact, it is evident that it is impossible to resolve objects having dimensions smaller than one wavelength. For pulsed beams the axial resolution depends on the bandwidth of the ultrasound pulse: this depends on the Q factor of the transducer. The quality factor Q of the transducer is defined as the ratio between the central frequency of the transducer and its band. In particular, in two-dimensional echo imaging, transducers with low Q values are used to have a better spatial resolution, short echoes = wide band. High Q values give a very narrow band whilst low Q gives a wide band. In diagnostics it is necessary to have a wide band, in fact, this means having short spatial pulses (SPL) which allow to resolve objects very close to each other.). It must be kept in mind that ultrasound undergoes attenuation in the tissues and that this attenuation is a function of frequency, i.e. high frequencies are attenuated more, this implies a narrowing of the frequency band of the spectrum, therefore, a reduction in resolution axial as the depth of penetration into the tissue increases.

So, axial (also called longitudinal) resolution is the minimum distance that can be differentiated between two reflectors located parallel to the direction of the ultrasound beam.

$$\text{Axial resolution} = \tfrac{1}{2} \times \text{spatial pulse length.} \tag{1}$$

The spatial pulse length is determined by the wavelength of the beam and the number of cycles (periods) within a pulse. Axial resolution is high when the spatial pulse length is short. Spatial pulse length is the product of the number of cycles in a pulse of ultrasound and the wavelength. Most pulses consist of two or three cycles, the number of which is determined by the damping of piezoelectric elements after excitation: high damping reduces the number of cycles in a pulse and hence shortens spatial pulse length. The wavelength of a pulse is determined by the operating frequency of the transducer; transducers of high frequency have thin piezoelectric elements that generate pulses of short wavelengths.

Supposing three cycles per pulse and using the equation $c = \lambda\nu$, you obtain the wavelengths and the consequent theoretical maximal axial resolution values reported in the following **Table 1**:

Frequency (MHz)	λ (mm)	Max (theoretical) axial resolution (mm)
2.5	0.616	0.924
3.5	0.440	0.660
5	0.308	0.462
7.5	0.205	0.308
10	0.154	0.231
12	0.128	0.192
15	0.103	0.155

Table 1.
Wavelength and theoretical maximal axial resolution supposing three cycles per pulse at different frequencies.

AAPM (American Association of Physicists in Medicine) suggests the following tolerance limits on axial resolution:

- ≤ 1 mm for transducers with a central frequency greater than 4 MHz
- ≤ 2 mm for transducers with a central frequency less than 4 MHz.

The axial resolution should remain constant over time. If changes occur, the intervention of the technical assistance service is advisable.

Generally, in commercially available phantoms the pins for axial resolution are slightly moved horizontally with respect to each other in order to reduce acoustic shadow effects. Also, this test is very simple to perform but the results are very variable, depending on the depth, the position of the pins and the quality of the scanner. For example, we found resolutions close to 0,3 mm for superficial pins at 10 MHz for high-quality scanners, and significantly higher values for low-quality ones.

3.3 Lateral spatial resolution

Lateral spatial resolution is defined as the ability of the instrument to resolve two adjacent point objects (pin in the phantom) arranged perpendicular to the beam axis. It is known that in ultrasound imaging the lateral spatial resolution is worse than the axial spatial resolution. The diameter of the beam is a quantity that crucially determines the lateral resolution: in fact, a single object smaller than the width of the beam produces an echo signal during scanning for the entire time it is inside the beam, therefore the object appears as if its size were equal to the width of the beam itself at that depth, so the lateral resolution of fixed focus transducers varies greatly with depth and frequency. However, systems with multiple focal zones or dynamic focus can produce more uniform lateral resolution over a wider range of depths. The lateral resolution can also deteriorate significantly due to side lobes and grating lobes and type of probe (best with linear, less with convex, sector).

As regards the lateral resolution values and their tolerance limit, there is considerable uncertainty regarding the solution that can be proposed for routine checks. In this regard, it is useful to refer to the −20 dB pulse echo focal length, W_{20} proposed by AAPM and proportional to the wavelength at the centre frequency and to the pulse-echo focal length and inversely proportional to the diameter of the active element (transducer).

		Lateral resolution	
Depth (cm)	ν (MHz)	Space between targets (mm)	FWHM (mm)
> 10	< 3.5	≤ 4	≤ 2
< 10	$3.5 \leq \nu < 5$	≤ 3	< 1.5
< 10	≥ 5	≤ 1.5	< 1

Table 2.
Theoretical lateral resolution for different depths and frequencies.

Frequency	Resolution (axial and lateral resolution)	
MHz	Axial resolution	Lateral resolution
3.0	1.1 mm	2.8 mm
4.0	0.8 mm	1.5 mm
5.0	0.6 mm	1.2 mm
7.5	0.4 mm	1.0 mm
10	0.3 mm	1.0 mm

This is for linear array transducers with parallel beams.

Table 3.
Desired axial and lateral resolution for different frequencies in real conditions.

In general, it is suggested by AAPM that the measured lateral resolutions respect the specifications reported in **Table 2**.

In real conditions (not mathematical), the optimization of spatial resolution will be obtained with the values proposed in **Table 3** for a medium-quality scanner.

3.4 Contrast resolution

The contrast in the object determines a modulation of the reflected echo that returns to the probe and produces a contrast in the grey scale of the image. Contrast resolution represents the smallest contrast in the object that determines a contrast detectable by the human eye in the image, in conditions that do not have problems related to spatial resolution.

Various test objects are available with circular areas of decreasing and known contrast (scatter relative to the background). **Figure 4** can be reproduced in the image. It is therefore possible to evaluate the response curve of the system (contrast in the image as a function of contrast in the object) and consequently the contrast resolution. First- and second-level checks can also be carried out for this parameter. In the first case, a visual inspection and the first insert not visible can give such an indication. In the second a plot with the grey scale contrast of the object in the image as a function of the actual one in the phantom can give a numerical assessment (contrast drop to zero). In our experience, the max contrast resolution is of the order of magnitude of 10 dB. No particular problems were found in performing this test.

3.5 Uniformity

Uniformity describes the ability of the ultrasound system to obtain a uniform image of a homogeneous object, i.e., to show echoes of the same amplitude and depth with equal brightness on the screen. Factors that influence uniformity are tissue attenuation and focus.

The attenuation of the beam in-depth, and therefore the underestimation of the echoes coming from the deeper layers, is corrected by the operator's adjustment of the time gain compensation (TGC), which determines the amplification curve of the signals as a function of their time of delay (and therefore depending on the depth of origin of the echo itself).

By using a phantom or a test object with a homogeneous tissue texture, i.e., made up of a homogeneous set of diffusers smaller than the wavelength of the beam and correct by means of an appropriate adjustment of the TGC, a "homogeneous" image is obtained and used to calculate parameters numbers that characterise uniformity. A first-level check is only a visual inspection and, if possible. A profile of the grey level as a function of depth. One way of the second level is to select a region of interest (ROI) which is moved deeper and deeper to be able to calculate the average and standard deviation of the grey levels of the pixels as a function of the depth itself. Ideally, ultrasound images are characterised by an irregular background (texture), in which, however, the average value of the grey levels calculated in areas of interest of dimensions much larger than the "grain" of the image, should not vary either with the depth nor with the angle of displacement from the centre of the image. In reality, however, there may be areas of the image with an average grey level value different from the rest. These can manifest themselves in the form of dark vertical and/or horizontal bands: these non-uniformities can be explained in different ways. Horizontal bands may arise due to multiple focusing. In some multiple focusing systems there may be a gap between some focal zones, this can cause the beam to be wider in this zone and therefore there is a reduction in the amplitude of the echoes, this "strip" of different values of the grey is therefore completely normal because it is associated with the technological "limits" of the equipment. The presence of vertical bands in the image, however, can be caused by a damaged transducer element, but it can also be due to a defect in the transmission circuit connected to a certain element of the transducer. Vertical bands due to these electrical causes appear in the upper part of the image.

However, the absence of vertical bands in the image is not sufficient to ensure that all elements of the probe are functioning; if there is only one non-functioning element, it is very difficult to realise this in the case of phased array probes that use all their elements to create each line of the image. In this case, the absence of a single transducer in a transmission line causes a weak effect that has repercussions in all lines of sight of the image. However, it is easier to recognise such a defect in the case of vector, convex and linear array probes. In these, in fact, it manifests itself in the form of small vertical bands.

An image grey scale histogram is normally used to quantify the uniformity.

As regards the reference limits, the AAPM recommends notifying the technical assistance service if significant non-uniformities occur. The term significant is, however, subjective; each user can identify an appropriate threshold; however, a reference value of 4 dB is suggested. No particular problems were found in performing this test.

3.6 Dead zone

By dead zone, we mean the distance along the direction of propagation of the beam between the front surface of the probe and the first reflector that produces a detectable echo on the image. This region, where no information can be collected, exists because the transducer cannot emit and receive ultrasonic pulses at the same time. This depends on the instrument and is the result of both the reverberations coming from the probe-object interface under examination and the length of the pulse train. As the frequency increases, keeping the other parameters constant, the length of the pulse decreases and consequently the depth of the dead zone.

The extension of the dead zone is controlled since its variations are linked to variations in the performance of the probe and therefore of the ultrasound equipment. For example, an increase in the extension of the dead zone can be caused by a long-lasting ultrasonic pulse (caused by the fracture of a crystal), the incorrect functioning of damping materials, or the breakage of a lens. However, it must be emphasised that the reverberations observed in phantoms can differ significantly from those in patients, especially when the impedance of the phantom surface is very different from that of the skin.

Several test objects and phantoms are available for first-level dead zone control (visual inspection). If the trough between the initial pulse wave and the first echo wave is 6 dB lower than the peak of the first echo wave, then we can distinguish them and this distance is the dead zone (second level). In our experience, the dead zone ranges from 0,5 to 1,5 mm. No particular problems were found in performing this test.

3.7 Focus

The focal point of the transducer is the point of the ultrasound beam at a given depth in the medium crossed, where the intensity is maximum and the width of the beam is minimum. By focal zone, we mean the region surrounding the focal point in which the intensity of the ultrasound beam is within 3 dB of the maximum. This area is clearly the region where lateral resolution is best.

The focusing can be fixed (in mechanical sector probes) or electronic (in all others); electronic focusing can also be dynamic (multiple focal zones at the same time).

Focusing control consists of verifying the correspondence between the focal zone selected by the operator and the actual one. It is possible to verify the position of the focal zone by evaluating that of the best lateral resolution (see spatial resolution) or by using special test objects that allow direct visualisation of the shape of the beam.

Focus control can be a first-level control or a second, measuring the FWHM of the pin in the position of focus selected. We found different results depending on the frequency and quality of the scanner.

3.8 Sensitivity (or depth of maximum penetration)

Sensitivity represents the system's ability to detect the weakest ultrasound signal that can be clearly visualised, originating from small interfaces located at a given depth in an attenuating medium. In practical terms, sensitivity represents the maximum display depth of the system under certain reproducible conditions. The weak signal in question is detected in the presence of noise, and its detectability is an indication of the signal-to-noise ratio. The factors that can influence the sensitivity are the frequency and intensity of the excitation pulse; the gain; the TGC; the focusing; the attenuation of the medium; the depth; the composition and geometry of the reflecting element; and finally, the noise due to the system electronics. The signal/noise ratio is maximum within the focal area and decreases on the sides of it.

By carrying out this type of check, it is necessary to be able to distinguish the echoes diffused from the background from those generated by electronic noise. The former, by keeping the transducer stationary on the phantom, will appear stationary, whilst the latter will present fluctuations.

For this parameter, it is possible to carry out first and second-level assessments.

In the first case, a vertical inspection of the pin in the phantom is performed and the depth of the last visible pin is measured. In the second a rectangular ROI around the pins is acquired and its profile analysed. No particular problems were found in performing this test. The depth of penetration (sensitivity) is generally limited to approximately 200 wavelengths, corresponding to a depth of 30 cm for a 1 MHz transducer, 12 cm for 2.5 MHz transducer, and 6 cm for a 5 MHz transducer.

3.9 Dynamic range

The dynamic range of an ultrasound scanner is defined as the ratio, expressed in dB, between the intensity of the largest echo and that of the smallest echo, which can be displayed simultaneously. The smallest echo is the one that is barely distinguishable from the system noise, whilst the largest is the one just below the saturation level. Therefore, the dynamic range provides a measure of the intensity (or amplitude) that the system can handle. All elements of the equipment, from the probe to the screen, influence the dynamic range. The dynamics of the signals reaching the probe are normally much wider than that which can be represented with a normal screen; a non-linear compression (usually logarithmic) of the intensity of the echoes is therefore necessary. In particular, most of the grey level range is reserved for weak signals, compressing very intense signals more.

The dynamic range can be evaluated with first and second-level controls with phantoms or test objects similar to those used for contrast resolution. No particular problems were found in performing this test. The typical value we found can reach 100 dB.

3.10 Size, shape and filling of pseudo-cysts

Cysts are fluid-filled, hypoechoic and normally poorly attenuating structures. The shape, size and consistency of the cyst must be represented correctly.

The first level of control of these parameters allows you to evaluate any changes in the system output, an incorrect selection of the TGC curve, an incorrect insonation non-uniformity, an incorrect focusing or an insufficient quantity of number of lines of sight in the image.

Various phantoms suitable for this purpose are available for the evaluation of the circular shape of the circular insert in the phantom. No limits are suggested.

No particular problems were found in performing this test.

3.11 Size, shape and shadowing of solid pseudo-masses

Solid masses are generally echogenic and normally with high attenuation. The shape, size and consistency of solid masses must be represented correctly.

The first level of control of these parameters allows you to evaluate any changes in the system output, an incorrect selection of the TGC curve that leads to an incorrect uniformity of the image, an incorrect focusing or an insufficient quantity of number of lines of sight in the image.

The result of the test is correct if all the scanner parameters are optimised and particular attention to the angles is taken.

4. Doppler velocity QC

To carry out periodic quality control on ultrasound equipment used as a Doppler velocity metre (continuous Doppler, pulsed, colour, power, etc.), it is necessary to define a minimum set of significant and measurable parameters to compare with parameters measured at the time of acceptance of the machine itself, or with levels of acceptability. To do this it is necessary to proceed with the measurement starting from a known machine state that can be reproduced in subsequent measurements. The standard configuration set for the test will be called the initial setup and consists of a series of settings and adjustments that must be recorded in the test protocol.

Some parameters, although important for image quality, are intrinsically determined by the setup of the equipment, such as the minimum measurable speed, determined by the choice of the wall filter once the angle has been fixed Doppler, the maximum measurable speed, determined by the choice of PRF, once the Doppler angle has been fixed and so on.

Therefore, no quality control is carried out on these parameters.

The parameters tested are the following:

- Flow sensitivity
- Accuracy of speed estimation
- Precision of speed estimation
- Accuracy of positioning of the sample volume
- Spatial resolution in colour
- Colour contrast resolution
- Coloured noise
- Flow registration
- Temporal resolution
- Motion discrimination

4.1 Flow sensitivity

It represents the minimum flow that produces a statistically different signal from the background and can be variable depending on the average velocity in the vessel. This parameter is measured by simulating a known, gradually decreasing flow in a phantom shown in **Figures 6–8** (in this case it is necessary to use synthetic blood), or in a test object (preferably decreasing the diameter of the pseudo-vessel and keeping the average speed constant). In fact, as is known, the laminar flow in a duct at a given instant is equal to the product of its section at a point and the average velocity of the particles passing through that section at that instant. It should be remembered that in laminar conditions, the speeds of the particles close to the axis of the vessel are greater than those placed near the walls and that this difference increases as the diameter of the duct decreases. This first level test is performed with the selection of the lowest wall filter and decreasing the velocity until the flow disappears in the image (spectral or Doppler).

In our experience, the lowest detectable velocity was 0.04–0.06 cm/sec in a 3.00-mm tube. Detection was about three to four times less sensitive in a 0.30-mm and 0.05-mm tubes. Amplitude Doppler US sensitivity is only slightly dependent on the angle of incidence and propagation medium.

4.2 Accuracy of speed estimation

It is obtained by comparing the true (known) average speed inside a pseudo-vessel with the one measured by the velocity meter, for the different speeds of interest.

The parameter in question depends, for a given flow, on the velocity profile inside the duct itself. If we assume that the flow is laminar, and therefore the velocity profile is parabolic, it is possible to estimate the peak velocity v_P using the following relationship:

$$v_p = 2\overline{v} \tag{2}$$

If the flow inside the vessel is turbulent, or has characteristics other than laminar ones, estimating the peak velocity is more problematic.

Generally, when a liquid enters a tube it does not present a laminar regime but reaches it after a certain distance whose length depends on the diameter of the tube, the speed and the viscosity of the liquid. In commercially available phantoms there are generally some sections of the pseudo-vessels in which the laminar regime conditions can be assumed to be satisfied. It is therefore necessary to measure the speed within these zones and compare it with that calculated with the following expression:

$$\overline{v}(cm/s) = \frac{\varphi(ml/s)}{A(cm^2)} \tag{3}$$

where φ is the flow rate and A is the vessel cross-sectional area.

In our experience, Doppler methods are capable of good absolute accuracy when suitably designed equipment is used in appropriate situations, with systematic errors of 6% or less (**Figure 7**, steady flow). This is a second-level test.

Particular attention must be taken in order to measure the parabolic region of the flow.

4.3 Precision of speed estimation

It consists of checking the reproducibility of a speed measurement and is obtained by carrying out subsequent measurements in the same conditions and checking the constancy of the measured value. The precision, taking attention at an angle is generally high. Particular attention must be taken in order to measure the parabolic region of the flow.

4.4 Positioning accuracy of the sample volume

The sample volume delimits the region of the image, and therefore of the object, from which Doppler signals are obtained. The check is carried out by positioning, at different points inside the vessel, the smallest sample volume available and carrying out a flow measurement in points from one wall to the opposite one. The highest speed value must be obtained when the sample volume is found at the centre of the vessel. It is a check equivalent to that of the accuracy of the speed estimate: a misalignment produces a variation in the average speed measured.

The results we obtained were very spread due to the quality of the scanners.

The test requires particular precision.

4.5 Spatial resolution in colour

It represents the system's ability to spatially discriminate two different flows, in conditions of high contrast, for example, two opposite flows, when they are very close to each other.

Typically, it can be checked by simulating two pseudo-vessels of decreasing diameter and increasingly closer (and parallel) to each other, in which two opposite flows flow (therefore one represented in red and one in blue).

It is necessary to evaluate the minimum conditions of flow velocity and distance that produce an image in which the flows begin to be unresolved (**Figures 9–12**).

If the phantom does not have two adjacent pseudo-vessels, it is still possible to carry out the check using a flow perpendicular to the ultrasound beam. In this case, the Doppler signal is affected by spectral broadening. If the flow is orthogonal to the beam, the broadening artefact must be the same for both positive and negative speeds and the amplitude of the spectrum must be the same. If the velocity spectrum was obtained in pulsed Doppler mode it must be symmetrical with respect to the axis. The values of vessel diameter and flow velocity that produce a flow not resolved are an index of the second level of spatial resolution in colour. Value obtained is very spread.

4.6 Colour contrast resolution

It represents the system's ability to discriminate different speeds, even in the same flow.

Typically, it can be checked by evaluating what is the minimum difference between two known speeds that determine two distinguishable speeds in the image. It is a second-level test. Value obtained is very spread.

4.7 Colourful noise

Coloured noise consists of the detection of flow in areas where there is no flow. Typically, it is sufficient to acquire a colour image using a phantom or a test object in conditions of no flow and gradually increase the colour gain until coloured noise appears. The higher the gain at which the coloured noise appears, the better the performance of the ultrasound. It is a second-level test. Value obtained is very spread.

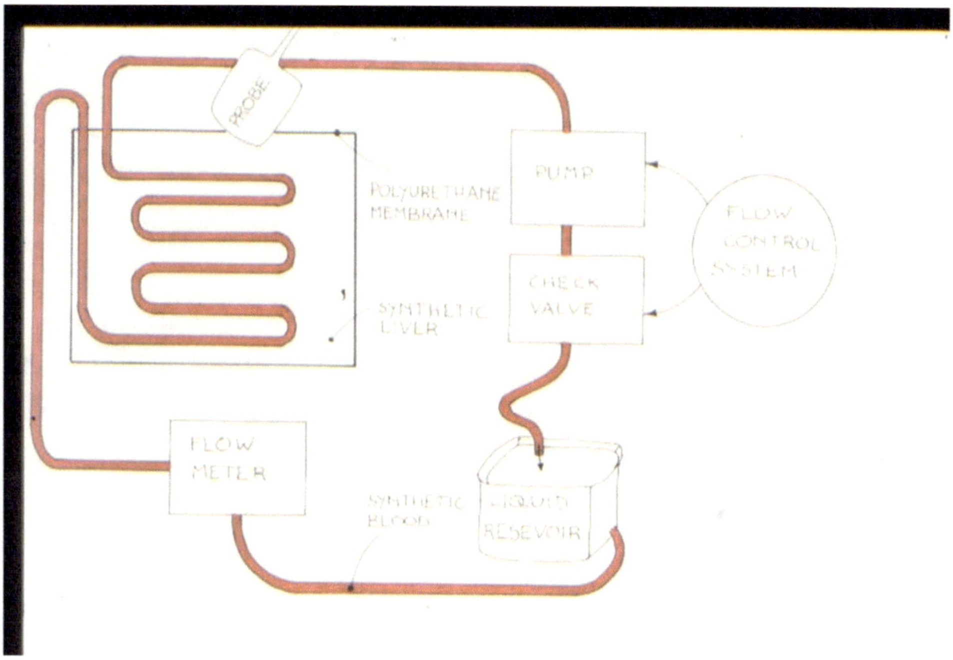

Figure 9.
Scheme of the homemade Doppler phantom for colour resolution.

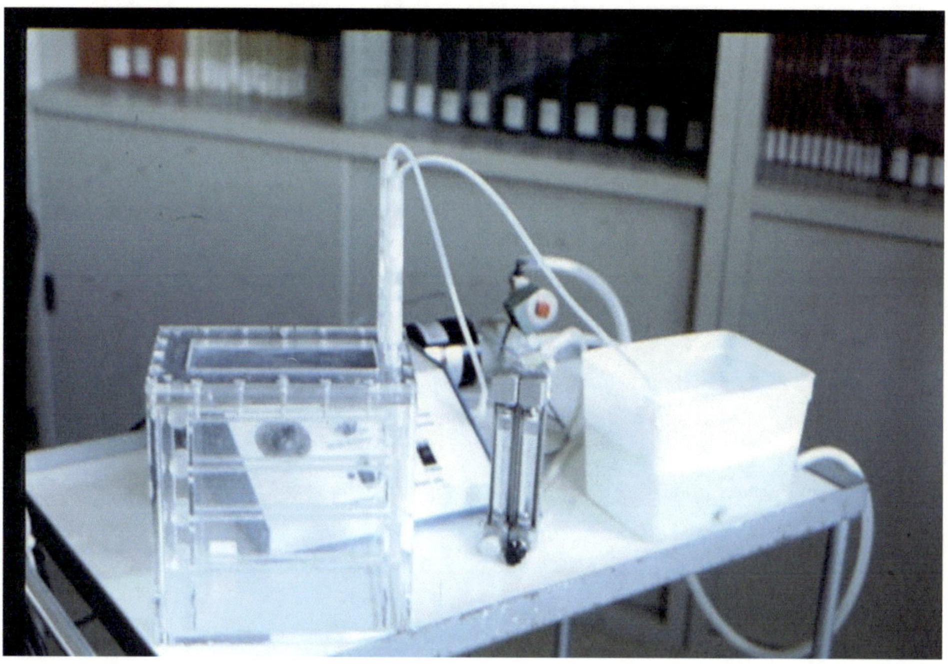

Figure 10.
Homemade Doppler phantom for colour resolution.

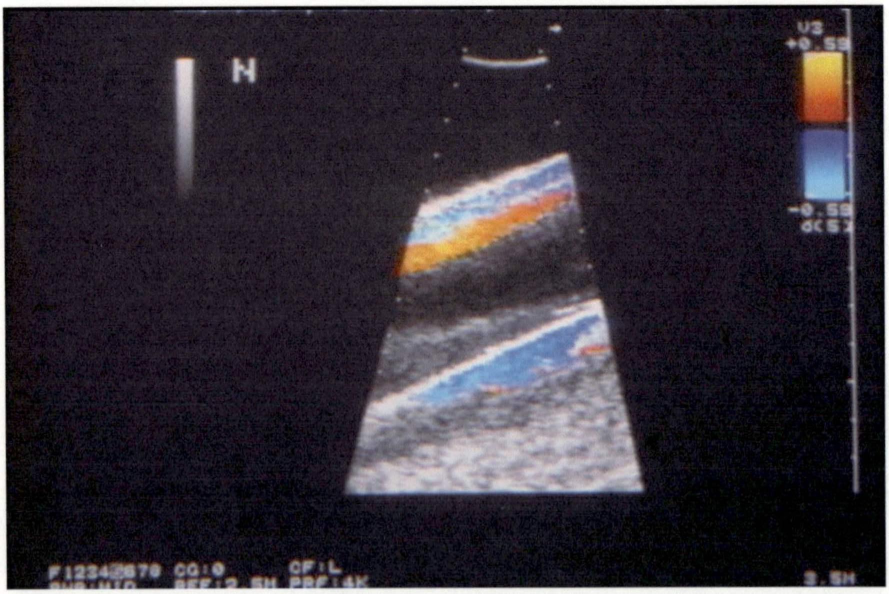

Figure 11.
Doppler phantom image with colour resolved.

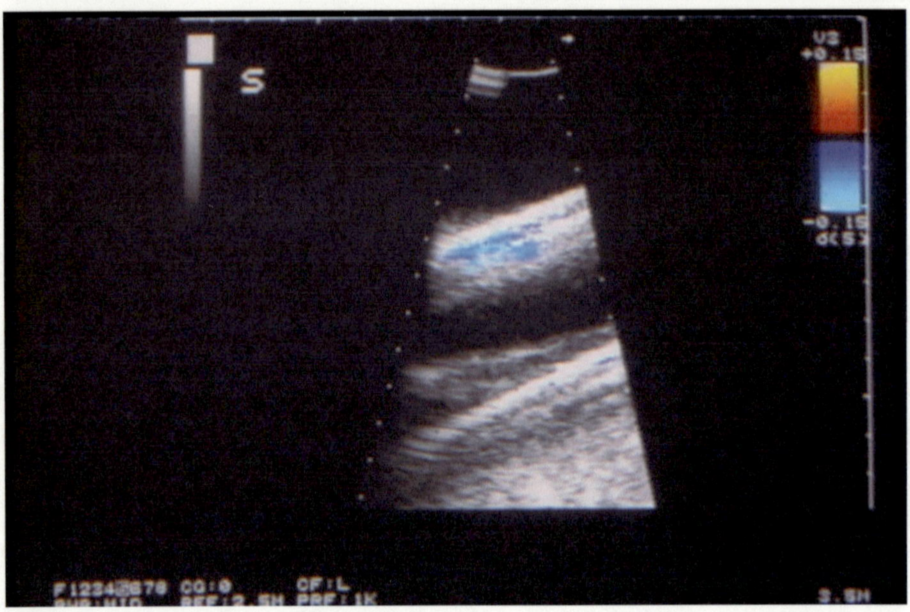

Figure 12.
Doppler phantom image with colour not resolved.

4.8 Flow registration

It represents the system's ability to give a colour signal exactly superimposed on the morphology of the pseudo-vessel. The check consists of carrying out a scan of a vessel using the Colour Doppler mode and verifying that the colour information on the flow is contained within the vessel, does not overflow but does not leave areas with an absence of colours, especially near the walls. It is a first-level test. Value obtained is very spread.

Particular attention must be taken to the overlap colour-B-mode.

4.9 Temporal resolution

It is related to the colour frame rate of the image. A dynamic event is simulated: speeds change rapidly and acceptable reproduction of dynamics in the image is verified. It's a first-level test.

4.10 Motion discrimination

It represents the system's ability to discriminate between the movement of liquid and solid masses. It is influenced by the choice of wall filter.

It is checked by simulating a known flow in a pseudo-vessel with very elastic and extensible walls and evaluating the signal of the walls, which appear displayed and which must not be in colour. It's a second-level test.

Referring to the frequency of evaluations, there are no indications whatsoever from international bodies or centres with particular experience.

5. Conclusions

It should be noted that a complete quality control programme requires some hours of machine time for acquisition only.

However, due to the very high number of ultrasound exams performed worldwide, we recommend at least a complete annual check for each piece of equipment, reserving the right to decide to carry out any simple checks at higher frequencies, which require little waste of time, and in any case whenever the radiologist deems it necessary.

Planning a quality control programme can reduce the number of repeated investigations, make the diagnosis more accurate and limit patient referral to other methods for diagnostic completion, making each ultrasound examination more accurate and informative.

It concerns real-time B-mode ultrasound and Doppler velocimetry, both spectral and colour mode.

It should follow a protocol that contains the parameters to be evaluated, the phantoms to be used, the measurement methods, the frequencies of the evaluations and, possibly, the tolerance limits.

A set of controls both for B-mode and Doppler are proposed. The parameters investigated were chosen following their clinical relevance and in the cases was possible, limits of tolerance suggested.

The tests were all used in commercial or homemade phantoms and the problems and general results were discussed. This can be a simple way to approach the physics of quality control in ultrasound. If someone wants to go into detail some useful reference is suggested. With the very high number of exams worldwide performed, it is mandatory to follow a QA programme like in the case of X-rays.

Author details

Raffaele Novario[1,2]

1 University of Insubria, Varese, Italy

2 Medical Physics Department, University Hospital of Varese, Italy

*Address all correspondence to: raffaele.novario@uninsubria.it;
raffaele.novario@gmail.com

References

[1] Gill R. The Physics and Technology of Diagnostic Ultrasound: A Practitioner's Guide. Sydney: High Frequency Publishing; 2012

[2] Scott C, Mount C. Ultrasound Physics and Instrumentation. Treasure Island (FL): StatPearls Publishing; 2023

[3] Cox B, Beard P. Imaging techniques: Super-resolution ultrasound. Nature. 2015;**527**(7579):451-452. DOI: 10.1038/527451a

[4] Novario R, Strocchi S. Ultrasound imaging and therapy. In: Keevill S, Padovani R, Tabakov S, Greener T, Cornelius L, editors. Introduction to Medical Physics. Boca Raton: CRC Press; 2022. pp. 235-263. DOI: 10.1201/9780429155758

[5] Wayne R, Hedrick R, Hykes D, Starchman E. Ultrasound Physics and Instrumentation. 4th ed. Amsterdam: Elsevier Mosby; 2020. ISBN 978-0323032124

[6] Grazhdani H et al. Quality assurance of ultrasound systems: Current status and review of literature. Journal of Ultrasound. 2018;**21**:173-182. DOI: 10.1007/s40477-018-0304-7

[7] Doyle AJ, King DM, Browne J. A review of the recommendations governing quality assurance of ultrasound systems used for guidance in prostate brachytherapy. Physica Medica. 2017;**44**:51-57. DOI: 10.1016/j.ejmp.2017.11.011

[8] Carson P, Nicholas J, Hangiandreou N, Lu Z. Clinical ultrasonography physics. In: Samei E, Douglas E, Pfeiffer M, editors. Clinical Imaging Physics: Current and Emerging Practice. Hoboken US: Wiley; 2020. pp. 249-260

[9] American Institute of Ultrasound in Medicine. AIUM Quality Assurance Manual for Gray Scale Ultrasound Scanners. Laurel: American Institute of Ultrasound in Medicine; 2014

[10] Sassaroli E, Crake C, Scorza A, Kim D, Park M. Image quality evaluation of ultrasound imaging systems: Advanced B-modes. Medical Physics. 2019;**20**:115-124

Chapter 12

Quality Assurance When Developing Software with a Medical Purpose

Jordy Coffa and Joost van den Berg

Abstract

In the field of development of scientific or medical software, questions may arise, such as how we define *if* software has a "medical purpose," what regulations may apply and how they influence the (projected) pathway. We may find ourselves embroiled with the new "In Vitro Diagnostics Regulation" (IVDR) and its implementation in organizations. In this chapter we will attempt to summarize and order key bits of information, as found in these standards and related publications, that seemed relevant along the way in our software development processes. After this we will try to expose possible pitfalls that could be encountered. We also reach out to (existing) methodologies that may aid in the endeavor to the realization of software. Top-down risk approaches consider hierarchical ordering of priorities based on process levels where context and meaning play a more significant role over content and documentation. To honor the different sources, we will seek to outline how this led to a form of understanding that allowed the development of software. Maintaining a high standard of risk control while keeping focus on product realization. Hopefully these outlines and referred source materials may bring slight relief to others on a similar quest.

Keywords: MLPA, digitalMLPA, DNA diagnostics, genomic research, next-generation sequencing analysis, data analysis, copy number assessment, software development, IVDR, medical devices, quality assurance, risk-based design

1. Introduction

Science may be seen as something that is involved in measuring what a thing is not. Science may be about hypotheses where it may be a form of communication, which is in a constant flux, challenging earlier stated definitions. Because the hypothetical does not exist, we set limitations and perform measurements according to systematic methodology, that may then lead to understanding of possibilities. When enough views are examined to concern that what is observed is certain within as set of specifications, then we may have a better understanding of what a thing may be. Techniques that investigate copy number aberrations may be no different, they typically measure that a test sample is not equal to a control reference. Controls to assure quality are also similar, in that they typically aim to measure that a sample it not according to certain standards. In practice, such standards may lead to empirically defined acceptance criteria, that could for instance be used to determine that the DNA concentration of a sample is not within the expected range. These ranges or acceptance criteria are often initially determined by predefined logic, but they may be adjusted following evaluation of discernable observations that via empirical logic may lead increased accuracy. Laws may be similar in origin as well; in that they typically focus on what should not be done, though they not as often may be adjusted based on empirical logic. Consequently, conscious law may invite its subordinates to examine "*what is too much*," but often fails to provide understanding and transparency about its purpose.

When developing software that may have a medical purpose, we may find ourselves embroiled in the practical application of regulations that are currently being enforced, such as the IVDR. We may also find ourselves in contact with these regulations, for instance via implementations of these standards into company policy. While standardization certainly has positive effects: to the quality of care, the consistency of outcomes and reducing costs, in other areas it may be experienced as impeding endeavor. Research may be such an area where regulations based on predefined logic block innovation. Innovation in science may leads to progressive leaps by discovery via intuitive understanding of discernable observations. Understanding of possibilities may for instance have led to the introduction of Multiplex Ligation-dependent Probe Amplification (MLPA) in 2002 [1]. When developing devices such as software, that may have a medical purpose, we may find ourselves stuck in the morasses of regulations as well. These standards and their implementation in organizations typically tell us what to do, or rather what we should not do, but they do not seem to focus on explaining how we should get things done. One-to-one implementations of regulations often lead to documentation cultures where the intended purpose of implemented regulations may have been lost altogether.

In this section, we like to discuss and consider strategies involved in the development of software that may or may not lead to a product with a medical purpose. We will try to use the new IVDR standard to provide guidance and consider obstacles that may be prevalent in commercial endeavor. Can we develop high quality products and fulfill the requirements of the IVDR harmoniously? By evaluation of the different pieces that make up the IVDR and observation in scientific writing of the effects of its implementation in practice, we will attempt a discussion to see if the IVDR really requires huge piles of documents and rigid structures. Perhaps the IVDR does value research and innovation, but it just takes time to understand the law, by gaining experience applying the law. We may then also learn how such standards are to be implemented in crossover design, where research and diagnostic are very near to each other such as clinical trials. Finally, we will discuss a methodology to quality assurance when developing software, that is designed to react quickly to changes, which tends to be essential in a dynamic field, while also upholding IVDR standards for quality assurance with patient safety as continuous mindset. We will then use this methodology and apply it in a few examples on a software project currently in development, that involves the analysis of digital Multiplex Ligation-dependent Probe Amplification (digitalMLPA) data to establish chromosomal aberrations [2].

2. Barriers to effective implementation of regulations

Since the release of MLPA, the technique has been become a widely used tool for copy number detection and methylation status analysis on DNA samples in both a research and diagnostic setting [3]. The increased use of MLPA in diagnostics, for instance where detection of copy number variation has prognostics significance [4, 5] has led to increased demands for more diverse products, as well as their certifications for In-Vitro Diagnostic (IVD) usage. MLPA has great properties to be used in routine applications and for screening [6] to detect, treat or prevent many more serious early-onset health conditions, such as when applied for newborn screening for Spinal Muscular Atrophy (SMA) [7]. Conventional MLPA reactions are however limited to about 40 targets, and thus broad-spectrum screening may need many different conventional kits, while some diseases may even need multiple kits for a single diagnosis, such as with Duchenne Muscular Dystrophy (DMD) [8]. Also, investigation of complex tumors, such as Multiple Myeloma (MM) often requires a more detailed mapping of Copy Number Alterations (CNAs). To respond to this demand, MRC Holland (Amsterdam, the Netherlands) developed digitalMLPA, a technique combining MLPA and Next-Generation Sequencing (NGS), to detect disease-related CNAs. This technique combines the advantages of NGS and MLPA and has the potential to include up to 1000 probes in a single reaction [2] requiring only a minimum of 20 ng. digitalMLPA already showed that the method is reliable in different studies and can provide comprehensive profiling of disease-related unbalanced genetic aberrations as well as to detect specific point mutations. Due to the ability to focus on a specific range of targets, relative low cost, and short turn-around time [9], digitalMLPA could represent a valuable addition to diagnostics and screening tests.

The increase in demand to product variation and product accreditation may be seen with more products in the biochemical arena and urges companies that are involved in manufacturing to upscale. The departments that are involved in production, support, testing, sales, and shipping for instance will all need to expand. Next to that, the demand for Conformité Européenne (CE) marked products requires accreditation, which often requires further expansion of departments that are involved in quality assurance, quality control and regulatory affairs, though we may find an overhead for nearly all departments, including software development. Today, patient safety is an increased priority to both patients, providers, payers, and policy makers alike [10]. This shift requires also changes to production systems that demand higher quality standards, as well as continuity in performance and delivery. For European countries the European Union rules enforces such standards by stating that medical devices or equipment intended for a medical purpose must undergo a conformity assessment to demonstrate they meet legal requirements to ensure they are safe and perform as intended. "Software is a medical device as well if the manufacturer has made the software for medical use...," states the website of the Dutch Government. For medical devices the Medical Devices Act (MDA) becomes applicable, and it is then important to determine in which class a device falls if one wants to continue in that direction. Depending on the class, certain requirements will apply for accreditation. This may for instance determine if the manufacturer may obtain a CE marking directly or whether an external audit need to be performed by a competent authority. Under the new Medical Device Regulation (MDR) the rules will be even stricter with regards to classification of a medical device and software has a high chance of being classified as a class IIa or higher. Though the European Union enforces such standards, there are no EU-harmonized rules governing the pricing and reimbursement of medical devices, member states are therefore responsible for establishing national social security schemes, including healthcare policies to promote the financial stability of their healthcare insurance systems [11]. One way of promoting social security systems is by expanding public health insurance aiming to reduce the risk of expenditures and improving health outcomes [12]. As a result, some healthcare providers in certain member states may only reimburse medical expenses in case the used products are CE marked. But in a commercial environment there may not always be interest to invest great amounts of time and money to obtain such markings, if a product is not economically viable to generate profit, for instance in the case of product intended to diagnose rare diseases. Even if a product for a rare disease is successfully developed, obtaining reimbursement from healthcare systems can be challenging. Pricing and reimbursement decisions often consider the cost-effectiveness of treatments, and with a limited patient population, it can be difficult to demonstrate cost-effectiveness compared to more prevalent conditions.

As regulations grow in scope and impact, understanding their precise role, effects, and limitations will become increasingly important for companies that make products and the hospitals using them. Especially for health providers there is a growing demand for transparency with the patient and the public in terms of how treatment decisions are made and when and why errors and unexpected outcomes occur [12]. So, while regulation may potentially provide a variety of benefits for instance in terms of maintaining safety and quality in patient care. At the same time, we may recognize that due to the extent, variability, and fragmentation of the regulatory system it is hard for regulations to act effectively, and it places a massive burden on provider organizations [13]. Design and production of biotechnological products and its related software may be an industry that is best understood by those who deliver it. It may thus be questionable whether there are competent authorities that have the expertise to provide dedicated investigations into quality systems that target specific areas where professionals are scarce to be found. Next to that with the new regulation being far more comprehensive than IVDD and it will require an estimated 80–90% of IVDs on the EU market to undergo a conformity assessment by a Notified Body (NB) [14]. Will there even be enough NBs to handle the increasing demands from the industry? When competent authorities cannot provide a dedicated system evaluation that provides insight into the effectiveness of such a system, then what will be their focus? Is it enough to merely evaluate that all generic protocols are in place and that version numbers on protocols and check forms match, or will they allow the profession to set the appropriate standards because it is best positioned to determined what is needed to improve safety.

Even if the profession itself may determine its protocol, then still the question arises if fast growing industries are equipped to effectively apply fast changing regulations internally while also expanding its internal machinery, especially when governments quickly apply new rules but do not have a framework yet in place to provide support and understanding on those legislations. The drive for companies to comply certainly exist, since care providers that use their products often may only reimburse their spendings in case the products used are CE marked. The drive to comply may therefore become a goal on its own and often is resolved within companies by expanding procedures and regulations. Clinical research may be an area that is being subjected by increasing regulation, research ethics committees and research and development offices are responding to social and political pressures that may lead to compromising clinical care [15].

Since progress must continue and patient care comes first, we may see in the industry that care providers find ways around the jungle of regulations to provide dedicated care for their patients. At the same time, we may see that in certain countries, individuals working in the industry completely shy away from care positions altogether where following regulations have overtaken the main priority above patient care. Shortages in dedicated trained staff are currently seen in all health care branches [16], subjecting patients to insufficient care, delays, and medical errors. Adequate staffing is essential for the delivery of quality care and safe nurse working conditions, which in turn are associated with better patient outcomes [17], A similar trend may be felt in research where overregulation is counterproductive, as it tends to discourage researchers. In other areas of endeavor such as software development or areas of technical production, companies may even outsource the required piles of papers needed to be sent to competent authorities. Outsourcing may have a negative effect, as it increases the gap between different entities involved in quality assurance, blocking harmonization of processes and feedback into design and production.

Next to that standards are often implemented by individuals that do not have an active role in the processes to which they apply. If we need to comply, then we may want to comply to patient safety and we would need to understand on an individual level how these standards may be implemented to further that goal. Enforcement to comply to regulation for its own sake may lead to atrophy of consciousness [18], as well as being a hinderance in intuitive progression via discovery. Why would this be any different when working in the medical sector where we are dealing with the health and lives of individuals? Caution comes from wisdom, not from fear (of non-compliance). Prudence requires awareness of the pitfalls to be avoided. Controls that are a threat to awareness, are a threat to patient safety and public health and therefore must be investigated on their effectiveness and improved or removed. Governments and managers ought to aid developers in providing the tools they need to properly perform the task at hand and to be established with minimal risks to patient safety. They may want to attempt to oversee and provide recommendations or key bits of information that leads to harmonization of different areas of endeavor rather than to disperse and divide. Consciousness itself may need to be our highest priority in establishing patient safety when developing products that are used as medical devices. Consciousness itself is self-regulating [19] and self-correcting which may often not be the case for ever-expanding systems that may become mostly self-serving. How can we then navigate through the morasses of regulations as developers while focusing on progress and prioritizing patient safety?

As an individual it seems irresponsible to attempt to vanquish the negative of regulation as it may lead to reductionism ignoring the social values which clearly may also play an important role in science [20]. Instead, we may apply the positive norm opportunely which practically exists only as spirit of the determining lawmaker, spirit which gives formal seriousness to some requirements which she/he considers rightful according to asset of values which contains the requirements and the interests historically determined [21]. As a scientist it may be easy to ignore the intangible and through creativity and logic deduction, dualistically reason that our intentions are good, thus as such we may avoid in clear consciousness, attempting to get deeply acquainted with new regulations. We may however also attempt to resolve the issue as a scientist and stimulate awareness and aim to create procedures that attends toward self-correction and allowing natural evolution of these procedures while considering both the product or goal and the human factor.

We could even draw this concern further and wonder why authorities themselves do not focus more on investigating legal consciousness, which is an area that can be defined as the attempt to make sense of the decisions of applicants. If even in areas of welfare, it may be clear that applicants 'know the law' and (ab-)use it [22], such may also be expected in other areas. In the end, when considering standards and regulations, as well as most integer companies, we may consider that they both aim for the same, with respect to patient safety. The new IVDR for instance, increases its focus on measuring effectiveness, accurate description and strengthening post marked research, all of which have been introduced for the benefit of the patient. We may also see that agencies are set in place to measure the real-world pertinence and relative clinical effectiveness of the IVDR [23]. We may further see that the IVDR as a key point aims to preserve innovation, safeguards quality, safety, and accessibility of innovative diagnostics.

When we want to design and develop medical software in a time of fast changing technologies and regulations, where demand often exceeds availability, whether related to development of products, or being knowledgeable about the exact meaning of new regulations, we may often just want to move forward and gain experience along the way if we keep our intentions always to act to the benefit of the patient. Manufacturers must strive to do the best as is possible with the information available at that time and evolve along the way as more information becomes available. Prudence, or acting in accord with what is real, is crucial when developing anything that is used in diagnosis, but we must also strive to have patience and understanding when working in teams. Stress, due to high workloads and unrealistic deadlines, rarely is an indicator that results in improvement of patient care [24] and productivity on the work floor. The virtues that we extent to patients, to some extend must as such also be applied to staff, disposing both parties to act well in relation to the end of medicine. Next to this, we may state that virtues of different people involved in medicine may be correlated with the goals of purposes of the medical encounter [25]. However, without a common agreement about goals and values, virtue theory cannot survive in health care [26]. In practice we may be more willing to get into a position of agreement, when the path ahead becomes clearer, and focus may be placed on well-defined shared objectives.

3. The pathway

The IVDR introduces a risk-based approach to classification, it is like the MDR in that it takes a lifecycle approach to IVD regulation. It also incorporates many elements of the current European guidance documents and concepts such as performance evaluation (with a major emphasis on clinical performance), vigilance and post market performance follow-up. The first step in the development of medical software, is to determine the class as it will highly influence the pathway to obtain accreditation which in turn influences how software should be developed and e.g., determines post market requirements. Besides, one may choose specific strategies that may make the progress easier and take steps that help to integrate the process into the organization. As such, the demonstration of conformity through technical files and records, quality systems and clinical evidence & performance may become a wide carried process and includes input from business executives, members of the data management team and workers throughout the whole line of development. By harmonizing the development of software and the required documentation they may work in synergy and time and effort may be spared.

3.1 digitalMLPA & software

To classify digitalMLPA software, it's important to understand its usage alongside digitalMLPA products. digitalMLPA probe mixes consist of probes targeting specific genomic sequences, comprising a Left Probe Oligonucleotide (LPO) and a Right

Probe Oligonucleotide (RPO). LPO contains a P7 Polymerase Chain Reaction (PCR) primer and a DNA hybridization sequence, while RPO includes a sequence for distinct digitalMLPA read identification and a DNA hybridization sequence. Hybridized probes that can bind to target sequences are ligated and amplified with a single primer pair, making MLPA robust for copy number assessment. The MLPA reaction involves mixing a minimum of 20 ng DNA sample with 2 ul of a unique barcode solution. After denaturation at 98°C for 10 minutes, 1.25 μL digitalMLPA probe mix and 1.25 μL buffer are added to each sample. Probes are then hybridized overnight at 60°C, saturating all target sequences in the DNA sample. Oligonucleotides are ligated enzymatically, incorporating a barcode oligo for sample identification. The ligase is inactivated by increasing the temperature to 98°C for 5 minutes and incubated at 65°C for 20 minutes to reduce background. Ligated probes are amplified using a single P7/P5 primer pair in a multiplex PCR, removing the need for unbound and non-ligated probes. The PCR-amplified products are quantified using an Illumina sequencer by utilizing the Rd1 sequence, which serves as the Illumina tag. The resulting data is stored in FASTQ files, a text-based format for biological sequences and quality scores. An average FASTQ file size is 23 Gb, containing approximately 25 million read sequences in a 7.5 Gb file. Specialized data analysis tools are required to analyze such data and provide a viable user experience that allows interpretation and give meaning to the results.

3.2 Classification

Here we describe possible classifications based on deductions using the provided information the EU has granted on the IVDR. Classification rules may vary based on regional regulations, intended use of the software and the different countries or regulatory bodies that may have their own classification systems. It is thus recommended to consult regulatory authorities or guidelines in the relevant jurisdiction for a more precise classification, we only wish to give an outline here. The IVDR uses seven rules for classi4ication along with guidance as provided by the Medical Device Coordination Group (MDCG 2020-16 Rev.2) on implementing the rules to help manufacturers to identify the risk class of the IVD devices. Assuming a risk-based position, classification starts by assuming the highest: Class D (detection of the presence of, or exposure to, a transmissible agent and/or working with infectious loads of a life-threatening diseases) or products that pose a high public health risk. The standard then tones down to Class A, products for general laboratory use which possess no critical characteristics (receptacles, buffer solutions, washing solutions, and general culture media). Guidelines to software classification follow more specific guidelines as laid out by MDCG 2019–2111. The starting point is setting out the devices, intended purpose, from which it can be determined whether it is a medical device and if it is, what the classification may be.

Software itself can be defined as a set of instructions, that processes input data and creates output data. Medical software is software intended to be used, alone or in combination, for a purpose specified in the definition of a "medical device" in the MDR or IVDR, regardless, of whether the software is independent, driving or influencing the use of a device. Medical devices are products or equipment intended for a medical purpose. This seemingly, simple question can be quite complex, and heavily related to the intended purpose that is assigned to the product, which a manufacturer should describe in detail to understand whether there is a medical action involved. In the European Union's Regulation (EU) they rely on No. 2017/745 on Medical Devices Regulations (EU MDR) and Regulation (EU) No. 2017/746 on in vitro diagnostic devices (IVDR). Software may be considered a medical device if it's intended by the manufacturer to be used alone or in combination, for human beings for one or more medical purposes such as: diagnosis, prevention, monitoring, prediction, prognosis, treatment or for providing information by means of in vitro examination of specimens derived from the human body, including organ, blood, and tissue donations. Not all software in healthcare is qualified as medical software, simple search and retrieval of records does not qualify as medical software, but software intended to process, analyze, create, or modify medical information may be qualified as a medical device software. We may also need to consider whether the software is a companion diagnostic or an accessory. Companion diagnostics only provide information for the use of a product and accessories indicate that the medical device can be used independently even in the absence of an accessory; neither seem to be true for software intended to analyze digitalMLPA data. MLPA data analysis software may be described as a molecular diagnostic tool as it aids in the interpretation of genetic data for diagnostic purposes. MLPA is often used in clinical genetics, cytogenetics, and research settings to identify genetic mutations or chromosomal abnormalities associated with various genetic disorders.

Once it is established whether the software is a medical device, then it needs to be considered whether it is an In Vitro Diagnostic Medical Device Software (IVD MDSW) or Medical Device Software (MD MDSW). IVD MDSW falls under the in vitro diagnostic medical devices Regulation (EU) 2017/746. In case the MDSW creates information based on data obtained by a vitro diagnostic medical device or if information provided is based on data, obtained solely from in vitro diagnostic medical devices, then it is an IVD MDSW. Software that takes raw data outputs and uses clinical algorithms for diagnostic and/or prognostic purposes also qualifies as a IVD MDSW. The IVDR requires that all existing and new IVDs are (re)classified based on a risk-based classification system [27]. Finally, the risk class may be considered that is applicable. This categorizes the risk of software, based on the combination of the significance of the information provided by the software to the healthcare decision and the healthcare situation or patient condition. We may for instance consider, that the software is used in combination with a medical device, that provides information on the statistical predisposition of Down syndrome (Trisomy 21), may be classified as Class IIb, since it may lead to an irreversible deterioration of a person's state of health. If the software is used to determine the etiology of intellectual disability and developmental delay to inform neurologists, geneticists, pediatricians, and patients' families, then the class would likely be Class IIa. The IVD class influences the requirements and assessment routes, and the required development process documentation needed to demonstrate that all requirements have been fulfilled, as for instance indicated by IEC 62304. A supporting approach to the documentation process that may be interesting for software development is Good Automated Manufacturing Practice (GAMP), GAMP 5, which is a science-based approach to understanding and managing risk for computerized systems that resembles the International Electrotechnical Commission (IEC), IEC 62304. These standards both suggest that for Class IIb and Class C devices should have detailed software design documentation while this is not necessary for Class IIa and Class B devices. Class I and Class A devices on the other hand only require planning, setup of user requirements and validation of those requirements and do not require architectural design specification, unit testing and integration testing, which is necessary for Class IIa and Class B devices and up.

If we want to make software for a product that is currently Research Use Only (RUO) oriented but as a future product has the intend to be a medical device, then one may argue that prudence entails both long-term preparation and short-term, goal-oriented planning. Quality assurance comes from due diligence in all stages of production. As such, we may choose to take up a higher classification initially as it will be easier to come down later, rather than to first built a high mountain and then attempting to climb it. Such decisions are however highly dependent on purpose, context, meaning, organizational structure, availability, priorities, and support by a wide range of members of the data management team, stakeholders, business users and business executives. If a team prefers to build a tool that will never be used for a medical purpose, then extensive documentation may be superfluous. Problems with documentation may become a complication that brings frustration and struggle. While documentation and code serve different goals, without documentation code is not a full-fledged product. Choosing an approach that is well thought through and supported through the whole line of development is therefore essential to the success of any project.

3.3 Innovative diagnostics

The IVDR recognizes the need for different types of products by admission of the clinical need for laboratory developed tests, In House IVD tests (IH-IVD), modified CE-IVDs, RUO tests, and off-label use, if their benefit for patients is obvious and can be proven. IH-IVD can be used by diagnostic laboratories if they meet several conditions and requirement as specificized by IVDR article 5.5. The IVDR thus explicitly states, that its requirement does not apply to IH-IVD, with exception of the relevant General Safety and Performance Requirements (GSPR) if several conditions are met [27]. Health institutions must justify that the specific needs of a group of patients cannot be met, at a desired level of performance by an equivalent device on the market that is a CE-IVD. Judgment of equivalence is up to health institutions, and they should have documented justification for their use of IH-IVDs, competent authorities only oversee compliance of health institutions notified bodies with respect to article 5.5. Justification, however, is a continuous responsibility and thus requires laboratories that have implemented IH-IVD to regularly monitor and evaluate new CE-IVDs. This extra work on its own, may be an incentive for replacing or marking IH-IVD through the industry. How equivalence is exactly determined, and what is expected in terms of comparison of performance, is open to speculation. In the absence of exact specifications what needs to be done, we may get started without delay. Epistemologically, for abstract ideas and beliefs to become more concrete, experience with the matter is necessary, which can only be obtained by doing the actual work. The implementation of the regulations may thus follow a similar iterative process where the final interpretation may only become clear while the law is applied. Innovative diagnostics are therefore typically expected to be first developed and applied locally, in healthcare institutions with specific expertise and then taken over by IVD manufacturers that produce above certain volumes, to make the tests cost effective.

When considering MLPA in its early development, we may see that this technique was first introduced in 2002 by MRC Holland as a research technique which then, among others obtained validity as a classifier [28] and as a technique to investigate copy number aberrations [29]. Through the years of usage and by experience via many validation studies it was able to confirm its performance. The IVDR also mentions that: "if a device is in a performance evaluation phase, it can be made available to institutions or laboratories to be subjected to one or more evaluations studies, with the intend to gather information on performance evaluation parameters" as mentioned in Annex I of the directive. This information can then be used for its conformity assessment. This would require the manufacturer to draw up a statement that the device conforms to the requirements of the directive, apart from those specifically itemized in the statement and that every precaution has been taken to protect the health and safety of the patient, users, and other persons. When IH-IVDs developed and used by the academic diagnostic sector make it into commercialization, other problems may arise, such as lack of interest by manufacturers due to economic viability and/or access patient-derived material for controls. In-house devices shall not be transferred to another legal entity, which thus entails that health institutions that manufacture in-house devices may only use those device within that same legal entity. Since a legal entity may be a single hospital or several hospitals that are part of one health institution, collective development of IH-IVDs may greatly reduce regulatory burden, as well as increasing access patient-derived material for controls. This may in turn lead to improvement of quality and standardization of tests. Access to patient-derived material may be one of the main factors preventing transfer of IH-IVDs to CE-IVDs, especially for rare disease and why it is nearly impossible for manufacturers to directly develop CE-IVD tests. Other factors include complexity, suitability for automation, algorithm dependence as well as a variety of national diagnostic and reimbursement practices that affect economic viability [23].

3.4 digitalMLPA RUO tests

Products designated for RUO are exempt from MDR and IVDR regulations, and if used solely for research purposes, they are not considered in-house devices. Common genetic disease research has already established knowledge on copy number variations using methods like conventional MLPA, array Comparative Genomic Hybridization Arrays (CGH), Single Nucleotide Polymorphism (SNP), SNP genotyping, and next-generation sequencing [30]. RUO products developed for research purposes typically fall into the category of general laboratory use, serving as elements in diagnostic testing plans to assess their potential future use in IVDs. digitalMLPA, as a screening technique, offers advantages such as replacing multiple MLPA assays with a single test, reducing costs and turnaround times, and targeting specific genomic targets without extensive NGS kit requirements. It ensures deliberate and purposeful testing to reduce ambiguity surrounding patient problems and prevents misleading disease associations. With its simplicity, low cost, target specificity, rapid detection of CNAs in up to 1000 target sequences, digitalMLPA is well-suited for diagnostic purposes. Furthermore, once development and quality control processes are established, new digitalMLPA products can be easily developed, including targeting niche diseases overlooked by commercial companies.

digitalMLPA data results cannot be interpret without specialized software, that can import FASTQ files and convert the sequence data into understandable results, that include meaning and significance for a variety of context dependent situations. MRC Holland, as the internal user, needs software for data analysis and quality assessments of digitalMLPA products and are therefore an important group of target user for which the software should be developed. We may expect that digitalMLPA will target areas for investigative research, such as the usage of copy number data for genomic profiling and their correlation to medicine effectiveness. The potential for molecular-guided therapy to affect the treatment of an individual patient has been recognized [31] and digitalMLPA may be a product that may aid in recognition of patients where such therapies may be beneficial. Especially in understanding cancer development and progression, next-generation sequencing has played an important role, and therapies that target a cancer's genomic dependencies has fueled the transition of genomic assays into clinical use [32]. Users of RUO products may therefore be, an important second line of users. While these products may initially be developed and used to investigate specific purposes, through usage and experience they may end up as IH-IVDs or as part of an IH-IVD device. We may also have users that already use conventional MLPA as a diagnostic tool or users that are in search of diagnostic tools and are evaluating how NGS devices may aid in lowering costs and improve turnover times. Typically, much of scientific software development, is carried out by scientists who create programs to support their investigations [33]. When software development is not done by scientists themselves, establishing requirements may be difficult considering that the goals in question are potentially unknown. In such cases it is particularly important to gather understanding of the environment, practices, technologies and wishes that potential and future users may have.

3.5 User surveys/pre-market research

Understanding of the user environment of potential customers is an essential step in an early phase of development since it may for instance lead to constraints to the resources a software program may depend upon. Software for digitalMLPA will need to work with FASTQ files which may very large files up to 100 Gb. If 8 GB is the standard amount of RAM for your average desktop computer in 2022, special operative processing capabilities will be necessary to support analysis of raw digitalMLPA data on an average computer system. These early requirements form the base for the operative and system requirements that need to be established early in development, as they may influence all secondary choices and design specifications. Failure in compliance to such requirements may lead to the development of software that either needs users to greatly invest in computer systems, or simply leads them to search into alternative choices. When we are dealing with a wide range of different user types, we also need to consider that different users may have different availability to computers systems. It is then thus imperative to design the software to support the group of users that have the lowest operative capabilities, though scalability is also a factor.

Another point that needs to be identified early in development is the mode of delivery of the software and the required configurability. Software for digitalMLPA eventually will preferentially be able to support all MLPA products over a longer period. Since digitalMLPA products are still under development, the exact requirements for the software are still changing. We may thus for instance recognize, that internal users at MRC Holland may have specific demands and want to use the software for quality control purposes. They also want to have an easy way to adjust and update the configuration, so its content matches the actual digitalMLPA products. While digitalMLPA product developers want flexibility in the software, users in a diagnostic laboratory often prefer products, software, and its configurations to be stable. This allows them to validate a technique which may then be used as an IH-IVD or part of an IH-IVD. Adjustments in software or its configuration will often an incentive for them to reperform the validation, completely or partially.

MRC Holland's internal users, who are primarily focused on researching digitalMLPA product development, can be considered as RUO users. Their preferred output should support adaptation, filtering, and manipulation, enabling comparison of results over longer periods. Unlike field users who value stable products and changing samples, Internal users anticipate that the digitalMLPA products will change while the samples remain relatively constant. RUO users typically handle larger sample groups simultaneously and require a data presentation format conducive to integrative review. On the other hand, prospective users in the diagnostic field concentrate on individual patients and prioritize extensive quality control. Eventually, diagnostic field users may prefer CE-IVD products and software due to the added burden on clinicians when using RUO products in an IH-IVD setting. To develop software that caters to all user groups, specific requirements from regulatory affairs must also be considered. For example, traceability of patient reports may be necessary in a diagnostic setting, while results should remain unalterable to ensure data integrity. Addressing these contrasting demands calls for a flexible framework with extensive configurability, along with modes that manage risks by limiting adaptability for diagnostic-focused users.

We may furthermore need to recognize that the success of the evaluation of RUO product by users influences the progressive move of those products into IH-IVDs, which may then stimulate the production of CE-IVD products. The transfer of research into diagnostics is also more likely to occur within institutions as a consequence of trust rather than availability alone. Within diagnostics we may also have different groups of target users. A small institute for instance, is not expected to have much automatization and is more likely to be interested in software that can be operated using a Graphical User Interface (GUI). A single user of a digitalMLPA kit will likely execute the MLPA procedure as well as analysis of the data. A large service lab on the other hand that performs hundreds of analyzes each day will probably have a specialized department for the analysis of data and automated pipelines that revolve around Laboratory Information Management System (LIMS). This type of user is unlikely to be interested in GUI applications and will rather integrate a technique and related software components into their existing pipelines. To be able to support these kinds of features specialized software architectures are necessary. Early in development, conceptual, logical, and physical data models need to be created in a sequential process that involves all members of the data management team and business users. Agreement on what is needed is essential to establish, otherwise, the data models may not fully capture the business context of data or meet an organization's information needs. Conceptual modeling may thus be an important part to get to a point of understanding of what is needed before building commences. Knowledge and understanding of a complex usage environment may result to the process of creating a more usable and useful tool. Another important part that needs to be considered early in development is the method of documentation, which may be the results of the risk classification, but proper documentation may also lead to transparency regarding evidence about its performance and algorithms. Choosing a fitting strategy early in development may make the development of software easier, as well as testing and integration.

3.6 Iterative behavior

To ensure a comprehensive and effective approach to software development, it is crucial to recognize and adapt to the varying needs of different elements throughout the development process. As depicted in **Figure 1**, attention to these elements should be iterative and correspond to specific stages of development.

Figure 1.
Elements that are expected to return on a cyclical basis.

By making well-considered decisions and obtaining team agreement on goals and timelines, significant time can be saved, and disappointment and frustration can be avoided.

It is important to acknowledge that these elements are interconnected and that the software's scope may evolve over time as more information becomes available about the goals. Therefore, making decisive choices regarding the prioritization of development components and understanding their mutual influences will significantly impact factors such as cost, agility, simplicity, scalability, fault tolerance, performance, and extensibility. In this context, the creation of data models and workbenches assumes a crucial role in the development process. They provide valuable insights into the possibilities and often inspire early users to envision software implementation approaches. Consequently, these investigations may lead to the identification of new requirements that will influence the establishment of the software architecture. Overall, by recognizing the dynamic nature of software development, adapting to changing requirements, and leveraging tools such as data models and workbenches, developers can ensure a more informed and effective development process.

3.7 Software architecture

The architecture style chosen for an application plays a crucial role in determining its characteristics and behavior. Certain styles are particularly well-suited for highly scalable systems, which are often necessary for accommodating a large volume of users, as seen in service labs. On the other hand, some architecture styles enable developers to respond quickly to changes in requirements. Each style possesses its own strengths and weaknesses, and selecting the appropriate one that aligns with specific business needs and goals will greatly impact the subsequent stages of application development, as well as its extensibility and usability. Prior to embarking on significant software projects, companies or institutes intending to develop software may find it beneficial to initiate modeling projects. During this phase, data modelers or architects interview stakeholders to gather comprehensive requirements about the processes involved. Business analysts may also contribute to designing conceptual and logical models, which can subsequently be employed to convey technical requirements to designers. Skipping these crucial steps can result in implemented software architectures and data models that fail to fully capture the organization's and users' needs. In recent years, the field of software architecture has witnessed significant advancements, leading to the emergence of new styles such as microservices and domain-driven design. Architecture styles provide an overview of the system's macro structure, while patterns offer structural building blocks that can be utilized within architecture styles to address specific problems. Well-designed patterns have the potential for reuse in applications and company infrastructure. This section aims to explore various considerations in architectural styles during the early stages of software development for digitalMLPA, with the objective of meeting project business needs. The choices made in architectural styles have far-reaching consequences, impacting the overall development process, including quality assurance and the pathway to certification.

3.8 Requirement drivers

3.8.1 Useable software without installation

Often some desired functionality will be the main drivers in choosing software architecture. When starting development of software for digitalMLPA it was seen as an advantage to create the opportunity for users to be able to use the computer that is used to run the NGS system. These computers are often powerful systems that could make the conversion of FASTQ files much faster as opposed to using their personal computers. However, users in institutes will often not have installation rights on such systems, as such one of the main requirements was to develop software that could be used without the need of installation.

3.8.2 Data sharing

The software system supporting digitalMLPA experiments requires the capability to handle data sharing among multiple user groups. To optimize costs, users can utilize a central facility for NGS runs, enabling the merging of digitalMLPA experiment products for sequencing. Consequently, the centralized unit must be capable of redistributing specific experiment-related portions of the data from a FASTQ file. To fulfill this requirement, the software should employ a file-based data storage system utilizing a customized format that allows for data updates throughout each phase of the analysis. In this setup, a FASTQ file can be divided into separate files containing data specific to individual samples. These files can then be grouped together within separate folders associated with each experiment, facilitating redistribution to the respective experiment owners. For example, on a MiSeq device from Illumina, it is possible to combine approximately 64 reactions for probemixes with around 600 probes, with single reads of 110 nucleotides or longer. On a Nextseq device, it is feasible to combine 192 reactions for probemixes with a similar probe count. Combining data in this manner can lead to significant cost reductions in NGS runs by minimizing the number of plates and reagents required. Automation, when coupled with data combination, further enhances sample processing efficiency while reducing variation and waste. While file-based data systems may have certain disadvantages, such as potential limitations in performance and scalability, they offer notable advantages in terms of data retrieval, backup, and sharing. The simplicity and efficiency of file-based systems make them an attractive option, eliminating the need for installing third-party components. Overall, the software's file-based data storage system, coupled with the ability to merge and redistribute specific experiment data, enhances collaboration and cost-effectiveness in digitalMLPA experiments, without the complexity of additional software installations.

3.8.3 Reusing of components

In the development of components for digitalMLPA analysis, it is worth considering their potential for reuse in future software programs, including updates to the analysis software for conventional MLPA data [34]. By accurately identifying the components that can be shared between conventional and digitalMLPA and designing them in a way that allows their availability as framework components within the company's infrastructure, economic benefits can be realized. By reusing validated and verified components, a significant portion of the work required for certification may already be completed even before the entire solution is finished. This approach not only saves time and effort but also improves the overall quality by standardizing procedures that are recognized as part of the general functionality for MLPA. For instance, techniques like normalization and population statistics are commonly employed in MLPA, and their standardization through the reuse of components can enhance quality and reliability. Moreover, by establishing a framework of reusable components, the company can leverage this infrastructure for other software programs and projects in the future. This approach promotes efficiency and consistency across different applications, reducing redundant development efforts and fostering a more streamlined development process. In summary, by identifying and designing components for digitalMLPA analysis in a way that enables their reuse and incorporation into a framework, economic benefits can be achieved. Reusing validated components saves time, effort, and potentially contributes to meeting certification requirements. Furthermore, standardizing common procedures enhances quality and establishes a foundation for future software programs within the company.

3.8.4 Continuous integration/continuous delivery

Continuous integration and continuous delivery (CI/CD) are important aspects of software development, especially when integrating software into an organization. In scientific software development, the end goal is often uncertain. A monolithic software architecture, which consists of a single deployment unit, may be cost-effective and easier to design and implement initially. However, it lacks scalability, fault tolerance, and elasticity. Domain-Driven Design (DDD) offers an alternative approach, focusing on the function of the domain rather than the workflow or technical components (**Figure 2**).

In DDD, all functionality related to a specific domain is grouped together, allowing changes within that domain to be self-contained within a specific area of the system. Microsoft has been promoting their version of domain architecture, which is based on developing microservices-based applications and managing them using containers. Microservices, as single-purpose, separately deployed units of software, excel in scalability, agility, and fault tolerance. They also facilitate risk management due to their

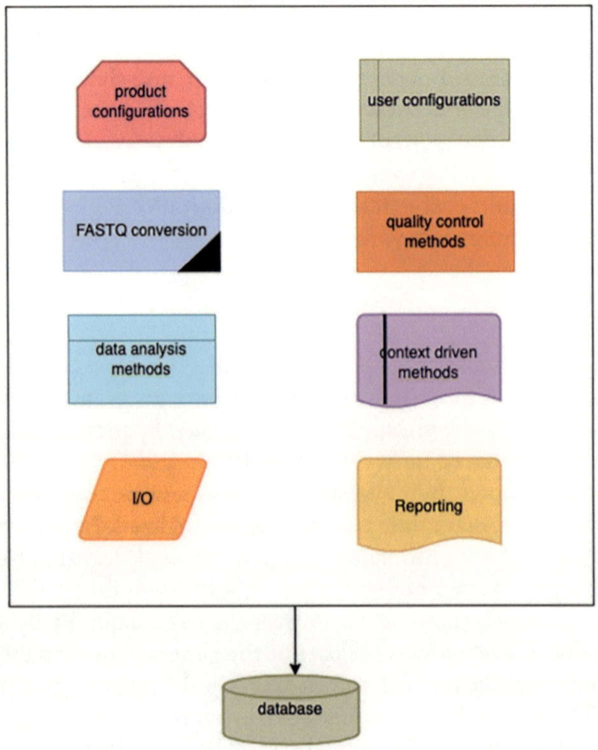

Figure 2.
Domain structures software architecture.

focused functionality. However, implementing and maintaining microservices in an environment with rapidly changing data sources and extensive database content across multiple functionalities can be challenging. Domain-driven design offers advantages in validation and verification testing, as separate components have well-defined inputs and outputs, enabling quality control and integrity testing. These components can be easily integrated into automated pipelines. DDD also allows close collaboration between development teams and domain experts, focusing on specific key parts of the system. Over time, the individuals involved in development may change, depending on the scope of the development cycle, which can limit team sizes while enabling specific testing and a focused approach. DDD also facilitates the gradual integration of components into the organization for evaluation purposes while development on other components continues. This iterative approach allows for feedback and evaluation at different stages, ensuring a smoother integration process. CI/CD is crucial for software development, and DDD offers a beneficial approach that emphasizes domain functionality. Microservices-based architectures provide scalability and fault tolerance, although they can be challenging to implement and maintain. DDD enables efficient validation and verification testing, close collaboration with domain experts, and gradual integration of components into the organization for evaluation.

3.8.5 Separation of concerns (design)

Separation of concerns in software design, particularly in layered architecture styles, ensures that each component within a specific layer focuses solely on the logic relevant to that layer. Each layer has its own role and responsibility in the application, which enhances its effectiveness and testability. For example, the presentation layer handles user interface and communication logic, while the business layer applies specific rules associated with the application's domain. Each layer forms an abstraction around a particular request, isolating its functionality from the other layers. The advantage of closed layers is that they provide independence to each layer, ensuring that updates or changes made in one layer do not affect the others. This isolation makes risk management easier and allows for specific scoping of development cycles or sprints, as changes are contained within isolated parts of the code. Understanding the impact of these changes enables better definition of associated risks and improves quality assurance. Additionally, the separation of layers enables the design of presentation layers to be deferred to a later stage while other components are being tested and integrated into the organization. In scientific software development, where functionality often takes precedence over usability, separating end-user presentation design features such as GUI and reporting options from the analysis modules proves beneficial. Analysis modules produce visual output primarily for verification purposes, allowing their development, documentation, and testing to be carried out independently, possibly even by separate groups. The output design can be optimized for easy automated integrity testing, promoting stability, and simplifying quality control while other parts of the software are being developed. Back-end specialists can focus on developing the functionalities of analysis modules, while design specialists can handle front-end design and report generation. Usability is an essential aspect that often requires close interaction with the intended user groups to ensure that the software efficiently and effectively achieves the set goals. By separating presentation design features and analysis modules, software developers can engage with design specialists and users to create a user-friendly and goal-oriented application. Separation of concerns, particularly in layered architecture styles, improves testability, risk management, and quality assurance. Separating presentation design features from analysis modules allows for independent development, testing, and documentation. Design specialists can focus on usability aspects while back-end specialists work on functional modules, leading to a more efficient and effective software application.

3.8.6 Interfaces

Interface design may be a process more akin to an art than a science and requires a different skill set and possible several rounds of iterations with communication between different end user groups and the developers to reach a satisfactory result. Challenges involved in usage of interfaces are often of a "more subjective order." The applications that a scientist works with are often less advanced, less secure and deliver an inferior user experience compared to the applications the same scientist uses on their smartphone after work [35]. Though a digital revolution is taking place biotechnology companies are slow to adapt and many companies still believe that software must be developed and maintained in-house, and that data is more secure on internal servers than on the cloud. Cloud services are often avoided being believed to be less secure than local storage which requires companies to spend both capital and operation expenses on in-house servers and in-house software development. By investing in components and codes that are encapsulated and independent, movement to a different platform in later stages may be made easier. Thus, when the overall organization has gained more trust in cloud services, much of the previous work can be easily transferred and documentation, testing procedures and risk assessment information are more readily available.

3.9 Life cycles

3.9.1 Principles of coding

1. Keep it simple: Write code that is easily understandable, self-explanatory, and divided into smaller, manageable pieces. Use clear variable names and separate bigger problems into smaller components.

2. Risk management: Prioritize patient and user safety by exercising risk management throughout all phases of development.

3. Epistemology: Invest in understanding the theory of knowledge especially regarding its methods, validity, and scope, and the distinction between justified belief and opinion and apply methods for testing and verifying conformance. Establish a shared nomenclature within the organization to ensure accurate descriptions and prevent confusion.

4. Research: Conduct thorough research before starting coding to identify user groups, their requirements, and constraints. Create separate components for evaluation and testing, allowing for early feedback and ease of reusability.

5. Meaning and context over content: Focus on implementing functionality that aligns with the intended use and minimizes patient risk. Emphasize the importance of context, intention, and meaning in the code.

6. Writing good requirements: Craft clear, unambiguous, testable, and correct requirements that serve as blueprints for the design. Good requirements facilitate understanding, minimize misunderstandings, and aid in creating acceptance criteria.

7. Tools development: Consider the development of tools that support various processes such as traceability, documentation, testing, configuration management, and deployment. Use available software tools to streamline testing and improve code quality.

8. Documentation: Maintain accurate and concise documentation tailored to the target audience. Keep code and documentation closely linked and updated together. Consider integrating code and documentation in the same file to enhance understanding.

9. Order of development: Start with macro structures and main functionality that can be easily validated. Then focus on implementing configurable options to meet specific user requirements. Finally, address user interfaces and user experience, which are subjective and require iterative development.

Adhering to these principles may promote effective communication, collaboration, and quality assurance throughout the software development process.

3.9.2 Agile

A life cycle is a process used to design, develop, and test high quality software. It consists of planning, requirement analysis, defining, designing, coding, testing, deployment, and maintenance. Agile approaches emphasize on continued collaboration and improvement. Tasks for a project are typically placed on boards to produce comprehensive overviews. The project is then divided into phases that each consist of planning, executing and evaluation. The advantage is that a large, complex project then consists of manageable chunks. Evaluation after each sprint improves collaboration and invites adjustments in the processes whenever this is necessary. As with the interpretation of the law, optimizations to processes are often recognized while executing them. Quality assurance departments recognize the need for corrective and preventive action but in practice only execute these after non-conformities are found. Though there is significant evidence that agile methods can be applied for developing software in the safety critical industry [36] and that this has significant benefits, organizations in highly regulated environments often lack suitability to accommodate fast response to changes. Processes may be dictated by standards rather than adapting methodologies to the company. The time and the energy that is needed to implement changes in processes is discouraging. So why would we still want to encourage agile methods? The obvious paradox is quality assurance itself.

Though the documentative culture caused by the need for compliancy to regulations and the fear for regulatory inspections [37] may prevent agile working, developing teams may still employ these methods to ensure the quality of the product. The idea is to create a process based on empirical logic that in the end fits into the defined logic of the company standards. This may create overhead for developers and lead to two track systems that then often requires syncing. Failure in doing so tends to make integration more difficult. The IVDR however is clear that we must attempt to take every action possible to minimize risks to the patient even if it means avoiding processes that are set in place due to that standard. Working agile in shorter sprints promotes quality because it limits the number of changes. If the software architecture has a strong domain-oriented architecture, it may prevent changes in one domain from affecting others. This focus minimizes the risks of introducing bugs because the scope of the changes is limited, and the effect are easily recognized. The scope of a cycle may for instance completely focus on interfaces, no changes in result data are then expected. In case discrepancies are found, they may be easily traced since the change set is limited and confined to a single domain. In comparison, if we choose to make a wide set of changes that affect analysis methods, configuration, and interfaces, then the complexity increases exponentially. The source of an issue may then be the result of changes in any of the domains or even be the result of unexpected interactions between these areas. Traceability may require rigorous investigation of possibly hundreds of changes to find the source of an error. Next to that, due to the increased length of a cycle, feedback from official testing may take such extensive periods of time that developers will need to invest time to get reacquainted with the topics at hand. Organization may also urge developers to continue working on a project before test results are produced. In those cases, changes may be placed on top of changes making traceability nearly impossible.

Shorter cycles may furthermore prevent scope creep, that is when new project requirements are added after the project execution has started. This has a major impact on project management and increases risks because these changes are often not properly reviewed, and developers are expected to complete the tasks with the same resources and in about the same time as the original scope. This in turn leads to stress and frustration, it crushes enthusiasm and invites quick-and-dirty programming. This tactic of programming leads to serious deficiencies also known as the technical debt. We want to obtain all objectives, but we do not have the resources needed so priorities are reassigned, and certain objectives are postponed. We may for instance assume that, like in traditional development, QA teams run their tests and code is expected to have bugs and that as such developers only need to ensure that the code runs. When developing software for a safety critical industry cutting corners should be avoided. If QA becomes agile as well, they can work closely with developers, and each group can focus on different aspects of testing. Harmony and balance may reduce work and stress for both groups leading to higher quality products. Limited cycles also make it more acceptable for new project requests to be executed in the next cycle of development. Another way to prevent scope creep is by defining the requirements for a project, making a schedule for their execution, and then verifying the scope with the stakeholders. In that way an agreement may be reached that is verified by team members in stakeholders, in practice we may notice that unless officially verified, agreements tend to be easily mended.

3.9.3 Software validation

An important part of ensuring patient safety is sound risk/hazard analysis. Also, in software development it is an essential and required process to logically process for risk management from risk identification through evaluation, rating, and mitigation. International Society for Pharmaceutical Engineering (ISPE), ISPE's GAMP5 is a framework that provides a structured approach to validation that can be applied to agile development processes, as well as more traditional development patterns. GAMP is an acronym for Good Automated Manufacturing Practices, suggesting a scientific approach to enable the development of high-quality software that meet regulatory requirement and ensure patient safety. Standards such CE and FDA describe what needs to be done, but they do not describe how this must be accomplished. Protocols that are placed into effect by companies based on these standards often are translated one-to-one leading to inflexible systems that are document based on predefined logic that are inflexible to new implementations of context. The protocols are often also not defined by those who use them, leading to operational deficiency. Methods that are not adapted and made suitable for those who work with them, may lead to conflict, prevents innovation, decreases efficiency, and takes focus away from the core essential of the regulations: patient safety and product quality. The problem often lies with the absence of knowledge and experience with the specific matter at hand leading to demarking everything with the same priority and the idea that everything must be documented. The evaluation of risk to quality may instead be based on scientific knowledge and linked directly to the risk to the patient. In GAMP5 the level of effort, formality, and documentation of each process commensurate with the level of risk associated with that process. One could argue that this is the same as the benefit/risk analysis for all risks and the overall residual risk as required by the IVDR. GAMP5 has also been endorsed by the FDA as an acceptable approach to risk-based software development.

GAMP5 has a top-down approach that looks at processes before systems and operates from the notion that the risks associated with the computerized system cannot be greater that the risk associated with the processes it supports. When considering software for digitalMLPA, then the risk category for the software is thus directly linked to the digitalMLPA product that has the highest risk classification. When considering the relative risk level within the software, an initial assessment should be done that is based on an understanding of the processes, typically derived from the user requirements, design specifications, operating procedures, regulatory requirements, and functional areas as well as understanding of the behavior of the users. This information is then used to identify the specific functions that have impact on patient safety, product quality and data integrity. These functions can be identified as possible hazards and may require extra controls to minimize their harm. The rigor of the risk can be based on the impact of the function, note that no function can have a higher risk than its related process. Risk assessment can typically be done using the Failure Mode and Effects Analysis (FMEA) derived risk assessment processes, where severity and probability are set out against each other to obtain a risk class. The risk class is then plotted against detectability to obtain the priority class. High impact functions will often require mitigation, for instance by adding requirements for control systems, but they may also be warnings or constraints to limit behavior. Correct functional risk assessment requires a strong knowledge of the system, as well behavior of end users. It is also advisable to communicate mitigations with the final users to identify if the controls are suitable in practice. For example, a warning message that out of habit is directly closed by users without reading the information may not be a suitable control. Such behavior can only be discovered by direct contact with final users.

3.9.4 Implementation and testing

Based on the identification of the functions that have a certain severity and risk, tests may be developed to evaluate if the implementations meet expectations. For testing low level impact functions, we may suffice with validation of a user requirement. For digitalMLPA software we may want to validate that a user is able to produce a log report, such low impact requirement meets compliancy if they match normal business expectations. Medium impact level functions need consideration of failure mode and typically involve the creation test case scenarios. These test scenarios describe use cases and their expected outcomes. For digitalMLPA we may for instance want to have a test case scenario where a user has specified incorrectly which digitalMLPA product was used. The digitalMLPA software has a control that matches the set digitalMLPA product to the automatically recognized digitalMLPA product. The test in the implementations when executed by a user must then show that a user was able to correctly identify the problem when using the software. Such outcome may involve subjective reasoning, since a user need to identify that there is a problem and what that problem entails. This may thus lead to the need to install extra controls or adjustments to the chosen implementation. For high impact functions specific risk scenarios must be tested. For digitalMLPA software, we may for instance expect that there are probes in probe mixes that are used to detect DNA mutations that are used when considering the diagnosis of a patient. Note again that the highest risk class cannot be higher than the risk class of the software. Under the IVDR that would automatically indicate that use cases that involve diagnosis of a patient are only appropriate if that software was distributed and intended to be used with a digitalMLPA product that has an intended purpose to influence the diagnosis. High impact level functions should be rigorously tested to ensure that the appropriate response is generated. For digitalMLPA this may involve creating test scenarios that involve evaluation of the appropriate response when reference signals are absent for these mutation probes or what happens if the settings in the configuration are incorrect. Based on the outcome of the tests, controls may need to be applied to increase robustness. Controls and their test results should be retraceable to their original risks. Risk assessment is typically repeated periodically and in case the risk class changes, for instance due to the installment of a higher class digitalMLPA product in the configuration of the software, then appropriate measures and controls need to be installed that match the impact level. The higher the risk level of the function, the greater the controls that should be applied. For digitalMLPA software when analyzing a mutation probe that has the highest impact level, we may for instance set in place a requirement that to analyze that probe in a test sample, we need in that same experiment both a positive sample and a negative reference sample. Such samples may then be used to establish the expected outcome under the same circumstances allowing higher confidence when calling a result in an unknown test sample.

An alternative strategy to estimate the necessary level of testing and investigation of the impact level of each function is by using the V-models (**Figure 3**).

In V-models often the depth is related to the risk class. As such low impact user requirement only need performance qualification. This usually entails requirement analysis, unit implementation and system testing (Class A – Level 1). Medium impact user requirements obtain functional specifications and perform both performance and operational qualification. Here design documentation is extended with architectural design and testing is extended with unit verification and integration testing (Class B – Level 2). High impact user requirements then also include the lowest layer of technical design specification and verification of these specifications (Class C – Level 3). Each step is followed by risk assessment, as well as evaluation of test results and installment of controls whenever needed.

It may be worth mentioning here that software designated for digitalMLPA cannot produce any data on its own, similarly a digitalMLPA product cannot produce useful output without specialized software. The product digitalMLPA, thus consist of the probe mix kit, the software and the service or support provided. A user requirement

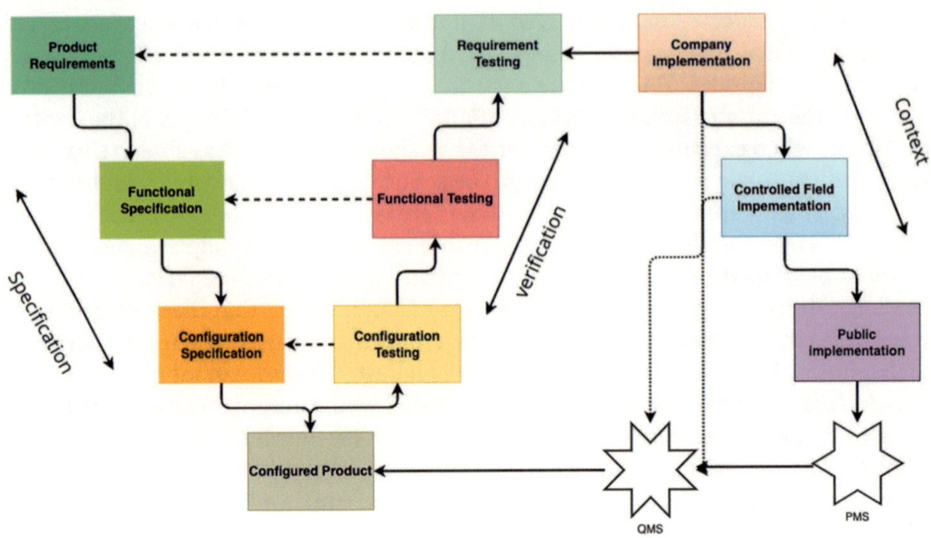

Figure 3.
Example of a product validation pattern.

of a product may then for instance be the ability to detect certain types of Duchenne muscular dystrophy. Associated risks when performing such an analysis may be that incomplete denatured DNA can show patterns that resemble deletions of copy numbers. Control mechanisms may thus be set into place to measure incomplete DNA denaturation by adding specific probes that are targeting GC% rich areas that will show early signs of denaturation issues. The functional tests to ensure such a control system works appropriately could be established by calibration series for that product using samples with increased salt concentrations. These salt concentrations will make it incrementally harder for the DNA to denature. We then want the control system to pass off a warning long before entering the process of copy number estimation related to disease diagnosis. However, the only way to measure this feature is to have specific software that can fulfill the needs related to the specific control mechanism. In this case the software is thus in direct association with the user needs of the overall product and the validity of the control mechanism can only be tested by the combination of the two elements. Controls should always be traceable to their original identified risks, and residual risk evaluation must be performed after installment of the selected controls for all high-risk functions. We may thus recognize that due to the associative actions between a digitalMLPA product and its related software, there may be many different V-models and that the depth of these models may be multilayered. Such multilayered systems can be executed in stages that have their own specific models for evaluation. The required depth should however be associated with impact level which is always related to the overall risk level associated with the intended use of the product as specified by the manufacturer. digitalMLPA software should fulfill the data analysis needs of many digitalMLPA products that may have different demands depending on the type of product. To do this, the system does not alter its original functionalities but rather allows adaptation of values that are assigned to certain variables to fulfill the needs of different users in the context of their processes. The risk level of the software itself should thus be related to the risk level of the digitalMLPA product that has the highest risk level. Please note that even though digitalMLPA products may use the software as a type of control that exists on a single level for the digitalMLPA probe mix, that we would still recommend multilayers testing and installment of controls for the software itself.

3.9.5 Review and monitoring

Once software requirements are verified and controls are implemented, they require monitoring. Implemented controls can reduce testing efforts, eliminating the need for calibration series with different salt concentrations for each digitalMLPA probe mix once we confirm that the installed controls function effectively across multiple mixes. Assuming the control configuration remains unchanged, residual risk analysis should be conducted to ensure the adequacy of controls. Periodic evaluation is necessary to identify new, unrecognized risks and determine if recognized hazards, their estimated risk levels, and controls remain appropriate.

4. digitalMLPA software

The analysis of digitalMLPA data involves assessing signals representing the relative number of target sequences in DNA samples. Conventional MLPA utilizes fluorescent tags connected to PCR primers for signal measurement [1]. Capillary electrophoresis separates and measures relative fluorescent units of probe PCR products based on length. In digitalMLPA, NGS devices sequence the probe products [2]. Analysis of resulting FASTQ files involves counting reads associated with specific probes in samples. Barcode sequences correspond to samples, while target sequences identify the probes. Normalization compares read counts to reference values, providing probe ratio values for copy number estimation. Probes can be clustered and sorted to identify genomic patterns for disease recognition, prognosis, and correlation with treatment effectiveness in precision medicine research.

4.1 The analysis workflow

The proposed software for digitalMLPA organizes the analysis of FASTQ files into different domains with specific tasks (**Figure 4**).

Each domain has its own executable subprogram, which can be accessed through a provided GUI that executes all steps sequentially. Alternatively, these subprograms can be integrated into existing pipelines and connected to LIMS. Incorporating LIMS can enhance automation, but it requires on-site experts familiar with the program's

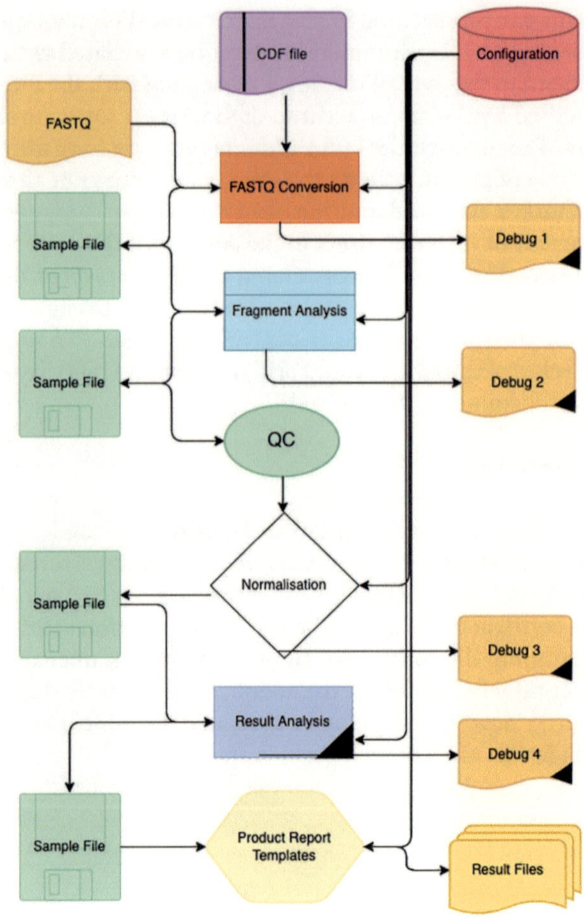

Figure 4.
digitalMLPA analysis processes flow.

input and output requirements. The analysis workflow begins with FASTQ conversion, which requires information about the NGS run (Coffalyser Definition File, CDF file) and a configuration containing digitalMLPA probe mix and barcode collection information. FASTQ conversion generates proprietary file formats for each sample barcode, serving as input for subsequent analysis stages. This allows users to demultiplex large FASTQ files at one location and continue experiment-specific analysis at a different location. In institutes where multiple groups share an NGS device, this behavior is often expected as each group may only be interested in a subset of the FASTQ file. FASTQ conversion can be automated through a tasked operation system that triggers the conversion when the FASTQ file is generated by the NGS device. As FASTQ conversion can be time-consuming, automation is beneficial. After conversion, the resulting sample files undergo fragment analysis, which performs sample-specific quality checks to determine their suitability for subsequent analysis steps. Several steps implement controls for higher processes, including NGS machine performance, DNA sample quality, sample identity, MLPA reaction quality, and context-related factors such as reference sample and probe quality factors. Quality control also ensures reliability of the overall results, aiding in quality evaluation when discrepancies occur. Following fragment analysis, the sample files are updated and used as input for data normalization, which applies specific instructions from the CDF file and configuration to normalize samples within the same experiment. After normalization, the sample files are updated again and serve as input for results analysis. This component sorts and clusters the results based on sample type for comparative statistical analysis, enabling determination of significant changes in sample probe results compared to reference collections. The results analysis facilitates confident result calling and customized recognition of genomic disease patterns. Results are stored in the sample files, which can generate various types of reports. The report format is highly configurable, allowing digitalMLPA probe mix designers to customize the products for customers. Each component produces debug files to facilitate validation and stability testing. This enables automated unit testing when updating a single component, ensuring stability in other components. This methodology supports quality assurance by facilitating estimation of the effects of changes within a single domain.

Domain-specific changes enable focused and specific verification, allowing scrutiny of individual software components while verifying the stability of others. This approach optimizes large software projects by integrating each part into the organization while other processes continue in parallel. Although domain-organized software may initially have a longer development path, it remains more agile throughout its lifetime. Documentation around quality assurance can benefit from this structure if organizations are prepared for this work style. Importantly, this structure lends itself well to a top-down risk approach that centralizes focus on the requirements of the highest processes, which may have an impact on patient safety. This approach allows for the implementation of dedicated controls to minimize risks effectively. The effectiveness of controls and regulations in minimizing risks to patient and user safety should be continuously verified through implementation and measurement in the field, encompassing the entire scope. These measurements are essential for establishing the effectiveness of controls and ensuring the usability of regulations.

In summary, the main requirements for digitalMLPA software involve the analysis of signals representing the relative number of target sequences in DNA samples, using NGS devices and FASTQ files. The analysis workflow encompasses several domains with specific tasks, allowing for automation, quality control, data normalization, and results analysis. The domain-specific approach facilitates focused verification and the integration of software components into the organization. Furthermore, this approach aligns with a top-down risk perspective, emphasizing patient safety and the implementation of effective controls. Continuous measurement and evaluation are crucial for assessing the effectiveness of controls and maintaining regulatory compliance.

4.2 Quality assurance when developing software

4.2.1 Initial development of digitalMLPA software

How do we approach quality assurance when developing software with unclear goals and intended purpose? Here are some considerations to guide us. The software's classification depends on the intended purpose of the supported probe mix, which may require us to anticipate future usage scenarios. During software development, we can incorporate configurable functionality to cater to various product needs. These configurations become a separate domain, requiring thorough testing and evaluation alongside other product aspects. When determining the risk class, we have two options: assuming the highest possible risk class throughout the software's lifetime or aligning it with the current risk level and upgrading when higher-risk products emerge. The choice depends on factors like project size, available resources, and documentation practices. Timely documentation, preferably done alongside coding, ensures synchronization and avoids outdated information. Identifying controls is often best achieved during method development. When the exact goal of the software is not completely defined, then we may search for guidance to quality assurance from the highest applicable authority. Authority comes from excellence, which may be defined by dedication to the highest standard. This may mean IVDR or FDA, or how its regulations are implemented in the organization, it may also involve some inner conviction and experience with the matter at hand. The main requirements of the company, such as MRC Holland's goal to provide affordable and user-friendly products for detecting chromosomal aberrations, can shape the software's objectives. Supporting input from measurement machinery and calculating MLPA probe ratios are key requirements. These ratios, in combination with specific MLPA products, assist in copy number estimation and disease diagnosis. The risk class depends on the potential impact on patients and its associated probability. To address uncertainty, initially assigning the highest risk class and gradually lowering it through control implementation and verification of validity can be pursued. Use cases that assess severity and probability help determine the initial risk class.

4.2.2 Use cases

4.2.2.1 User case one

Conventional MLPA as a technique focuses on the detection of genetic diseases, that are typically tested for when a phenotype is observed. The intended use may then be described as a function that is employed to know something that could lead to understanding of its genetic underpinnings. Genetic test kits could also be used to understand the probabilities of certain inherited genetic traits being passed on to offspring. For instance, we may have a scenario where a user has a genetic test result and the intended purpose of the probe mix is to provide understanding in underlying conditions of a patient, but not treatment. Positive (false) results may be expected to be confirmed by a second method (as stated per IFU), but even without that confirmation, the product is not expected to harm the patient directly. In case of false negatives, patients are typically expected to remain being monitored due to their already observed phenotype and to be subjected to alternative types of testing. The severity to the patient is therefore low as well as the probability to do harm, the risk class may therefore be expected to be a IIa risk class.

4.2.2.2 User case two

We may also consider a use case where product has the intention to test for genetic diseases on newborn. In this case, we may deal with specific syndromes, where an outcome may be used in a chain of a decision-making apparatus, that ultimately leads to application of medicine, for instance to delay progression of disease. In that case, a false negative may lead to harm to the patient, but not death. A false positive would still be expected to be confirmed, especially since the pharmaceutical industry is expected to adhere to higher risk classes under FDA or the IVDR. The severity in this case may also be expected to be higher than with the previous case since an extended delay influences the patient sovereignty. The initial risk class for such a product may be IIb. We need to keep in mind however, before such product obtains such an intended purpose, they usually must pass extensive rounds of testing by external users. The post market survey data from these tests are important to understand how products are implemented and what role they play on different locations. This data may result in changes to risk class as well as help to provide information about its performance in the field. Such data may be generated by allowing a new genetic test to be used for an extensive period alongside current implemented technologies or so-called gold standards. Information about its application, context of usage and performance in the field may then lead to better insights that can change the severity and probability leading to a lower or higher class. Developers are usually limited to verification and unit testing, that have their own importance, but do not grant validation of a devices multi operational implementation. To determine to the risk class of medical software we thus must adhere to the risk class of the known process that lies above its application. For software for digitalMLPA it's the product itself. Processes above that may be the design and production of the product, but without data about its usage in the field, no verifiable validity can be determined. Verifiable validity may mean that the intended purpose has compliancy by operation and evaluation of its performance in the field. Please note that we do not intend to dictate an importance to a certain risk class here, merely to examine and invite discussion to the outcome of certain case scenarios.

4.2.2.3 User case three

If we would consider digitalMLPA in its current employment, we may state that its focus relates mostly to research in cancer genomics [2]. It's very suitable to investigate the copy number changes that investigate specific genomic locations and combinations of changes that lead to effects in biochemical pathways. Such understanding can lead to better insight into prognosis and patient specific medicine. Software intended for research may be classified as a general laboratory purpose device and does not need to undergo external validation by notified bodies. However, if we also consider the commercial interest in such products, then there is also a need to develop trust with customers in the field that a product including its software version and configuration results in a stable element in a process they use to investigate cancer diagnostics. Considering commercial endeavor, then a positive outcome of a research trial may lead to the establishment of a diagnostic product for a manufacturer. This may be a screening test or a device that is part of standard operating procedures in clinical diagnosis. Data that may lead to insight into performance in the field also requires a stable supply of a product and its specific software and related configurations. The risk class that may be considered in this category under the IVDR may be IIb, since the deployment of the product may influence diagnosis related to cancer genetics. The obvious problem here is that data that may lead to insight to its implementation, does not exist before it is implemented. When developing software, we may thus better be prepared for the road ahead and invest in accuracy and stability to attempt conformity as best as possible to the company's main requirements. To create software that may remain stable over periods of 10 years broader considerations need to be evaluated. We may for instance consider to the decisions I want to be accountable for, when that period has passed. We may then conclude that we should be prepared for its ultimate usage in practice. If we accept a high-risk classification early, we may have the option to choose a suitable documentation style that matches its purpose and is an accepted company standard. We could for instance choose to comment in such a way in code, that it leads to automated description of processes and/or understandable diagrams and workflows. It may be considered that using such data in combination with Artificial Intelligence (AI) linguistical models could amount to great potential in this area, as well as leading to an easier flow of data considering support to QA documentation and support material.

Note that risk class is often dependent on its verifiable validity of the software implementation in the world, and that there is a dependency on that data to establish and optimize controls. This however does not mean that we cannot attempt to mitigate risks, via case scenarios that lead to controls that allow adjustment of the probability. Controls also require testing and evaluation of these results to understand whether a mitigation has been properly implemented and that the final risk is acceptable. Sometimes we may also choose to mitigate risks with lower probabilities, because their detectability is high and easy to implement. Such situations are typically based on verifiable data and consider the detectability of a failure mode. We may also imagine failure of the highest possible processes that may influence the overall intended use and highest risk class to which applicable detection controls may be set in place. Risk priority is determined by the risk class combined with the detectability. That thus means that via the severity, it may be relevant to also work on low probability failures, due to availability of the mitigations. It's not in the scope of this piece to consider all scenarios, but we may attempt to evaluate a few that may rank the high.

4.2.3 Use scenarios

Prioritization by combining detectability with the risk class allows focusing on implementation of controls to ensure its endured overall intended purpose, while considering safety to patient and user alike, in every step of the process. While the severity may be directly linked to the intended purpose of the product, the probability to different failure modes may be a different component altogether. Probabilities may be adjusted by increasing the detectability of failure of the main processes. As such, in discussion we adhere to the highest severity imaginable, within confines of the context at the time of implementation of the controls, to the highest processes that are available for the software to investigate or support. In this section we discuss some examples of failure modes that may exist in practice, and the types of control that have been implemented to mitigate risks.

4.2.3.1 Case scenario one

Failure of NGS machine measurement device. *Failure mode*: The NGS machine measurement device fails to obtain an accredited accuracy rate. Very low read quality results in changes in probe ratio, as reads are not recognized or assigned incorrect identities.

Severity: Complete failure of the device may lead to false positive results. Verification: The validity of the software's methodology should be verified by provable specificity and sensitivity in real-world implementation. Logical reasoning and aiming for the highest degree of accuracy can also be employed. *Specificity and sensitivity*: Read recognition is a critical requirement of the software. With digitalMLPA, read identification is simplified as it only needs to match against a limited probe set. By implementing a method that distinguishes barcode sequences and links them to specific sample barcodes, probe mix identification is established, increasing specificity. Further segmentation and setting of thresholds enhance read recognition. *Controls*: Establishing control systems to measure data outside standard limits is important regardless of severity and probability. Controls for NGS data may include sequence quality, detection range, signal amplitude, sample signal bias, and more. New controls can be developed based on observed measurements, which may impact the probability of risk. *Priority and detectability*: NGS run quality and library preparation are prioritized due to their effect on outcome and high detectability. Improving NGS run quality can enhance the detectability of controls in later stages of data analysis. Rerunning digitalMLPA products can improve data quality without repeating the experiments. *Adjusting probability and detectability*: The effectiveness of implemented controls determines whether adjustments can be made in risk classification. Actual measurement under configured conditions is necessary to evaluate the impact of controls.

4.2.3.2 Case scenario two

Copy number estimation failure. *Failure mode*: High salt concentration in certain samples negatively affects the specificity to discern chromosomal aberrations from poorly denatured DNA, resulting in incomplete probe hybridization. This behavior may introduce patterns originating from chromosomal locations, increasing the probability of a false positive result. *Severity*: High salt concentration can mimic a false deletion pattern, leading to a high severity level. *Probability*: The strain on discerning copy number estimation via probe ratio and genomic pattern recognition increases due to the influence of high salt concentration, affecting the probability of accurate estimation. *Detectability*: digitalMLPA products incorporate smart controls using specially designed probes to test for different quality aspects, such as proper denaturation and DNA fragmentation, enhancing detectability. *Priority*: Supporting the requirement of proper denaturation and minimizing false positives becomes a high priority. It is prioritized due to the increased probability of false positives, the high detectability provided by the product, and its potential impact on later processes. The software should exclude such results from copy number estimation processes.

4.2.3.3 Case scenario three

MLPA reaction failure for copy number estimation. *Failure mode*: Poorly closed tubes in the MLPA reaction may result in sample volume evaporation, leading to changes in salt concentration that negatively affect probe hybridization completeness. *Severity*: Similar to Case Scenario Two, this scenario affects larger groups of probes and leads to poorer distinction of chromosomal aberration patterns, resulting in false positives. The severity is therefore high. *Probability*: Since this scenario can mimic genomic aberration patterns, there is a higher probability of an end user incorrectly calling a result aberrant while observing a false positive. *Detectability*: The digitalMLPA product includes specially designed control fragments that measure the extent of the hybridization reaction, enhancing detectability for identifying such issues. *Priority*: Given the high severity and probability of false positives, along with the high detectability facilitated by the product, addressing this scenario becomes the highest priority. The software should aim to prevent end users from attempting copy number evaluation on identified cases related to this failure mode.

However, the exact implementation of the control system's effects cannot be solely attributed to the software itself. It should align with the intended purpose and risk class of the overall product. For example, a risk IIb class product should not produce probe ratio results but instead issue warnings and error reports to eliminate the chance of false positives. In contrast, if the software is distributed with digitalMLPA products intended for research purposes only, a less strict control system may be applied, and end results could include warnings. The behavior and effectiveness of these control systems can be observed in practice, providing insights on how to improve the accuracy of the product to achieve its intended purpose. Adjustments to the control systems can be made based on their effectiveness, ultimately enhancing the overall product's accuracy through harmonious implementation of all processes and control systems.

4.2.3.4 Case scenario four

Controls on software processes for compliance and measurement. In this case scenario, we focus on implementing controls on the main software processes used to establish compliance with company design, specifically related to signal measurement and probe ratio estimation. It is important to recognize and prioritize these processes based on their influence and significance in achieving the intended purpose of the product. *Failure mode*: One possible failure mode is when part of an experiment becomes unreliable due to structural variations, such as insufficient master mix during the MLPA procedure. This failure can introduce changes in the biochemistry, leading to aberrant patterns and false positives. *Severity*: Such changes may lead to aberrant patterns and thus false positives. The severity of this failure mode is determined by the severity of the product and its specific configuration. It is important to address this issue to prevent misleading results. *Probability*: The probability of this failure mode occurring may be considered low since laboratory staff is trained to avoid such situations, and the usage of multichannel pipettes is recommended. *Detectability*: To establish control systems for such scenarios, specific recommendations can be provided through the information for usage (IFU) that consider different experimental contexts. For example, recommending the inclusion of reference samples at the start, end, and intermediate positions of an experiment sample population. The software can then measure the variation over these reference samples, although the effectiveness of this control may be context-dependent and influenced by assumed limitations of higher processes. *Priority and effectiveness*: The priority of implementing control systems should be determined by the severity and detectability of the failure mode, with a focus on higher processes that have a greater influence. While the potential effectiveness of measuring the variation over reference data can help prevent false positives, its actual value can only be established through field experience and data analysis. Acceptance criteria and definable logic can be implemented in the software to prevent false positives, and adjustments can be made based on data in the field to improve the control's effectiveness.

Harmonization and collaboration: The effectiveness of control systems relies on the harmonization of all underlying processes and collaboration between departments, manufacturers, and users. Sharing data and experiences can lead to improvements in control systems. The increased focus on Post-Market Surveillance (PMS) in the IVDR is understandable considering the importance of ongoing improvement. It is important to note that these examples do not cover all possible failure modes but provide an examination of a few scenarios and their hierarchical order. Controls can be established based on machine manufacturer guidelines or by adhering to controls designed by the digitalMLPA product itself. A top-down risk-based approach should prioritize implementing controls for higher processes with higher probability and detectability. The severity of the software processes should align with the severity level of the product, and comprehensive testing of the product, including software, configuration, and support material, is necessary to ensure conformity to the intended purpose. By implementing appropriate control systems, such as evaluating variation in reference data for copy number estimation in MLPA, the accuracy of probe ratio measurements can be improved. Strict control criteria should be initially researched for effectiveness, and adjustments should be based on empirical logic and observable data. Stricter controls may be applied for specific probes or arbitrary borders can be tightened to reduce false negatives. However, it is essential to always prioritize the effectiveness of controls on higher processes over lower processes.

5. Discussion

When working on scientific software, encountering the new IVDR regulations can initially seem rigid and defining, while the implementation regarding patient safety and effectiveness may appear vague and transitory. Standards are often based on existing devices and knowledge, and while they set limitations to minimize outcomes, they rely on comparing and examining observable data. However, even with definitions in place, the practical implementation of patient safety measures can be challenging.

There seems to be a correlation between the focus on patient safety and the level of connection between the provider and the recipient of care. Direct patient contact allows tangible correlations, but in other cases, the ethical implications remain a question. It's important to reflect on how our day-to-day actions and decisions influence patient safety, regardless of our level of direct interaction.

Standardization has its advantages, such as uniform results, lower product costs, and predictability. However, it can also invite commercial interests and potentially lead to monopolization. Conformity to standards can become a goal, conflicting with personal integrity, and potentially causing resignation. In the context of patient care, caregivers should prioritize providing care rather than simply conforming to definitions. In diagnostics, focusing on delimitation of factors of categorization may be a strength leading to a more accurate result. In research this limitation may result in a blockage to discovery, that may lead to innovation.

While it may be easy to differentiate between these two categories in theory, the status of products and their implementations can be more fluid in practice. Implementing regulations in a way that ensures patient safety often lacks harmonization and may create discord and separation. Harmonization is crucial for effectively achieving the intended purpose of medical products while maintaining the highest standards, which should ultimately serve the well-being of the patient. To achieve this, there needs to be a personal understanding of how regulations translate into improved patient safety in day-to-day actions and decisions. In practice such an approach may be admirable, but vain.

A top-down risk approach therefore prioritizes the highest processes and their intended purpose. For companies, improving patient care should be aligned with the goals of the IVDR. It is therefore surprising that the EU does not invest more in improving communication between medical institutions and manufacturers to exchange information on standards' implementation and their effectiveness. They are in a prime position to advise on implementation since they can measure the effects of standards and could consider endorsing a legal entity for products and their users. Such collaboration would enable the placement of patient safety controls at a high-level process and allow PMS to focus on measuring product accuracy and the effectiveness of implemented controls.

Empiric logic, however, does require a vantage point. Lawmaking on the other hand often seems necessary and is often met with deadlines. In an urge to control content there may be a drive to document everything. It could also be so that the extensive nature of the regulations invites to make a set of procedures that are of equal size making sure that every delimiting box is checked to showcase conformity to the standard. Enforcing controls to promote patient safety is great but we need to take an active stand, taking interest to measure its accuracy and effectiveness in practice.

The extensive nature of regulations may tempt the creation of equally extensive procedures to ensure conformity, potentially overshadowing the goal of promoting patient safety. It is essential to take an active stance, measuring the accuracy and effectiveness of enforced controls in practice.

The extent to which the IVDR will be implemented and actively practiced remains unknown. In this piece, we discussed how applying such standards to software development in the intersection of research and diagnostics can have implications. We explored how software architecture choices contribute to quality assurance and how domain-based programming can establish quality control points. While we can attempt to safeguard quality through predefined logic and controls, the proof of a product's intended purpose can only be measured through discernable observations. Methods and controls should be subjected to empirical logic and adjusted based on newly obtained information to improve accuracy. However, it's important to recognize that harmonious implementation of products and controls often comes with its own set of limitations. Using a top-down risk-based approach, we can prioritize and establish hierarchical logic to ensure quality. These controls remain subject to change if they stimulate the accuracy of the intended purpose or increase the detectability of failure modes. However, for the sake of patient safety, we must acknowledge our limitations and recognize that processes and context beyond our own, offer higher value in effectiveness but require harmonious implementation.

Acknowledgements

We thank MRC Holland for its support and goal to provide affordable and user-friendly products for detecting chromosomal aberrations as well as for its ongoing support to the MRC Holland Foundation with the aim to contribute to the development of a better world.

Notes/thanks/other declarations

Thanks to Jan Schouten for his ongoing support in the medical and science world and providing a pleasant a comfortable working environment. Also thanks the MRC Holland staff for aiding in development, providing feedback and support. A special note to all referenced authors, their valuable work provided insights that helped to try to make sense of it all, without that it was not possible. Thanks to Joost van den Berg for many years of excellent programming work and providing understanding into the world of coding. Very grateful for Lydia de Burger, her unwavering support and unending understanding have made a great difference. If any references were left out, sincere apologies, this work was intended to summarize and attempt to provide some understanding and describe different considerations we went through when developing scientific software. This is also intended to invite discussion, share experiences to implementation of regulations to move forward.

Author details

Jordy Coffa[1]* and Joost van den Berg[2]

1 MRC Holland Amsterdam, The Netherlands

2 Berg IT solutions, The Netherlands

*Address all correspondence to: jordy.coffa@mlpa.com

References

[1] Schouten JP, McElgunn CJ, Waaijer R, Zwijnenburg D, Diepvens F, Pals G. Relative quantification of 40 nucleic acid sequences by multiplex ligation-dependent probe amplification. Nucleic Acids Research. 2002;**30**(12):e57. DOI: 10.1093/nar/gnf056

[2] Benard-Slagter A, Zondervan I, de Groot K, Ghazavi F, Sarhadi V, Van Vlierberghe P, et al. Digital multiplex ligation-dependent probe amplification for detection of key copy number alterations in T- and B-cell lymphoblastic leukemia. The Journal of Molecular Diagnostics. 2017;**19**(5):659-672. DOI: 10.1016/j.jmoldx.2017.05.004. Epub 2017 July 19

[3] Stuppia L, Antonucci I, Palka G, Gatta V. Use of the MLPA assay in the molecular diagnosis of gene copy number alterations in human genetic diseases. International Journal of Molecular Sciences. 2012;**13**(3):3245-3276. DOI: 10.3390/ijms13033245. Epub 2012 March 8

[4] Yu CH, Lin TK, Jou ST, et al. MLPA and DNA index improve the molecular diagnosis of childhood B-cell acute lymphoblastic leukemia. Scientific Reports. 2020;**10**:11501. DOI: 10.1038/s41598-020-68311-9

[5] Hömig-Hölzel C, Savola S. Multiplex ligation-dependent probe amplification (MLPA) in tumor diagnostics and prognostics. Diagnostic Molecular Pathology. 2012;**21**(4):189-206. DOI: 10.1097/PDM.0b013e3182595516

[6] Madrigal I, Rodríguez-Revenga L, Badenas C, et al. MLPA as first screening method for the detection of microduplications and microdeletions in patients with X-linked mental retardation. Genetics in Medicine. 2007;**9**:117-122. DOI: 10.1097/GIM.0b013e318031206e

[7] Romanelli Tavares VL, Monfardini F, Lourenço NCV, da Rocha KM, Weinmann K, Pavanello R, et al. Newborn screening for 5q spinal muscular atrophy: Comparisons between real-time PCR methodologies and cost estimations for future implementation programs. International Journal of Neonatal Screening. 2021;**7**(3):53. DOI: 10.3390/ijns7030053

[8] Lalic T, Vossen R, Coffa J, et al. Deletion and duplication screening in the DMD gene using MLPA. European Journal of Human Genetics. 2005;**13**:1231-1234. DOI: 10.1038/sj.ejhg.5201465

[9] Kosztolányi S, Kiss R, Atanesyan L, Gángó A, de Groot K, Steenkamer M, et al. High-throughput copy number profiling by digital multiplex ligation-dependent probe amplification in multiple myeloma. The Journal of Molecular Diagnostics. 2018;**20**(6):777-788. DOI: 10.1016/j.jmoldx.2018.06.004. Epub 2018 August 8

[10] Kachalia A, Mello MM, Nallamothu BK, Studdert DM. Legal and policy interventions to improve patient safety. Circulation. 2016;**133**(7):661-671

[11] Covington & Burling LLP. Spotlight: Medicine and Medical Device Pricing and Reimbursement in European Union. NW, Washington, DC: European Union, Covington & Burling LLP; 2023. pp. 20001-4956

[12] Erlangga D, Suhrcke M, Ali S, Bloor K. The impact of public health insurance on health care utilisation,

financial protection and health status in low- and middle-income countries: A systematic review. PLoS One. 2019;**14**(8):e0219731. DOI: 10.1371/journal.pone.0219731. Erratum in: PLoS One. 2019 Nov 7;14(11):e0225237

[13] Oikonomou E, Carthey J, Macrae C, Vincent C. Patient safety regulation in the NHS: Mapping the regulatory landscape of healthcare. BMJ Open. 2019;**9**(7):e028663. DOI: 10.1136/bmjopen-2018-028663

[14] The new EU IVD Regulations and CE Marking Invitro Diagnostic Medical Devices, by Jacques du Preez—Director Psephos Biomedica. Available from: https://www.psephos.com/newsandarticles

[15] Warlow C. Over-regulation of clinical research: A threat to public health. Clinical Medicine (London, England). 2005;**5**:33-38. DOI: 10.7861/clinmedicine.5-1-33

[16] Chervoni-Knapp T. The staffing shortage pandemic. Journal of Radiology Nursing. 2022;**41**(2):74-75. DOI: 10.1016/j.jradnu.2022.02.007. Epub 2022 February 28

[17] Han X, Pittman P, Barnow B. Alternative approaches to ensuring adequate nurse staffing: The effect of state legislation on hospital nurse staffing. Medical Care. 2021;**59**(Suppl. 5):S463-S470. DOI: 10.1097/MLR.0000000000001614

[18] Thoreau HD. On the Duty of Civil Disobedience. Lehi, UT: Libertas Press; 1849

[19] Posner MI, Rothbart MK. Attention, self-regulation and consciousness. Philosophical Transactions of the Royal Society of London. Series B, Biological Sciences. 1998;**353**(1377):1915-1927. DOI: 10.1098/rstb.1998.0344

[20] Habets MG, van Delden JJ, Bredenoord AL. The social value of clinical research. BMC Medical Ethics. 2014;**15**:66. DOI: 10.1186/1472-6939-15-66

[21] Ignatescu C. Letter and spirit of the law- an interpretation based on principles of law. Revista Romaneasca pentru Educatie Multidimensionala. 2013;**5**:75-82. DOI: 10.18662/rrem/2013.0502.07

[22] Cowan D. Legal consciousness: Some observations. The Modern Law Review. 2004;**67**(6):928-958

[23] Dombrink I, Lubbers BR, Simulescu L, Doeswijk R, Tkachenko O, Dequeker E, et al. Critical implications of IVDR for innovation in diagnostics: Input from the BioMed alliance diagnostics task force. Hema. 2022;**6**(6):e724. DOI: 10.1097/HS9.0000000000000724

[24] Herraiz-Recuenco L, Alonso-Martínez L, Hannich-Schneider S, Puente-Alcaraz J. Causes of stress among healthcare professionals and successful hospital management approaches to mitigate it during the COVID-19 pandemic: A cross-sectional study. International Journal of Environmental Research and Public Health. 2022;**19**(19):12963. DOI: 10.3390/ijerph191912963

[25] Drane JF. Becoming a Good Doctor: The Place of Virtue and Character in Medical Ethics. Kansas City, MO: Sheed and Ward; 1988

[26] Thomasma DC. Virtue theory in philosophy of medicine. In: Khushf G, editor. Handbook of Bioethics. Philosophy and Medicine. Vol. 78. Dordrecht: Springer; 2004. pp. 89-120. DOI: 10.1007/1-4020-2127-5_5

[27] Lubbers BR, Schilhabel A, Cobbaert CM, Gonzalez D, Dombrink I, Brüggemann M, et al. The

new EU regulation on in vitro diagnostic medical devices: Implications and preparatory actions for diagnostic laboratories. Hema. 2021;**5**(5):e568. DOI: 10.1097/HS9.0000000000000568

[28] Gross E, van Tinteren H, Li Z, Raab S, Meul C, Avril S, et al. Identification of BRCA1-like triple-negative breast cancers by quantitative multiplex-ligation-dependent probe amplification (MLPA) analysis of BRCA1-associated chromosomal regions: A validation study. BMC Cancer. 2016;**16**(1):811. DOI: 10.1186/s12885-016-2848-2

[29] Lips EH, Laddach N, Savola SP, Vollebergh MA, Oonk AM, Imholz AL, et al. Quantitative copy number analysis by multiplex ligation-dependent probe amplification (MLPA) of BRCA1-associated breast cancer regions identifies BRCAness. Breast Cancer Research. 2011;**13**(5):R107. DOI: 10.1186/bcr3049

[30] Slatko BE, Gardner AF, Ausubel FM. Overview of next-generation sequencing technologies. Current Protocols in Molecular Biology. 2018;**122**(1):e59. DOI: 10.1002/cpmb.59

[31] Dhir M, Choudry HA, Holtzman MP, Pingpank JF, Ahrendt SA, Zureikat AH, et al. Impact of genomic profiling on the treatment and outcomes of patients with advanced gastrointestinal malignancies. Cancer Medicine. 2017;**6**(1):195-206. DOI: 10.1002/cam4.992. Epub 2016 December 28

[32] Berger MF, Mardis ER. The emerging clinical relevance of genomics in cancer medicine. Nature Reviews Clinical Oncology. 2018;**15**(6):353-365. DOI: 10.1038/s41571-018-0002-6

[33] Sloan D, Macaulay C, Forbes P, Loynton S. User research in a scientific software development project. In: Proceedings of the 2009 British Computer Society Conference on Human-Computer Interaction, BCS-HCI 2009, Cambridge, United Kingdom. 1-5 Sept 2009. pp. 423-429. DOI: 10.1145/1671011.1671066

[34] Coffa J, van den Berg J. Analysis of MLPA data using novel software Coffalyser.NET by MRC-Holland [Internet]. In: Modern Approaches to Quality Control. London, UK: IntechOpen; 2011. DOI: 10.5772/21898

[35] Scheitz CJF, Peck LJ, Groban ES. Biotechnology software in the digital age: Are you winning? Journal of Industrial Microbiology & Biotechnology. 2018;**45**:529-534. DOI: 10.1007/s10295-018-2009-5

[36] McHugh M, Cawley O, McCaffcry F, Richardson I, Wang X. An agile V-model for medical device software development to overcome the challenges with plandriven software development lifecycles. 2013 5th International Workshop on Software Engineering in Health Care, SEHC 2013 - Proceedings. 2013. pp. 12-19. DOI: 10.1109/SEHC.2013.6602471

[37] Hajou A, Batenburg R, Jansen S. An insight into the difficulties of software development projects in the pharmaceutical industry. Lecture Notes on Software Engineering. 2015;**3**:267-274. DOI: 10.7763/LNSE.2015.V3.202

Section 4

Applications of QA/QC in Education

Chapter 13

Course Delivery Evaluation as an Instrument for Quality Enhancement in Higher Education Institutions

Kayode Samuel Adekeye

Abstract

Course delivery evaluation of programmes by students in higher education institutions has become a global phenomenon for improving the quality of teaching and learning. From the past, it has been discovered that the quality of effective teaching and learning in Higher Education is dwindling and thus, the relevance of graduates to the development of national growth becomes an issue to be addressed. To solve this challenge, quality control and enhancement of course delivery comes to the fore in attacking the menace. The data used in this chapter consist of six thousand, eight hundred and fifty-eight responses from four hundred and seventy-six courses. The data were clustered first according to departments and then according to the lecturers. The thematic areas studied are lecture delivery, lecturer's attitude to learners, coverage of curriculum, use of library and internet for research, healthy lecturers-learners' relationships, and feedback system. A simple but powerful statistical analysis was deployed to analyse the data. The results reflected areas that have critical effect on teaching and learning. The results showed that 24% of the factors considered in this study have significant effects on effective teaching and learning. Furthermore, about 20% of the lecturers assessed were lacking in the determined factors.

Keywords: course delivery, evaluation instrument, quality improvement, teaching, learning

1. Introduction

Quality assurance of teaching and learning procedures has become an important tool in ensuring the production of competent and relevant graduates from higher education institutions of learning. It is, therefore, important to review, examine and critique higher education processes and procedures such as teaching and learning to ensure fitness of purpose and for purpose [1]. Course evaluation instruments are frequently used in higher education institutions to address the relevant and crucial aspects that are required for effective teaching and learning to take place. Course evaluation instruments should be used to assess some key factors that are believed to have significant influence on teaching and learning. The factors such as communication skills, organisational skills, enthusiasm, flexibility, attitude towards the student, teacher–student interaction, encouragement of the student, knowledge of the subject, clarity of presentation, course difficulty, fairness of grading and exams and global student rating have been listed in the literature [2]. Course delivery evaluation instrument can be used to harvest feedback from students for the assessment of quality of instruction. It can also be for gathering information about the impact of learning and teaching, analysing and interpreting information gathered and responding to and acting on the obtained results [3]. Therefore, the information gathered can be used by administrators, along with other input, to make decisions on staff for promotion, tenure, salary increases and identification of areas of improvement [4]. In the university where the data used in this study was collected, course evaluations are combined with annual performance evaluation report by supervisor and the grades of students to determine the overall performance of academic staff. Course evaluations can be implemented, either as summative or formative. Student evaluations are said to be formative when the purpose is to assist faculty members in improving and enhancing their teaching skills [5]. Another model that has been shown to be highly effective for getting feedback for course improvement is the 'Stop, Start, Continue' (SSC) format (see [6–8]). However, some of these methods have their own disadvantages, which thus do not make them all sufficient approaches.

Student ratings of teaching have been used, studied and debated for almost a century [9]. According to Lakeman et al. [10], student evaluation of teaching (SET) has become a ubiquitous feature of higher education. Intentional design, structured administration processes and transparent reporting of result of evaluation have been advocated by Melissa et al. [11] to be adopted to better ensure that student ratings of instruction are used to improve teaching. However, student evaluations have limitations in informing improvements to the quality of teaching and effective learning [12].

The focus of this paper is to determine the areas of weakness of instructors based on the perspective of students of a higher education institution. Apart from using simple statistical tools to unravel the unknown in the raw data, the Pareto analysis was conducted to determine the most influential variable that affects course delivery to achieve effective teaching and learning in higher education institutions.

2. Methodology

A questionnaire was designed for the purpose of evaluating course delivery and lecturers' performance. The sample of the questionnaire is in the appendix. The questionnaire contained twenty-nine factors, which were carefully selected to assess the course delivery and performance of lecturers.

The twenty-nine factors were coded Q1, Q2, Q3, … and Q29 for the purpose of analysis and they were clustered into five subgroups, which are: lecturer's behaviour in the lecture room, lecturer's adherence to teaching etiquettes, students/lecturers' relationship, mode of course delivery and use of smart electronic board. The four-Likert scale instrument was used because of the nature of the study. It is our belief that it is not possible for a student who is the respondent to be neutral in the choice of options available for the assessment of a course taken over a period of 14 weeks.

The completion of the questionnaire was done online by each student evaluating each lecturer that taught in the semester. The period for the exercise was well communicated to the students through different channels such as email, public announcement and department notice board. Furthermore, an electronic slip was generated for the student once the questionnaire was completed. This approach ensures high degree of compliance and reduction of non-response rate.

Two major statistical tools used in this study are the median and percentage scores. Furthermore, the Pareto analysis was used to determine the factors that have more impact on the academic performance of the students out of the twenty-nine variables captured in the questionnaire.

3. Analysis

The Data used in this paper is six thousand, eight hundred and fifty-eight (6858) responses, which account for 98% of response. Thus, the reliability and validity of the instrument is reasonably high.

The following indices were determined for each course taught by a lecturer:

a. Percentage of respondents (i.e., percentage of students that completed the form for the course and lecturer)

b. Area(s) for improvement. This is the percentage score less than 62.5% (2.5 out of 4)

c. Overall percentage scores. This parameter was used to summarise the overall performance of each lecturer. The parameter takes into consideration the total number of courses taught in the semester under study.

d. To further strengthen the index and measure students' performance per course, the above-listed indices were aligned with the scores of students in the semester's examinations.

4. Results

The obtained results were summarised to capture for each lecturer the number of registered students, percentage of respondents, overall performance, weak areas (average score of less than 2.5) and percentage of students with scores Grade A to F.

From the results, the following information was inferred:

i. There was positive correlation between lecturers' ratings and examination results.

ii. Staff with the highest overall score per programme were determined for appreciation.

iii. The weak areas were determined across board out of the twenty-nine factors using Pareto analysis. The Pareto chart reflects that Factors Q29, Q28 and Q27 are the most prevalent factors that most lecturers that are rated weak lack substantially (see **Figure 1**).

iv. **Figure 2** presented the frequency of weak areas for the twenty-nine variables used to evaluate each academic staff and courses taught. The idea was to

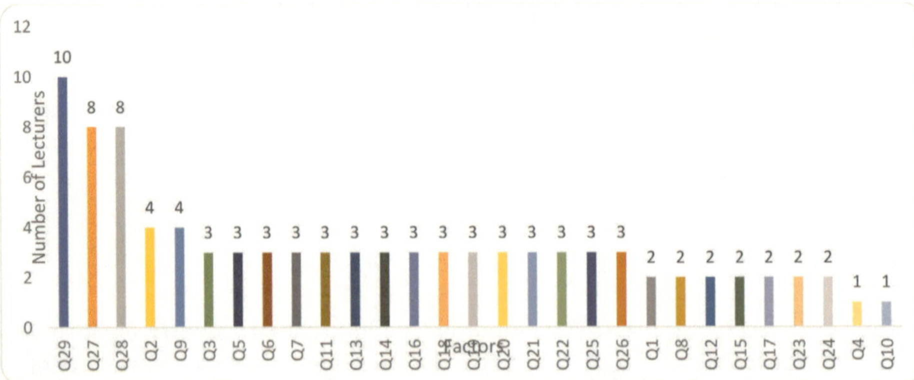

Figure 1.
Pareto chart on number of lecturers rated weak.

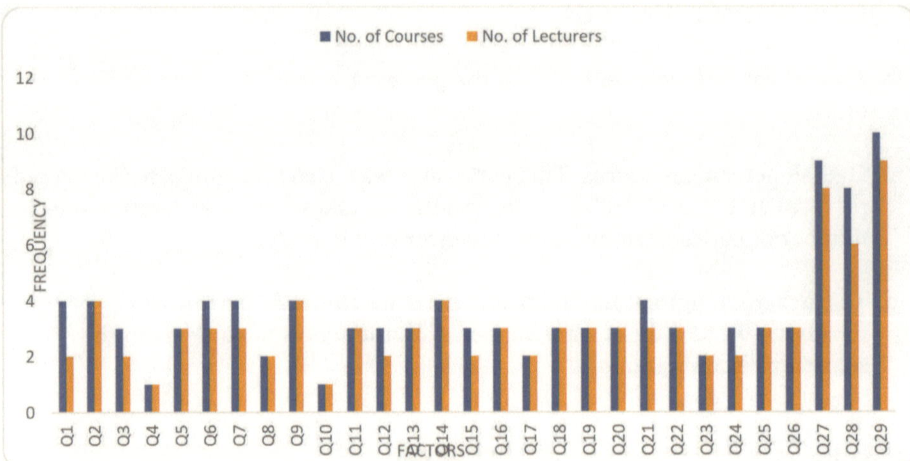

Figure 2.
Frequency of courses and lecturers with weak areas.

know the areas that need urgent intervention of the management of the higher education institution for effective teaching and learning to take place. Furthermore, the pattern of the information, in **Figure 2**, reflected a close relationship between the number of courses and the number of lecturers affected by each of the factors considered in this study.

v. Twenty per cent of the lecturers assessed were rated weak in the following variables:

a. *An appropriate course description/course outline was not provided at the beginning of the course;*

b. *Course outline was not structured and detailed enough;*

c. *The course curriculum was not adequately covered;*

d. *Lecturer do not make effort to explain difficult topics to students;*

e. *Marked tests and assignments were not returned to students regularly;*

f. *The lecturer's attendance at lectures was not usually punctual and regular;*

g. *The lecturer does not usually provide adequate examples and illustrations.*

From the obtained results, it was observed that higher percentage of weak areas has no significant effect on the performance of the students in the courses. This is a reflection that the course evaluations of lecturers might not be a yardstick to measure performance of students. A course with 85% weak areas has 63% of the students that have C grade and above while a course with 38% weak areas has 50% of the students that have grade and above. Whereas, a course with 55% weak areas has 100% of the students that have grade C and above. Thus, there is no definite pattern that can be used to describe the relationship between student's performance in a course and the performance of the lecturer that taught the course. However, the beauty of the course delivery evaluation instrument is in the x-ray of the behaviour of lecturer in the lecture room, lecturer's adherence to teaching etiquettes, students-lecturer relationship while lecture is ongoing and mode of course delivery.

5. Conclusion

The course delivery evaluation instrument has been shown to be a useful instrument to significantly improve the commitment of faculty members to teaching and learning and by effect, the performance of students in examination over time. Furthermore, the Pareto analysis reveals that academic staff needs more training on the use of the smart board for the delivery of instructions for courses at higher education institutions. It should be noted that the data used in this paper was collected prior to the COVID-19 era. The results of the study were useful for the implementation of blended teaching during and post Covid-19 era. The university was relatively prepared for online teaching and learning, though majority of lecturers were not fully equipped with the use of the smart board as reflected in the Pareto analysis. It is recommended that there is the need for more training of academic staff in the use of smart board in the new norm where the hybrid model and AI (artificial intelligence) in higher education teaching and learning is trending. Furthermore, the embracement of course delivery evaluation of lecturers by students has become an exercise that should not be discarded if graduates from higher education institutions must be relevant in the labour market.

Appendix: course delivery and lecturer's evaluation questionnaire to be Completed by Students.

Please complete the form below:

A. **Information on Course**

a. Course Title: … … … … … … … … … … … … … … … … … … …

b. Course Code: … … … … … … … … … … … … … … … …

c. Course Units: ... ...

d. Name of Instructor: ... ...

e. Program of Study of Student: ... ...

f. Semester/Session: ... ...

B. **Assessment/Evaluation**

Dear Student, this questionnaire is a valuable tool for improving the quality of teaching. Please tick the appropriate response and be as objective and honest as possible as the Lecturer will be looking forward to fair criticism as well as praise.

Your response will be of immense help to improve the quality of teaching as the result has the purpose of returning the lecturers' feedback on their teaching skills and ability and it will be used for joint discussion concerning the quality of the courses.

The four-Likert scales used are Strongly Agree, Agree, Disagree and Strongly Disagree.

B1 Instructor's Behaviour in Lecture Room

1. Coursework, continuous assessment tests and assignments were given during the course.

   Strongly Agree ☐Agree ☐ Disagree ☐ Strongly Disagree☐.

2. Marked tests and assignments were returned to students regularly.

   Strongly Agree ☐ Agree ☐ Disagree ☐ Strongly Disagree☐.

3. The instructor usually arranged make-up classes for missed lectures.

   Strongly Agree ☐ Agree ☐ Disagree ☐ Strongly Disagree ☐.

4. The instructor usually explains complex subjects to the students.

   Strongly Agree ☐ Agree ☐ Disagree ☐ Strongly Disagree ☐.

5. The instructor has enthusiasm for teaching.

   Strongly Agree ☐ Agree ☐ Disagree ☐ Strongly Disagree ☐.

6. The instructor demonstrated fair temperament in handling students in class.

   Strongly Agree ☐ Agree ☐ Disagree ☐ Strongly Disagree ☐.

7. The instructor did not tolerate distractions during lectures.

   Strongly Agree ☐ Agree ☐ Disagree ☐ Strongly Disagree ☐.

8. The instructor displayed adequate command of accurate and simple English in class.

   Strongly Agree ☐ Agree ☐ Disagree ☐ Strongly Disagree ☐.

B2 Instructor's Adherence to Teaching Etiquettes

1. The instructor hour (period) was always fully utilised.

   Strongly Agree ☐ Agree ☐ Disagree ☐ Strongly Disagree ☐.

2. The instructor was usually punctual and regular in class.

   Strongly Agree ☐ Agree ☐ Disagree ☐ Strongly Disagree ☐.

3. The instructor encouraged students' participation in class.

   Strongly Agree ☐ Agree ☐ Disagree ☐ Strongly Disagree ☐.

4. Adequate examples and illustrations were provided by the instructor.

   Strongly Agree ☐ Agree ☐ Disagree ☐ Strongly Disagree ☐.

5. The instructor always ensured mode of presentation was used to advantage.

   Strongly Agree ☐ Agree ☐ Disagree ☐ Strongly Disagree ☐.

6. The instructor was usually quite audible during lectures.

   Strongly Agree ☐ Agree ☐ Disagree ☐ Strongly Disagree ☐.

7. Attendance at lectures was regularly taken by the instructor.

   Strongly Agree ☐ Agree ☐ Disagree ☐ Strongly Disagree ☐.

B3 Students-Instructor Relationship

1. The instructor was concerned that students understood his/her lectures.

   Strongly Agree ☐ Agree ☐ Disagree ☐ Strongly Disagree ☐.

2. The instructor was always ready to offer individual assistance.

   Strongly Agree ☐ Agree ☐ Disagree ☐ Strongly Disagree ☐.

3. Would you want this instructor to teach you again? Yes ☐ No ☐.

B4 Mode of Course Delivery

1. Lectures were regularly delivered as scheduled on the timetable.

   Strongly Agree ☐ Agree ☐ Disagree ☐ Strongly Disagree ☐.

2. Lecture materials were presented in a way that students could readily assimilate.

   Strongly Agree ☐ Agree ☐ Disagree ☐ Strongly Disagree ☐.

3. An appropriate course description/course outline and expected learning outcomes were provided at the beginning of the course.

   Strongly Agree ☐ Agree ☐ Disagree ☐ Strongly Disagree ☐.

4. Course outline was structured and detailed enough.

   Strongly Agree ☐ Agree ☐ Disagree ☐ Strongly Disagree ☐.

5. Textbooks, reading lists and reference materials were recommended.

   Strongly Agree ☐ Agree ☐ Disagree ☐ Strongly Disagree ☐.

6. Students were encouraged to use the main library and the electronic library regularly.

   Strongly Agree ☐ Agree ☐ Disagree ☐ Strongly Disagree ☐.

7. The course curriculum was adequately covered.

   Strongly Agree ☐ Agree ☐ Disagree ☐ Strongly Disagree ☐.

B5 Use of Electronic Smart Board

1. The instructor always used the smart board to deliver his lecture.

Strongly Agree ☐ Agree ☐ Disagree ☐ Strongly Disagree ☐.

2. The instructor used the smart board appropriately to deliver his lecture.

Strongly Agree ☐ Agree ☐ Disagree ☐ Strongly Disagree ☐.

Thank you for your time and sincerity, please look forward to the best improvement based on your response.

Author details

Kayode Samuel Adekeye
Department of Mathematics and Statistics, Redeemer's University, Ede, Nigeria

*Address all correspondence to: adekeyek@run.edu.ng

References

[1] Netshifhefhe L, Nobongoza V, Maphosa C. Quality assuring teaching and learning processes in higher education: A critical appraisal. Journal of Communication. 2016;7(1):65-78

[2] Kim C, Damewood E, Hodge N. Professor attitude: Its effect on teaching evaluations. Journal of Management Education. 2000;**24**(4):458-473

[3] Rahman K. Learning from your business lectures: Using stepwise regression to understand course evaluation data. Journal of American Academy of Business, Cambridge. 2006;**19**(2):272-279

[4] Dunegan KJ, Hrivnak MW. Characteristics of mindless teaching evaluations and the moderating effects of image compatibility. Journal of Management Education. 2003;**27**(3):280-303

[5] Mohanty G, Gretes J, Flowers C, Algozzine B, Spooner F. Multi-method evaluation of instruction in engineering classes. Journal of Personnel Evaluation in Education. 2005;**18**(2):139-151

[6] Hoon AE, Oliver E, Szpakowska K, Newton PM. Use of the 'stop, start, continue' method is associated with the production of constructive qualitative feedback by students in higher education. Assessment and Evaluation in Higher Education. 2015;**40**(5):755-767. DOI: 10.1080/02602938.2014.956282

[7] Raza MA, Bilal M, Rasheed MR, Chandio BA, Ahmad N, Sawand FA. Quality assessment in higher education. International Letters of Social and Humanistic Sciences. 2015;**50**:162-171. DOI: 10.18052/

[8] Stark PB, Freishtat R. An evaluation of course evaluations. ScienceOpen Research. 2014:1-7. DOI: 10.14293/S2199-1006.1.SOR-EDU.AOFRQA.v1

[9] Tang TL-P. Teaching evaluation at a public institution of higher education: Factors related to the overall teaching effectiveness. Public Personnel Management. 1997;**26**(3):379-391

[10] Lakeman R, Coutts R, Hutchison M, Massey D, Nasrawi D, Fielden J, et al. Playing the SET game: How teachers view the impact of student evaluation on the experience of teaching and learning. Assessment and Evaluation in Higher Education. 2022;**48**(6):749-759. DOI: 10.1080/02602938.2022.2126430

[11] Medina MS, Thomas Smith W, Kolluru S, Sheaffer EA, DiVall M. A review of strategies for designing, administering, and using student ratings of instruction. American Journal of Pharmarceutical Education. 2019;**83**(5):753-764. DOI: 10.5688/ajpe7177

[12] Laundon M, Cunningham S, Cathcart A. Institutional approaches to evaluation of learning and teaching: A sector scan of Australasian universities. Journal of Higher Education Policy and Management. 2023;**45**(5):511-528. DOI: 10.1080/1360080X.2023.2196646

Chapter 14

Quality in Higher Education: Navigating the Academic Business Nexus

Terry Twomey

Abstract

Universities are powerhouses of knowledge and innovation where staff are hailed as their greatest assets. Engaging the commitment and innovation of staff is a stated aim of higher education management. However, universities are increasingly challenged by the need to serve many masters, and in particular, to bridge the academic and business culture aspects of the organization. Tensions emerge between the perceptions of staff role groups, the goals of academic endeavor and the business needs of the organization. Organizational and cultural ambiguity can hinder rather than enhance academic quality. This chapter explores the Academic Business nexus, the resulting perceptions and tensions between "tribes and territories" in a higher education institution. The chapter provides insights into orientations towards quality assurance in higher education, with examples from different countries and how they are navigating the Academic Business nexus. The chapter explores the question of the importance of organization culture, values and attitudes as drivers of quality in higher education. Cognizant of the diversity of university profiles, organization structures and culture, the chapter explores changing staff and management views on university quality. The chapter moves beyond the existing focus on management, professional and academic tribes, to address the changing models of universities that bridge the Academic Business nexus and cultivate quality through different approaches.

Keywords: quality, academic, business, culture, professionalization, higher education

1. Introduction

Universities strive to promote their reputation as centres of learning, opportunity, enterprise, innovation and social development. Yet the business and financial operations of universities are inseparable from these lofty objectives and aspirations. Hence, the business prowess of universities is often conflated with academic quality. The application of proxy criteria for academic quality to benchmark higher education at national and international level can be problematic. The criteria and techniques used to measure and compare institutions impact benchmark performance. For example, among the top 50 national universities in the *US News and World Report Best*

Colleges rankings there is an estimated 70% correlation between level of university funding and university benchmark performance [1]. When examined further we find these rankings include the following proxy criteria for quality or "Best":

- Opinions of university leaders
- Alumni donations to the university
- Size of the university endowment

Government and state support, strong research funding, high student fees, large endowment funds and philanthropy form mutually supporting characteristics that enable the "best" universities to trade on prestige and exclusivity. Yet there remains a question mark over this perception of quality and the quality of the student experience measured against specific educational criteria. One might also ask how the other 30% of universities outperform their funding-based rankings within the self-perpetuating wealth centric model.

In contrast to the USA, universities in the European Union (EU) operate a somewhat different model, with measures of quality that value social capital as well as economic drivers. In both the USA and Europe there are sufficient aspects of academic quality at the student experience level that can be controlled and assured to raise concerns regarding the use of proxies for academic quality. In acknowledging the academic and business aspects of university identity and mission, it is important not to conflate these as one and the same.

Within the European Higher Education Area (EHEA) the philosophy of quality control and quality assurance has evolved significantly since the Bologne Declaration in 1999 [2]. In addition to quality control and quality assurance, quality enhancement is now central to the EU academic quality philosophy. Acknowledging that different states in the EU are at different stages in higher education quality development, the European Standards and Guidelines [3] guide the development and enhancement of higher education quality across the EU. At the micro level, periodic institutional review, critical self-study and peer review are the key techniques and tools used to underpin institutional quality.

2. The academic business nexus in higher education

In light of the complex and distinctive mission of universities to inspire learning and nurture new ideas, definitions of academic quality based on single stakeholder views are often contested [4–6]. To capture the essence of inspiring students, nurturing learning and of creating new knowledge, concepts of higher education quality limited to inputs and outputs or to extra-curricular experience or to value adding, do not in themselves provide sufficient insight to the participants in an academic institution [7]. This research based on one university in Ireland, looked inside the black box of quality control and assurance in higher education.

The quality journey in higher education goes beyond the framework of policies and procedures to quality as a lived and meaningful experience from which knowledge and innovation can arise. This dynamic paradigm of quality control and assurance addresses the constantly changing nature of higher education organizations.

Shifting job roles, changes in organization mission and higher qualifications among professional staff have given rise to the concept of the "third space", reflecting the heightened interdependence of different staff roles emerging in Australia and Singapore [8].

Tam [7] contends that in universities, "*Quality is a highly contested concept and has multiple meanings to people who conceive higher education and quality differently*". Tam concludes that the varying models of measuring quality are underpinned by explicit and/or tacit differences in understanding and assumptions. The wide range of internal and external stakeholders, with often quite varied vested interests, has resulted in HE quality becoming contested territory of national, industrial, academic, non-academic and student perspectives.

The academic business nexus is central to the organizational tensions in universities. In the EHEA the dominant contemporary view of universities is as an instrument of the state, social cohesion and economic development [9]. Universities could be characterized as finding themselves in a tug of war between academia on the one hand and external stakeholder interests on the other [10].

A study of staff views on academic quality assurance (AQA) was carried out in a university in Ireland to discern and understand the views of the different categories of university staff on AQA. The methodology of the study employed a series of progressively refined survey questionnaires, known as the Delphi Method, and a focus group of the university's quality assurance managers and executive managers, to illuminate underlying perceptions and tensions between different staff role identities [11]. Over 250 staff participated in the study, self-identifying as 72% academic, 11% administration, 9% student services and 8% management.

3. Staff views on quality control and assurance

The study provided an evidence-based understanding of staff identities and their impact on AQA within the university. Structured around six themes, the study initially documented staff views on quality and the application of quality control and assurance techniques and tools. The study found a range of views on quality management in general, yet with a high level of consensus among the different staff role groups on the value of the key quality techniques and applications in higher education. There was general endorsement across the university for the inclusion of all staff in quality management. In the words of one quality manager:

> *"Yes, because it reflects all the stakeholders involved in ensuring that what we do provides quality. So, you are getting the delivery side, the administration side in ensuring that all the service areas, and the management thinking in terms of the policy side of what we decide in terms of quality. It's very much reflective of all the audiences that need to be represented".*

The quality experts in the study confirmed that the focus of quality management is the provision of quality service to students. Responsibility for education quality management was attributed to academic staff by 78% of respondents, with 19% attributing primary responsibility to management and 3% referencing quality management by external bodies. These differing views were considered to arise mainly from the specific organization culture of collegiate participation and to a lessor extent from staff role group identities. While external oversight is an internationally accepted aspect of higher education quality management, over reliance on external responsibility can reflect an internal abdication of quality responsibility.

Different staff role groups held overlapping views on academic quality control and assurance: 63% of management staff and 64% of administration staff viewed academic quality as the primary measurement of a university, while 53% of academic staff and 64% of student support staff expressed the view that education quality goes beyond academic quality as the primary measurement. We see a contrast here in views of quality between staff who interact with students directly and those who operate organization systems.

On the nature of the quality control and assurance system in place, 57% of staff viewed the system as *"A collegiate system of excellence"* while 38% of Staff viewed the quality system as *"Operational policies & procedures"*. A danger for quality control and assurance in people-based systems, is that it can be experienced as box ticking exercises. To avoid this pitfall in higher education it is important for staff to embrace quality control and assurance as an essential part of the organization's commitment to excellence, by continually raising consciousness of the nature and importance of quality. In this study, 78% of staff acknowledged the effectiveness of the quality system in helping to improve quality, an important measure of credibility for any quality control and assurance techniques and applicants. In addition, 66% of staff expressed the view that the quality system also impacted positively on the wider student experience. From the university perspective, allowing for the systemic importance of contrarian views to innovation and education, both management and quality experts were satisfied with this level of support for quality control and assurance, a position that might not be so reassuring from a purely business perspective.

While staff groups expressed some differing views of quality assurance, there was none-the-less a consensus across the staff as a whole. Where staff groups held differing views, leading at times to tensions between groups, the differences were often superficial or even perceptions of difference. The collaborative approach to quality helped to reduce these perceptions of difference and tensions between staff identity groups.

4. Quality tools and applications in higher education

Quality control and assurance in higher education employs policies and procedures to standardize operating practices. There is general agreement that this approach provides clear quality implementation, enforces checks and balances and helps to promote a quality culture among staff. An additional complexity in universities it that these policies and procedures need to be more than a rule book, open to enhancement from internal debate, periodic reviews and external stakeholders. Hence, quality policies and procedures are required to reflect the underlying philosophy of innovation and the overarching framework of university education. This tension between academic and business management perspectives is documented in the literature in Newton's [12] study of the differences between academic and management perspectives and the follow-up study on these differences by Cartwright [13].

A distinctive feature of quality systems in higher education is the reliance on collegiate academic judgment. In this study, 58% of academic staff and 67% of management were highly supportive of the importance of collegiate academic judgment as a technique in academic quality control and assurance. Education is not a one size fits all in terms of curriculum and delivery, so needs collegiate academic judgment. This feature of quality systems in higher education differentiates the academic orientation from the business orientation of other actors within the university. Closely related to the above is the identification of critical self-reflection on teaching by academic staff as an important technique in quality management in higher education [14].

Seventy percent of staff surveyed agreed that management monitoring of quantitative outputs is important for university quality control and assurance. The reservation expressed was that management monitoring of quantitative outputs in higher education needs to strike the correct balance between standard business measures and those relevant to quality education.

It is unusual to find universal agreement on quality techniques and applications in higher education. However, in this study there was universal agreement that external examiner monitoring of assessment is important for quality control as a type of safety valve. The external examiner systems for assessment and for programme validation were highlighted as two of the best techniques for quality control, assurance and enhancement.

Similarly, student feedback on programmes is a valuable and valued technique for quality management in universities. Despite the subjective aspect of this measurement, it supports the personal nature of education and the student-centred pedagogical approach. It reflects the importance of the relationship between student and teacher for learning. An interesting paradox in the study was both acknowledging the importance of student feedback and yet seeing it as a problematic area leading to tensions. This paradox exemplifies the business and academic cultures touching off each other. Academics apply student feedback to review and improve their module and provision, while remaining mistrustful of student feedback as an instrument of management monitoring, a blunt instrument of performance management. Nevertheless, 94% of all staff agree that student feedback on assessment is an important measure of quality. Interrogating this level of agreement further, identified a managerial culture that interpreted findings in a very specific and rather narrow, procedural way and different to a more reflective academic culture.

Industry stakeholder feedback was considered important for quality, reflecting a feature of higher education quality that external perceptions and market profile impact heavily on business performance, irrespective of the de facto measures of quality within the university. Understanding this business dynamic, university management were slightly more reserved in their support of industry feedback (88%) as a measure of quality, compared to other staff groups.

An interesting finding that emerged from the study was that the greater the level of staff participation within the quality system, the more critical the staff become towards the quality control and assurance techniques and applications. Perhaps this should not be surprising in a university setting where critical thinking is important and encouraged. Critique may be the price of staff participation. Greater staff participation provides a better understanding of quality techniques and applications among staff. This in turn provides valuable critique for quality enhancement and mitigates against risks of superficial measurement and box ticking.

Management commitment itself is viewed as a key element in establishing a viable quality culture that supports the techniques and applications in the quality system. In the words of one quality manager:

> *"If senior management do not walk the talk, in terms of quality assurance and the culture of quality assurance, it does not happen".*

The importance of management commitment to quality may seem obvious from a business perspective and attracted 91% support from staff in this study. However, it is not just an unquestionable affirmation for management, as it also designates responsibility.

5. Assessing quality in higher education

This study identified four main strengths of quality control and assurance in higher education:

1. *Academic staff quality, commitment, motivation, professionalism, integrity and self-reflective were returned as attributes of the academic staff. This response reflects on staff in general, but academic staff is repeatedly mentioned specifically.*

2. *Quality Standards in operation were also considered as a particular strength, with repeated positive references to policies, procedures and documentation.*

3. *External Examiners and External Reviews were noted as a strength of a higher education quality system.*

4. *Student feedback and student involvement in the quality systems were considered important.*

These metrics provide a strong assessment reflecting the lifecycle of delivering academic services to students. Management takes a wider view reflecting their responsibility for quality of the three pillars of teaching, research and engagement, adding engagement with industry and community as an important metric of quality.

Many academic staff self-identified as both teacher and researcher, research active teaching staff. Others see themselves primarily as teachers or as researchers. While these identities contribute differently to university quality, comparison of the result of the UK Research Assessment Exercise and the UK Teaching Excellence Framework suggest a close correlation between teaching and research excellence. This finding suggests the academic identity as both teacher and researcher is particularly valuable in supporting academic quality. The relationship between teacher and student, particularly in terms of access to teachers, was also identified as supporting the quality of the learning experience.

The study identified five main weaknesses of quality control and assurance in higher education, including some overlap with the strengths:

1. *Academic staff*: disinterested; resistant to change; underperforming; too busy to be reflective; not monitored and not supported to do research.

2. *Students*: the student experience; student unwillingness; lack of awareness among students; plagiarism and unequal treatment of students were given as examples of where students represent a weakness.

3. *Quality System*: too removed from teaching; emphasis on efficiency conflicting with a quality focus; lack of communication and training for staff.

4. *Management*: overbearing management structure; managerialism; micro-management; lack of commitment; disregard for staff; short-term focus; loose management practices; management based on pass rates, were stated as contributing to quality management weakness.

5. *Teaching*: when teaching quality not the highest priority; disconnect of quality assurance from teaching practice; limited internal oversight of teaching quality; and a focus on quantity versus quality, were identified as reasons why teaching can be a weakness.

Exploring the apparent contradiction that academic staff are seen as both a strength and a weakness of quality control and assurance, 81% of staff attributed this to the quality of teaching varying widely. The emphasis on teaching as a determinant of quality was somewhat surprising in the Irish context as teaching quality is not a criterion of the national cyclical review evaluation or internal institutional systems. This incongruence in quality control and assurance is significant, suggesting as it does the need to evaluate and enhance teaching as a criterion of higher education quality.

The business dimension impacts university ability to deliver a quality student experience. Commenting on the potential for quality enhancement:

- 90% of staff agreed that resourcing is needed to improve quality.
- Two-thirds of staff agreed that "speedier reaction times to changes in outside influences" and emphasis on external benchmarks would improve QA.
- 82% of staff agreed that more student evaluation and feedback on all aspects of the student experience improve quality.
- 78% of staff agreed that stronger links with employers strengthen quality.

The above indicators of quality strength and weakness identify the criteria for quality assessment of higher education. Benchmarks focused on other measures of quality or that use proxy measures are less reliable indicators for assessment of education quality. Quality control and assurance techniques and applications need to reflect these criteria to be effective. In the view of 84% of staff quality control and assurance in higher education extends to quality improvement and enhancement of the wholistic student experience. Similarly, 79% of staff identified the quality system as presenting opportunities for the improvement of academic staff performance and wider organization processes, while emphasizing that the quality system should maintain the central focus on the quality of research, teaching and learning.

6. Quality and the academic business nexus

The proposition that higher education is both an academic and a business endeavor is not widely accepted in Ireland. All Irish universities operate within the state funded public sector. In this study 84% of staff viewed Higher Education as primarily the Pursuit of Knowledge. There was little support for higher education being a public service, let alone a business. Responses to this question uncovered deeply held views on the nature of education, particularly among academic staff.

The manager interviewees were more supportive of the concept of higher education as a public service, but not as a business. One manager commented:

> *"Yes, it is a public service because it is publicly funded almost in its' entirety. And the fact the students pay a contribution, that's actually public funding but it's funding from the public except its direct funding from the public as opposed to via the State. So it is a public service. And the Pursuit of Knowledge can be quite abstract when we are trying to be practical and industry focused and focused on employability".*

Within the study survey, 80% of staff disagreed with the view that higher education is a business. This view carried over to the view of 64% of staff that higher education focuses on training for employment. The differences in views on this question of the nature of higher education were explained by staff as stemming from different role-identity staff groups and differences in understanding or perceptions. This finding underlines the significance of the role-group perspectives that are largely unrecognized within quality control and assurance in higher education. The finding suggests the potential of explicitly addressing the different perceptions through use of collaborative and integrative techniques and applications.

For example, 68% of academic staff believe that academic quality is best achieved with a flat management structure. Administration and student services staff take the contrary view that a hierarchical management structure is best to achieve quality. A quality expert commented that this structure issue has little to do with quality. Structure here is used as a pseudo measure for who is responsible for quality control and assurance. Administration and student Support staff are used to operating within a defined hierarchical management structure, while academic staff value their autonomy and academic freedom as a balance to management hierarchy. While not diametrical opposites in practice, there is a tension here between academic collegiate management and administrative bureaucratic management. The equality culture among academics is based on professional knowledge, experience and qualifications, which distinguishes them from the more hierarchical culture in terms of grades and promotion.

While the management interviewees understandably responded somewhat defensively to critique of hierarchical management, they were accepting of the value and importance of engaging the diversity of staff and their professional knowledge. The majority of staff (89%) supported the assertion that quality is best achieved where standards are legitimate, and all staff buy into the virtues of the quality system. There was near total agreement among staff to the need for balance in management focus to support collegiate culture. One manager commented:

> *"A good manager will create that collegiate focus and will be part of the collegiate focus".*

Similarly, the study found a consensus around the view that higher education quality is best managed through a mix of management and academic measurements. Agreeing in general with this view, the quality management professionals pointed to the complexity in deciding the type of management and academic measurements to employ and how to assemble a balanced mix of measures. For example, student results provide a very convenient, quantitative data set for performance measurement. However, they are a weak quality measurement tool, without context of how the pass rates and results relate to the added value to the student cohort, as students vary considerably by university intake policy and in learning ethos. This question surfaced the tensions that arise between two distinctive views, a business management view that quantitative measures are good performance indicators and an academic perspective that performance measurement in higher education is more complex and nuanced.

The importance of staff ownership and responsibility in universities encourages a management authority balance within the culture of universities. While directive and decisive management are needed at times, balanced management is needed to encourage staff ownership and participation. In this study 91% of staff confirmed that quality is driven by staff ownership. Management commitment to this balance is itself a driver of quality in higher education, supporting a collegiate culture of ownership. In this research, 75% of staff agreed that management commitment is a driver of quality. Management interviewees fully supported this understanding of their role of management in higher education to support quality control and assurance. This presents a tension for managers between asserting management authority and nurturing collegiate ownership to achieve this balance that supports quality ownership across the organization.

The orientation towards quality control and assurance in universities involves tensions to be navigated by the staff and the organization culture. The balance in management of front-line staff is important to the quality of the teacher-learner relationship, with a direct impact on education quality and the student experience. Similarly, 66% of staff agreed that their independence and ownership of quality requires external scrutiny and accountability to provide the checks and balances that underpin staff ownership.

7. Higher education quality culture

While quality control and assurance are at the heart of quality culture, 61% of staff surveyed referred to the importance of a focus on quality enhancement. The rationale here is that within a learning organization the outputs of quality control and assurance experience should generate ideas and opportunities for quality enhancement. Acknowledgement of the importance of quality monitoring by 54% of staff, contrasted with a view that quality control and assurance in higher education can become bogged down in paperwork, reporting and over-reliance on statistics, such as pass rates, retention rates and completion rates. In the words of one quality manager interviewed:

> *"I suppose it may be the perception that we focus too much on statistics... and that we do not actually take a qualitative view rather than a quantitative view from time-to-time. I could see why some staff would feel that way".*

Some skepticism was expressed with the value of assessments of quality in higher education. Management interviewees suggested that higher education quality is more about creating a quality framework for ongoing review of quality, which in their view is not the same as assessment. One manager commented:

> *"We're not constantly rigorously testing education quality and we probably should not because you'd tie yourself in knots trying to find the metrics on how to do it."*

The predominant view of 78% staff was that the focus of quality assurance in higher education is a combination of quality enhancement, quality monitoring and some assessment of quality. In this context the quality professionals referred to the circle of quality – plan, organize, control, measure.

The business and competitive environments in which universities operate have increasingly given credence to business like responses to the market. Nonetheless 61% of staff disagreed with the suggestion that the focus of quality assurance is on impression management. Similarly, there was almost universal agreement that traditional industrial models of quality control and assurance do not fit well to higher education. For instance, discipline and technology-based quality, that is central to ISO methodology, does not fit well in universities, though there may be something to learn from those models.

8. Addressing the academic-business nexus

The growing importance of technology and the knowledge economy has established higher education is a driver of innovation and national economy. However, with this important status of universities has come the desire for greater external control on universities through the legal and financial powers of the state. Striving to protect their independence, universities around the world are adapting to external controls, including quality control and assurance systems. As learning organizations, universities for the most part have managed to integrate these external controls, adapting them to the existing internal quality policies and procedures that achieve what the external quality policy may also be aiming to achieve. Similarly, universities internally have to some extent resisted market forces, so that their quality systems remain predominantly focused on substance rather than form.

In this study, 78% of staff confirmed that the focus of quality assurance was on adapting external quality policies, rather than on adopting external quality policies. Universities generally have pivoted to make external policies their own and fit for purpose. A quality professional commented:

> "*Yes, you must make them your own and see how they actually fit in. Because you may already have policies and procedures that achieve what the external quality policy may be trying to achieve as well*".

Three senior managers offered their views on how universities respond to external quality policy:

> "*Yes, we don't really tend to blindly copycat whatever we pick up elsewhere. We do have a culture of tailoring it and having our own debate. That's good*".

> "*Yes I think we adapt, we localize*".

> "*Yes, that would be a practice I would have seen here over the years. Of maybe using templates of other institutions but adapting them to our conditions*".

There remained a concern among 54% of staff that the focus of quality control and assurance in higher education can change to accountability as a priority over improving the quality of operations. In discussion, it was clear that this is not a binary choice, yet business managers displayed a preference for accountability while academic and quality managers were more committed to quality improvement. The multi-factorial, multi-faceted, multi-view nature of higher education operates across the academic-business nexus.

9. Concluding remarks

The complex nature of quality control and assurance in higher education revealed itself in this study. While standard business techniques and tools apply to the purely business aspects of the organization, academic quality assurance was most supported by an integrated, collaborative approach to define the quality system and measures that match the organization culture and the education mission.

Faced with multiple demands in a rapidly changing world university models continue to evolve. One such evolution is repositioning at the centre of economic and technological development that has brought universities closer to business models and needing to straddle the Academic Business nexus. Universities have been largely successful in doing so. In this study, 73% of staff confirmed they have a generally positive view of quality control and quality assurance.

Universities pride themselves on their different cultures, ethos and values as well as their education offerings. However, in education essentials they are often more the same than different. Similarly, there are different staff role grouping in higher education with somewhat different academic and business perceptions of the organization. These group differences are in reality less than the commonly held views of difference and tension. Participation, collaboration and integration of staff groupings within quality processes and systems can overcome these tensions.

The apparent and actual differences between academic and business views of higher education are reconcilable. Quality control and assurance systems, tools and applications create natural opportunities for different staff groups to collaborate and integrate.

Quality control and assurance in higher education goes beyond the understanding and implementation of policies and procedures. Higher education quality at its best is a lived and meaningful experience from which knowledge and innovation can arise. This research study was motivated by a need to address the changing nature of higher education organizations. The Academic Business nexus in higher education is at the root of current changes. A widening business focus requires a shift in staff roles and identity, and changes in organization mission. Higher qualifications among profession staff have given rise to the concept of the "third space", reflecting the heightened interdependence of different staff roles.

Author details

Terry Twomey
Technological University of the Shannon, Limerick, Ireland

*Address all correspondence to: terry.twomey@tus.ie

References

[1] US News and World Report. Best Colleges 2022-2023. Washington, DC: US News and World Report; 18 Sep 2023. Available from: https://www.usnews.com/best-colleges

[2] Joint Declaration of the European Ministers of Education. Ministerial Conference Bologna 1999: The Bologna Declaration of 19 June 1999. Italy: University of Bologna; 1999

[3] ENQA. ESG 2015: Standards and Guideline for Quality Assurance in the European Higher Education Area. Brussels: EURASHE; 2015

[4] Elassy N. The concepts of quality, quality assurance and quality enhancement. Quality Assurance in Education. 2015;**23**(3):250-261

[5] Warner CK. Contested definitions of excellent teaching: An analysis of the discourse of quality. Journal of Thought. 2016;**50**(1-2):20-36

[6] Yorke J, Vidovich L. Learning Standards and the Assessment of Quality in Higher Education: Contested Policy Trajectories. Vol. 7. Switzerland: Springer International Publishing; 2016

[7] Tam M. Measuring quality and performance in higher education. Quality in Higher Education. 2001;7(1):2001

[8] Veles N, Boon H, Carter M. The university third space phenomenon: Investigating perceptions of professional staff working across boundaries in an Australian university and its Singapore campus. In: Australian Association for Research in Education (AARE) Conference 2017. Canberra, Australia: AARE; 2017

[9] Karseth B, Solbrekke TD. Curriculum trends in European higher education: The pursuit of the Humboldtian University ideas. In: Higher Education, Stratification, and Workforce Development. Switzerland: Springer International Publishing; 2016. pp. 215-233

[10] Wright S, Greenwood DJ. 'Universities run for, by and with the faculty, students and staff: Alternatives to the neoliberal destruction of higher education', in learning and teaching. The International Journal of Higher Education in the Social Sciences. 2017;**10**:42-65

[11] Twomey T. Identity nexus in higher education: Perceptions and tensions. In: European Conference on Education Research. Hamburg, September 2019 Paper 2457. Berlin, Germany: European Education Research Association; 2019

[12] Newton J. Feeding the beast or improving quality?: Academics perceptions of quality assurance and quality monitoring. Quality in Higher Education. 2000;**6**(2):153-163

[13] Cartwright MJ. The rhetoric and reality of "quality" in higher education: An investigation into staff perceptions of quality in post-1992 universities. Quality Assurance in Education. 2007;**15**(3):287-301

[14] Brookfield SD. Becoming a Critically Reflective Teacher. 2nd ed. New Jersey, USA: Jossey-Bass; 2017

Chapter 15

Quality as a Sustainable Tool against Institutionalized Corruption

Pablo Dávila and Diego Mantilla

Abstract

Quality management systems are crucial for controlling and administering public institutions, promoting a culture of continuous improvement and reducing costs. This research evaluated three critical aspects in the fight against corruption in public procurement: risk management, compliance, and the implementation of ISO 37001 standards. Factors such as collusion between officials and companies, repetitive contracts, solvency checks, supplier experience, and certification were considered. The study found that collusion, procurement requirements, and supplier experience formed a pattern of fraud and corruption in companies. Compliance planning, internal procurement guidelines, training, commission monitoring, and contract evaluation were effective, but areas like planning and contract evaluation needed improvement. Low scores in implementing ISO 37001 standards indicated recognition but incomplete adoption, highlighting the need for stronger anti-corruption policies, ethical codes, and corruption identification practices.

Keywords: quality system, anti-bribery, compliance, risk management, ISO 37001

1. Introduction

Corruption in Latin American countries gains strength year after year, accentuating both public and private enterprises. For this reason, managers and public officials must adopt a preventive attitude to establish mechanisms that allow for the implementation of control at all stages. In this sense, their application will achieve the fight against corruption and fraud, thus becoming a complex challenge that requires a combination of effective measures and tools. Therefore, the following aspects highlight some key tools for control and the fight against corruption:

1.1 Legal and regulatory framework

A solid legal framework is essential to prevent and sanction corruption. This involves enacting clear, comprehensive, and up-to-date laws that prohibit and penalize acts of corruption, as well as establishing control bodies responsible for their implementation. In this context, compliance goes beyond mere fulfillment; it also requires evaluating the timely compliance with the regulations, which allows for establishing the foundations on which a company or institution develops. "The results show that Integral Audit is the multidisciplinary, independent, and systems-focused evaluation of the degree and manner of achieving an institution's objectives." This position from [1] highlights that not only the legal and regulatory framework of organizations allows for multidisciplinary evaluations, but also the fulfillment of institutional goals and objectives through management indicators.

1.2 Transparency and access to information

Another essential element is transparency in management, establishing technological mechanisms for accessing information as powerful tools to prevent and expose corruption and fraud. Transparency involves publishing documents, methodologies, and actions carried out by the company or institution on their portals, including the activities and budgets used. Additionally, governments should promote active disclosure of information about their activities, budgets, contracts, and important decisions, facilitating citizen participation and public scrutiny. As stated by [2], "to improve processes and have good management of the company," actions of transparency and access to information are considered a priority for timely controls.

1.3 Accountability

It is necessary to establish effective mechanisms and procedures to hold public officials accountable for their actions and decisions. This involves strengthening internal and external control systems, promoting auditing and evaluation of policies and programs, and ensuring that those responsible for acts of corruption are investigated and fairly punished. Additionally, in the case of companies, the contributions of a professional known as a compliance officer ensure that control processes are periodically fulfilled, as it becomes a constant evaluation of the company by applying legal regulations and promoting training with updated information, thus establishing a compliance culture throughout the organization to prevent and mitigate risks.

1.4 Citizen participation

The participation of civil society, citizens, and non-governmental organizations is crucial in the fight against corruption. Mechanisms such as implementing an anonymous reporting system, integrity systems, and dialog spaces between the government and society can help prevent corruption, identify corrupt practices, and promote transparency and accountability. All this becomes a commitment of the community, considering that institutions and companies are part of it, and therefore, participation becomes a commitment of citizenship.

1.5 Institutional strengthening

It is important to strengthen the institutions responsible for preventing, investigating, and sanctioning corruption. This involves ensuring their autonomy, providing them with adequate resources, improving the training of their staff, and promoting cooperation and coordination between different organizations and levels of government. This strengthening starts with leaders who actively position control in key processes of the organization, allowing the use of this mechanism to guide the actions of the company, to have timely information and make the best decisions for the organization. "A process audit procedure to verify the effectiveness of strategic planning" [3], this activity is part of institutional strengthening and represents an excellent managerial control strategy.

1.6 Technology and information systems

The use of information and communication technologies can be a powerful tool in the fight against corruption. Electronic public procurement systems, expense tracking platforms, online reporting systems, and data analysis tools can help detect corruption patterns, increase transparency, and facilitate accountability. These activities focus on evaluating control; most control models currently used by organizations promote the immediate availability of information and ensure its quality and transparency through evaluation and treatment, comparing processes and systems over time. "Applying artificial intelligence techniques in the field of auditing promises to be an effective strategy for improving efficiency and productivity in these processes and therefore in the organization" [4]. This position is part of a technological transformation where information quality becomes a decision-making tool based on the analysis of adequate information systems managed by technology.

1.7 Culture of integrity

Promoting a culture of integrity and ethics in both the public and private sectors is essential to prevent corruption. This involves fostering values such as honesty, responsibility, and justice, and promoting training and awareness programs on the risks and consequences of corruption. It is important to note that the effective implementation of these tools requires a comprehensive approach and collaboration among different actors, including governments, civil society, the private sector, and international organizations. Furthermore, it must be considered that specific strategies and tools may vary according to the context and needs of each country or community. This is about guiding activities through codes of ethics and conduct, which represent the path that organizations must follow as part of corporate social responsibility. It is here that values and principles are defined, which all employees of a company or organization must embrace. It is essential that companies do not have an ethical instrument and leave it unused or only adhere to the basics.

2. Elements of quality-related internal auditing

The search for effective tools for internal auditing is the first step in taking actions that allow for an effective evaluation of an organization. Currently, technological transformation, reproductions of processes through artificial intelligence, and controlling the lack of ethical values in professionals, who are the major perpetrators, are key aspects. Auditing, applied to fraud and corruption, is a fundamental tool for detecting and preventing illicit acts within an organization. The following are some common tools and techniques used in auditing to address fraud and corruption:

2.1 Risk analysis

Auditing begins with the identification and evaluation of risks associated with fraud and corruption in the organization. Factors such as organizational culture, internal controls, procurement and purchasing processes, and financial transactions are examined to determine the areas of greatest vulnerability. For fraud analysis, the risk of fraudulent activities in an organization is evaluated. Areas of repetitive risk are identified, existing controls are evaluated, transactions and unusual patterns are analyzed, and fraud indicators and other risk factors are considered. Risk is quantified, risk treatment is performed, mitigation strategies are developed, and additional controls are implemented for subsequent application. The process is continuous and requires periodic review and updating. "Audit risks focus on material misstatement," as stated by Cruz [5], emphasizing that risks focus on material misstatement, which is part of the analysis identifying and generating indications of fraud because of failed processes.

2.2 Compliance testing

Auditing focuses on verifying compliance with policies, procedures, and relevant laws to prevent fraud and corruption. Documents, contracts, accounting records, and other records are reviewed to identify possible irregularities. To achieve compliance, it is necessary for corporate governance to have the willingness to establish a policy that allows, if necessary, the hiring of a compliance officer who trains personnel and ensures the periodic enforcement of deficient processes that repeatedly lead to fraud. Without this commitment, little can be achieved by solely evaluating control. This compliance activity is part of the responsibility of how managers promote control as a policy for change.

2.3 Substantive testing

Substantive testing is conducted to detect unusual or fraudulent transactions. This may involve a detailed review of invoices, receipts, bank statements, and other documentary evidence to identify possible deviations or manipulations. Substantive testing is a tool used in fraud detection. It is applied to identify transactions or events that could indicate the existence of fraudulent activities. These tests are based on the identification of significant deviations, anomalous patterns, or unusual discrepancies in financial or operational data. By performing these tests, the goal is to highlight elements that cannot be explained by normal circumstances or random fluctuations. Careful interpretation of the results of these tests can help focus on more detailed investigations and reveal possible hidden frauds.

2.4 Data analysis

The use of data analysis tools can help identify suspicious patterns and trends that may indicate the existence of fraud or corruption. By analyzing large volumes of financial and operational data, anomalies, duplications, fictitious transactions, or other indications of irregularities can be detected. It is important to focus on standards in the case of processes and on the organization's historical data. Data analysis is a powerful tool in auditing that allows for the efficient and effective examination of large volumes of information. By applying data analysis techniques such as statistical sampling, trend analysis, anomaly detection, and data mining, auditors can identify patterns, irregularities, and potential frauds. These tools help improve the accuracy and thoroughness of audits while enabling a more proactive and data-driven approach. Data analysis in auditing provides greater confidence in detecting errors and fraud, and facilitates timely, informed decision-making.

2.5 Training and awareness

Auditing may include training and awareness programs on fraud and corruption targeted at the organization's audit staff, becoming an integral part of the control activity. These preventive educational programs help promote a culture of integrity and ethics and provide guidance on the detection and prevention of fraudulent behavior. Training and awareness are fundamental tools in auditing to strengthen internal controls and prevent fraud [6]. Through training, employees are provided with the necessary skills and knowledge to recognize warning signs, understand associated risks, and comply with established policies and procedures. Additionally, ethics and compliance awareness promote an organizational culture based on values and transparency [7]. This encourages early detection of irregularities, fosters collaboration with internal and external auditors, and promotes greater accountability and commitment from all members of the organization in fraud prevention and detection. Finally, it is important to emphasize that auditing applied to fraud and corruption should be carried out by professionals specialized in forensic investigations and regulatory compliance. Additionally, it is essential to have an organizational environment that supports the auditing process and does not hinder its work. Similarly, transparency and accountability should be promoted within the organization.

3. Quality in auditing

3.1 Internal investigations

A commonly used technique within preventive auditing is conducting investigations through the technique of inquiry in cases of specific suspicions of fraud or corruption. Auditors can conduct internal investigations to gather additional evidence [8]. This involves employee interviews, reviews of electronic communications, forensic computer analysis, and other methods to obtain relevant evidence. Internal investigations are important tools in auditing, especially in fraud detection and prevention [9]. These investigations involve thorough examination and analysis of an organization's activities, transactions, and records to identify possible irregularities or non-compliance. Internal investigations can be driven by internal or external reports and become particularly useful when there is an anonymous reporting process. By verifying data, reviewing documentation, and conducting employee interviews, these investigations provide a more detailed insight into potential fraud or misconduct, enabling auditors to take corrective actions and establish more effective control measures to prevent future incidents. The auditor relies on their intellectual capital, judgment, and experience to determine these actions.

3.2 Evaluation of internal controls

Auditing also evaluates the effectiveness of existing internal controls within the organization to prevent and detect fraud and corruption. Authorization processes, segregation of duties, supervision, and monitoring are reviewed, and recommendations are made to improve and strengthen them [10]. The evaluation of internal controls is an essential tool in auditing. It involves analyzing and assessing the systems and procedures established in an organization to ensure the effectiveness of its operations, the reliability of financial information, and fraud prevention. Through comprehensive testing and reviews, auditors assess the adequacy and effectiveness of internal controls, identifying potential weaknesses or deficiencies. This evaluation allows auditors to recommend improvements, strengthen risk management, and safeguard the organization's assets. It also provides greater confidence to users of financial reports and ensures compliance with applicable regulations. This is accompanied by a commitment from corporate governance to design internal control policies that enable the proper functioning of internal controls [11].

3.3 Ética (Ethics)

Ethics refers to the principles and values that guide human behavior. In the context of public management, ethics becomes an important tool for preventing institutionalized corruption, as it promotes integrity, honesty, and responsibility in the performance of public functions. "The purpose of ethics is duty, duty towards oneself or towards others. It is primarily individual or personal, even when it refers to duties and obligations towards others. The quality of human life is related to both solitude and sociability" [12].

Therefore [13], describes ethics as the part of philosophy dedicated to reflecting on morality. Ethics is a type of knowledge that seeks to be constructed rationally, using conceptual rigor and methods of analysis and explanation inherent to philosophy [14].

Ethics is a fundamental part of any organization as it clarifies concepts and brings arguments that allow us to understand the moral conduct of individuals [15]. Ethics is the moral philosophy that weighs on each human being in their behavior, or it can also refer to the good customs that allow for living in a healthy and balanced coexistence among people. Ethics enables individuals to differentiate between right and wrong and apply it to specific practices or societies [16]. It is more about human behavior based on personal commitment to the realization of common well-being, based on the social and psychological study of individuals or groups, and imposing the rules or customs of society or the environment in which they are found.

3.4 Access to information

Access to information is a fundamental right of citizens and a key tool for preventing corruption. The availability of clear and timely information about the functioning of public institutions, decision-making processes, and public resources can contribute to transparency and accountability in public management. "The recognition of access to information implies, correlatively, a guarantee of a free public opinion, indissolubly linked to political pluralism and the constitutional rule of law adopted by our constitution" [17]. It is considered a positive act that the right to access public information guarantees transparency so that citizens can oversee and hold public servants accountable at every step of the process and at any time. Furthermore, it constitutes an effective means against corruption [18].

The Swedes discovered that access to information is a solvent for patrimonialism, discretionary, illegal, or blatantly corrupt practices. And that solvent accompanied the economic takeoff process of that country, marking it forever. "Having access to official certificates was initially intended to support the land distribution process but later spread to other areas of government, becoming part of the administrative culture of the State and its citizens" [19].

Access to information benefits society, without any distinction, because by having sufficient information on a topic of interest, members of a society acquire the ability to engage in debates with others. This is an important aspect of freedom of expression and a fundamental instrument for promoting the rule of law and other individual rights to foster trust.

The right to access information as a fundamental right is entirely necessary in any country, depending on who or what requests the information and the type of information involved. Clearly, having access to information would prevent corruption, shed light on decision-making processes, budget allocation, and management that has been carried out to demonstrate the proper fulfillment of obligations, leading to public measures and policy design that help the population.

In this regard, we pose the following research question: Does the application of quality tools to combat corruption influence their effectiveness? To explore this, we will analyze three influential tools within management auditing: i) risk control, ii) compliance, and iii) ISO 37001 (anti-bribery).

3.5 Managerial implications

The study reveals critical managerial implications for effectively combating corruption and promoting quality management practices:

1. Prioritizing quality tools is a foundational step. Management must commit to implementing proven approaches like Six Sigma and Lean, which enhance processes and reduce susceptibility to corruption.

2. Proactive implementation of preventive measures is essential. Neglecting anti-corruption actions within organizational processes equates to permitting corruption to persist. Management should establish robust policies, promote a culture that encourages reporting corrupt acts, and protect whistleblowers.

3. Fostering a culture of accountability and transparency is vital. Leaders must cultivate an environment where responsibility and openness are core values.

4. Continuous improvement is paramount to stay ahead of evolving corruption risks. Management must regularly assess and update anti-corruption policies, processes, and quality management practices.

5. Ensuring legal compliance is a non-negotiable requirement. All measures must adhere to legal and regulatory standards.

6. Ongoing employee education and training play a crucial role in corruption prevention. Training programs should cover ethics and the significance of reporting corrupt behavior.

7. Visible commitment from top management in anti-corruption and quality management initiatives sets the organizational tone.

8. Regular monitoring and auditing: Uncover and rectify vulnerabilities to corruption. Managers should establish clear procedures for reporting and addressing corruption.

9. External collaboration with experts and organizations can provide valuable resources for enhancing organizational integrity. These managerial implications underscore the necessity of adopting a comprehensive, proactive approach to tackle corruption and ensure effective quality management.

4. Methodology

The depth of this methodology is evident in the thorough approach taken to address the research question. It encompasses a detailed examination of the causes outlined in the conceptual framework related to combating corruption through the application of quality tools. This method involves the creation of 5 specific criteria for each tool, focusing on the key tools mentioned earlier. Additionally, it involves consulting with 8 expert auditors to determine the average rating for each tool. The resulting rating scale, ranging from 1 (not influential) to 10 (totally influential), categorizes and describes the direction of the responses. The methodology goes on to highlight the most influential answers, indicating a comprehensive and in-depth analysis of the research question.

4.1 Results

4.1.1 Risk management

For this variable, we considered the following criteria within public procurement: repetitive contracts, collusion at all levels, review of requirements and solvency, supplier experience, and supplier certification. The results are as follows:

In **Figure 1**, we can observe that the audit experts rated collusion between officials and companies with a score of 8. This element is consistently identified as a significant driver of corruption. If collusion occurs at all levels, it becomes difficult for control systems to detect it. The experience of auditors indicates that this factor, combined with requirements and the experience of procurement participants, becomes the pattern for fraud and corruption in companies.

4.1.2 Compliance

The criteria we highlight in this tool regarding public procurement are as follows: contract compliance planning, internal contracting guidelines, guideline training, contract commission monitoring, and contract management evaluation.

In **Figure 2**, we can observe that the highest average is for the commission monitoring and training of officials in compliance, with a rating of 6 and 5, respectively. This implies that this quality tool, created for a commission to evaluate and proactively train officials to prevent acts of corruption in procurement, is effective. However, it is noteworthy that there is a low average for planning, guidelines, and

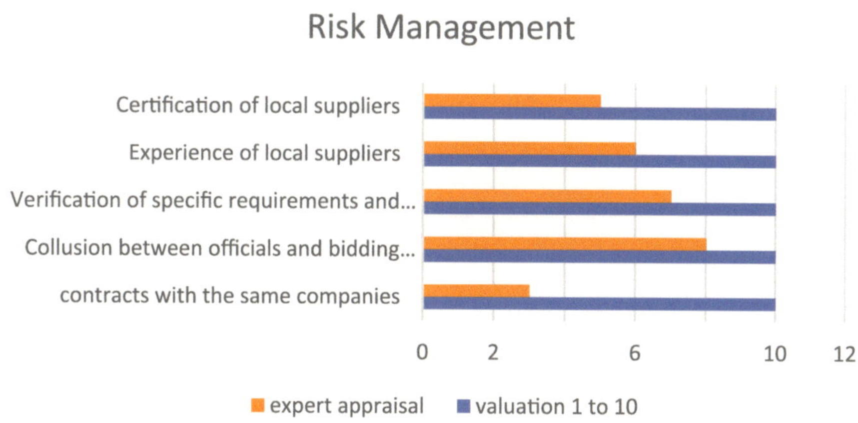

Figure 1.
Risk management analysis in public procurement.

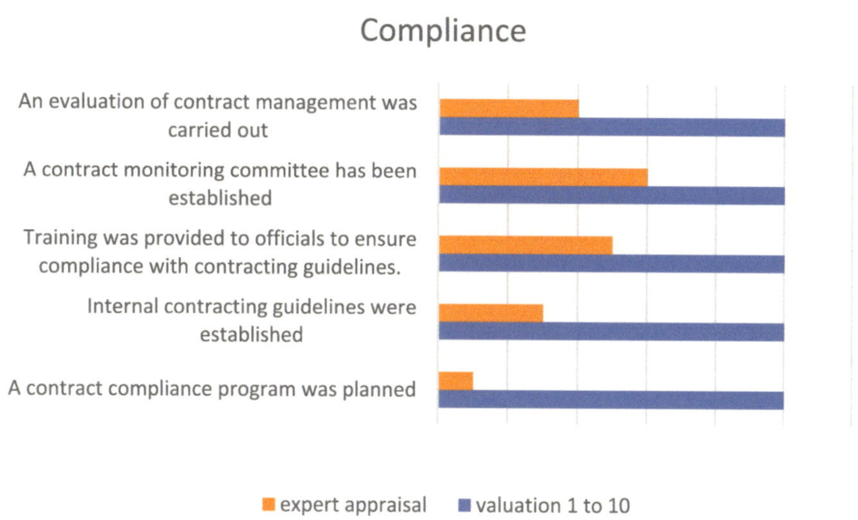

Figure 2.
Analysis of compliance in procurement.

contract management evaluation, with ratings of 1, 3, and 4, respectively. This suggests that compliance in these areas is very low or that there is a lack of commitment from corporate governance.

4.1.3 ISO 37001

In this process, the ISO 37001 anti-bribery standard was considered a preventive cross-cutting framework to combat corruption. By applying this standard, the conditions involve conducting an organizational diagnosis and addressing its weaknesses in a specific process. This implies implementing a quality condition to meet compliance requirements. The variables to consider are as follows: level of commitment to corporate governance, availability of a compliance officer, utilization of the code of ethics, identification of anti-bribery practices, and evaluation and sanctioning of corrupt suppliers and officials.

Figure 3.
Analysis of elements of the ISO 37001 anti-bribery standard.

In **Figure 3**, we can observe that the average ratings are low. The variables of the availability of a compliance officer and the evaluation of corrupt officials received a rating of 2. The remaining variables, such as the anti-bribery commitment policy, the code of ethics, and the identification of anti-bribery and collusion practices, currently serve as essential mechanisms for preventing corruption but have a rating of 1. This suggests that many companies have not fully implemented this ISO standard, although they recognize the importance of having a compliance professional and implementing an ethical code to prevent crimes and fraud.

5. Discussion

Risk management reveals that collusion between officials and companies stands out as a critical driver of corruption. This underscores the importance of enhancing control and supervision systems to detect and prevent this practice at all levels. Auditor experience highlights that collusion, when combined with procurement requirements and expertise, establishes a pattern of fraud and corruption within companies.

Compliance demonstrates effectiveness in commission monitoring and official training, but planning, guidelines, and contract management evaluation require significant improvements. This suggests a lack of commitment in critical areas of corporate governance.

Regarding ISO 37001, the low scores indicate that many organizations recognize its importance but have not fully implemented this standard. This underscores the need for greater commitment to establish robust anti-corruption policies, ethical codes, and practices for identifying corruption and collusion.

6. Conclusions

Combating corruption and fraud in Latin American countries requires the application of key tools. A solid legal framework, transparency in management, accountability, citizen participation, institutional strengthening, technology and information systems, and a culture of integrity are essential elements. When these tools are implemented comprehensively and collaboratively, they can contribute to preventing and combating corruption in both the public and private sectors. It is crucial for leaders and organizations to promote ethics, responsibility, and justice and foster a culture of integrity at all levels.

By identifying causes related to quality in internal auditing, the following elements stand out: i) Risk identification: examining factors such as organizational culture, internal controls, procurement and purchasing processes, and financial transactions to determine areas of greater vulnerability, ii) Compliance testing: reviewing documents, contracts, accounting records, and other records to verify compliance with relevant policies, procedures, and laws to prevent fraud and corruption, iii) Substantive testing: conducting detailed tests on invoices, receipts, bank statements, and other documentary evidence to detect unusual or fraudulent transactions, iv) Data analysis: utilizing data analysis tools to identify suspicious patterns and trends that may indicate the existence of fraud or corruption, v) Internal investigations: conducting internal investigations, which may include employee interviews, reviews of electronic communications, and forensic computer analysis, to gather additional evidence in specific cases of suspicion, vi) Evaluation of internal controls: assessing the effectiveness of existing internal controls in the organization to prevent and detect fraud and corruption, identifying potential weaknesses and deficiencies, vii) Training and awareness: implementing training and awareness programs on fraud and corruption for audit personnel to promote a culture of integrity and ethics.

This set of processes allows for the identification of causes for fraud and corruption crimes, and its application should be continuous and compliant with regulations and laws. Additionally, it can be developed as a policy enforced by corporate governance. Furthermore, transparency and accountability should be intelligently promoted within the organization, and specialized professionals in forensic investigations and regulatory compliance should be present.

Risk management in public procurement presents significant challenges. Our analysis reveals that collusion between officials and companies is a persistent problem and represents the primary driver of corruption. The lack of detection by control systems is due to its pervasive nature. Additionally, auditors' experience highlights that this factor, along with requirements and participants' experience, are common patterns of fraud and corruption in contracting companies.

On the other hand, our analysis reveals that, in terms of compliance in public procurement, the commission monitoring and training of officials receive positive ratings, with average scores of 6 and 5, respectively. These quality tools aim to prevent acts of corruption. However, the low ratings for planning, guidelines, and contract management evaluation, with scores of 1, 3, and 4, respectively, are concerning. This indicates low compliance or a lack of commitment from corporate governance.

The analysis of elements of the ISO 37001 anti-bribery standard reveals low average ratings. The availability of a compliance officer and the evaluation of corrupt officials received a score of 2. The variables related to the anti-bribery commitment policy, the code of ethics, and the identification of anti-bribery and collusion practices are considered essential mechanisms.

Citizens have a fundamental right to access information, which serves as a crucial tool for preventing corruption. This suggests that many companies have not fully implemented this ISO standard, although they recognize the importance of having a compliance professional and applying an ethical code to prevent crimes and fraud.

In this way, we observe that the research question is fulfilled, and we emphasize that the use of effective quality tools enables the prevention and detection of fraud and corruption. Not implementing measures within organizational processes means allowing them to continue without taking any action, which also implies a certain complicity with administrations that do not take steps to change, and that is also part of corruption.

Conflict of interest

The authors declare no conflict of interest.

Author details

Pablo Dávila and Diego Mantilla*
Universidad Central del Ecuador, Quito, Ecuador

*Address all correspondence to: dmmantilla@uce.edu.ec

References

[1] Herrera H. Corporate Social Responsibility and Business Ethics. Colombia: Ecoe Ediciones; 2011

[2] Medina A, Crespo M. Checklist to Evaluate Strategic Planning as a Process Audit Tool. Magazine de las Ciencias. 2019. pp. 1-15. Available from: https://revistas.utb.edu.ec/index.php/magazine/article/view/671

[3] Merrill JC & Barney RD. Press and Ethics. Essays on the Morality of Mass Media. Buenos Aires, Argentina: Universidad de Buenos Aires. Available from: https://catalogosiidca.csuca.org/Record/UNACHI.5485

[4] Law of Transparency and Access to Public Information. 2004. Available from: https://books.google.com.ec/books?hl=es&lr=&id=AZ3FAQAAQBAJ&oi=fnd&pg=PP1&dq=Herrera+H.+Ecoe+Ediciones%3B+2011&ots=vK409t3eZD&sig=M1IIcnmmpTJTzAM0KO0r8aHiNlY#v=onepage&q=Herrera%20H.%20Ecoe%20Ediciones%3B%202011&f=false

[5] Cruz J. Auditing in the COVID-19 environment. Use of technology and risk approach. Podium. 2020;**38**:1-15. DOI: 10.31095/podium.2020.38.5

[6] Descalzo A. Good Governance: Scope of Application, General and Action Principles, Disciplinary Offenses, and Conflict of Interests. Transparency Access to Public Information and Good Governance. Study of the Law. 2013. Available from: http://www.scielo.edu.uy/scielo.php?script=sci_arttext&pid=S2301

[7] Zorro M. Development: Perspectives and Dimensions. First ed. Bogota: University of the Andes; 2007. DOI: 978-958-695-432-7

[8] Aguilera Izaguerra G. Good Governance and Human Rights in Mexico. 2020. Available from: http://portal.amelica.org/ameli/jatsRepo/187/1872095005/index.html

[9] Cortina A, Martínez E. Ethics. (1st edition). 1996. Available from: https://books.google.es/books?hl=en&lr=lang_en&id=_HtC1_tTjfUC&oi=fnd&pg=PA9&dq=ethics+&ots=hFrFbP-aGJ&sig=RbookDoAR5MuzhuehCixlfD_3I#v=onepage&q=ethics&f=true

[10] María L, Clarita G, Galo C. Importance of management audit as a continuous improvement tool in companies. Contemporary Dilemmas: Education, Politics, and Values. 2021;**42**:52-72. DOI: 10.46377/dilemas.v8i.2723

[11] Fernández JA. Total quality management. The EFQM Model of Excellence. Medifam. 2002;**12**:41-54

[12] Guajardo E. Total Quality Management. Concepts and Teachings of the Great Quality Masters. 1996. Available from: https://books.google.es/books?hl=en&lr=&id=9zYyYc6i9JwC&oi=fnd&pg=PA1&dq=total+quality&ots=2bzZ65J7Th&sig=XtlTK5AmkcDCaCUfZw0fsMX6ANY#v=onepage&q=total%20quality&f=false

[13] Ferrando Sánchez M, Granero Castro J. Total Quality: EFQM Model of Excellence. Spain: Editorial; 2005. DOI: 84-96169-68-5

[14] Cantos M. Comprehensive audit as a validation tool for institutional management. Telos: Journal of Interdisciplinary Studies in Social Sciences. 2019;**21**:422-448. Available from: https://dialnet.unirioja.es/servlet/articulo?codigo=6923471

[15] Guajardo Garza E. Total Quality Management. 2019. Available from: https://books.google.com.ec/books?hl=en&lr=&id=9zYyYc6i9JwC&oi=fnd&pg=PA1&dq=total+quality+scientific+article&ots=2bzZ52O1Ul&sig=msaHQooNzQoUpNmB7tRcX6rIkEY#v=onepage&q=total%20quality%20scientific%20article&f=false

[16] Iñiguez ES. Good Governance: Political Studies. Mexico City: Planeta Num; 2016. p. 26

[17] Basterra M. The Right of Access to Public Information. Buenos Aires, Argentina: Universidad de Palermo; 2012. Available from: http://hdl.handle.net/10226/853

[18] Hernández J, Duque F. Artificial intelligence at the Service of Auditing: A systematic literature review. Iberian Journal of Systems. 2020;27:213-226. Available from: https://www.proquest.com/openview/8a2868ccf43245be9a642a31d5454ca4/1?pq-origsite=gscholar&cbl=1006393

[19] Ackerman M. Laws of Access to Information in the World. Mexico: National Institute of Transparency, Access to Information, and Personal Data Protection; 2008. Available from: https://ru.iis.sociales.unam.mx/handle/IIS/5301